Radiation Detectors
for Medical Imaging

Devices, Circuits, and Systems

Series Editor
Krzysztof Iniewski
CMOS Emerging Technologies Research Inc.,
Vancouver, British Columbia, Canada

PUBLISHED TITLES:

Atomic Nanoscale Technology in the Nuclear Industry
Taeho Woo

Biological and Medical Sensor Technologies
Krzysztof Iniewski

Building Sensor Networks: From Design to Applications
Ioanis Nikolaidis and Krzysztof Iniewski

Circuits at the Nanoscale: Communications, Imaging, and Sensing
Krzysztof Iniewski

CMOS: Front-End Electronics for Radiation Sensors
Angelo Rivetti

Design of 3D Integrated Circuits and Systems
Rohit Sharma

Electrical Solitons: Theory, Design, and Applications
David Ricketts and Donhee Ham

Electronics for Radiation Detection
Krzysztof Iniewski

**Embedded and Networking Systems:
Design, Software, and Implementation**
Gul N. Khan and Krzysztof Iniewski

Energy Harvesting with Functional Materials and Microsystems
Madhu Bhaskaran, Sharath Sriram, and Krzysztof Iniewski

**Graphene, Carbon Nanotubes, and Nanostuctures:
Techniques and Applications**
James E. Morris and Krzysztof Iniewski

High-Speed Devices and Circuits with THz Applications
Jung Han Choi

High-Speed Photonics Interconnects
Lukas Chrostowski and Krzysztof Iniewski

**High Frequency Communication and Sensing:
Traveling-Wave Techniques**
Ahmet Tekin and Ahmed Emira

PUBLISHED TITLES:

Integrated Microsystems: Electronics, Photonics, and Biotechnology
Krzysztof Iniewski

Integrated Power Devices and TCAD Simulation
Yue Fu, Zhanming Li, Wai Tung Ng, and Johnny K.O. Sin

Internet Networks: Wired, Wireless, and Optical Technologies
Krzysztof Iniewski

Labs on Chip: Principles, Design, and Technology
Eugenio Iannone

Laser-Based Optical Detection of Explosives
Paul M. Pellegrino, Ellen L. Holthoff, and Mikella E. Farrell

Low Power Emerging Wireless Technologies
Reza Mahmoudi and Krzysztof Iniewski

Medical Imaging: Technology and Applications
Troy Farncombe and Krzysztof Iniewski

Metallic Spintronic Devices
Xiaobin Wang

MEMS: Fundamental Technology and Applications
Vikas Choudhary and Krzysztof Iniewski

Micro- and Nanoelectronics: Emerging Device Challenges and Solutions
Tomasz Brozek

Microfluidics and Nanotechnology: Biosensing to the Single Molecule Limit
Eric Lagally

MIMO Power Line Communications: Narrow and Broadband Standards, EMC, and Advanced Processing
Lars Torsten Berger, Andreas Schwager, Pascal Pagani, and Daniel Schneider

Mobile Point-of-Care Monitors and Diagnostic Device Design
Walter Karlen

Nano-Semiconductors: Devices and Technology
Krzysztof Iniewski

Nanoelectronic Device Applications Handbook
James E. Morris and Krzysztof Iniewski

Nanopatterning and Nanoscale Devices for Biological Applications
Šeila Selimović

Nanoplasmonics: Advanced Device Applications
James W. M. Chon and Krzysztof Iniewski

Nanoscale Semiconductor Memories: Technology and Applications
Santosh K. Kurinec and Krzysztof Iniewski

Novel Advances in Microsystems Technologies and Their Applications
Laurent A. Francis and Krzysztof Iniewski

PUBLISHED TITLES:

Optical, Acoustic, Magnetic, and Mechanical Sensor Technologies
Krzysztof Iniewski

Optical Fiber Sensors: Advanced Techniques and Applications
Ginu Rajan

Optical Imaging Devices: New Technologies and Applications
Dongsoo Kim and Ajit Khosla

Organic Solar Cells: Materials, Devices, Interfaces, and Modeling
Qiquan Qiao

Radiation Effects in Semiconductors
Krzysztof Iniewski

Semiconductor Radiation Detection Systems
Krzysztof Iniewski

Smart Grids: Clouds, Communications, Open Source, and Automation
David Bakken

Smart Sensors for Industrial Applications
Krzysztof Iniewski

Soft Errors: From Particles to Circuits
Jean-Luc Autran and Daniela Munteanu

Solid-State Radiation Detectors: Technology and Applications
Salah Awadalla

Technologies for Smart Sensors and Sensor Fusion
Kevin Yallup and Krzysztof Iniewski

Telecommunication Networks
Eugenio Iannone

Testing for Small-Delay Defects in Nanoscale CMOS Integrated Circuits
Sandeep K. Goel and Krishnendu Chakrabarty

VLSI: Circuits for Emerging Applications
Tomasz Wojcicki

Wireless Technologies: Circuits, Systems, and Devices
Krzysztof Iniewski

Wireless Transceiver Circuits: System Perspectives and Design Aspects
Woogeun Rhee

FORTHCOMING TITLES:

Advances in Imaging and Sensing
Shuo Tang, Dileepan Joseph, and Krzysztof Iniewski

Analog Electronics for Radiation Detection
Renato Turchetta

FORTHCOMING TITLES:

Cell and Material Interface: Advances in Tissue Engineering, Biosensor, Implant, and Imaging Technologies
Nihal Engin Vrana

Circuits and Systems for Security and Privacy
Farhana Sheikh and Leonel Sousa

CMOS Time-Mode Circuits and Systems: Fundamentals and Applications
Fei Yuan

Electrostatic Discharge Protection of Semiconductor Devices and Integrated Circuits
Juin J. Liou

Gallium Nitride (GaN): Physics, Devices, and Technology
Farid Medjdoub and Krzysztof Iniewski

Ionizing Radiation Effects in Electronics: From Memories to Imagers
Marta Bagatin and Simone Gerardin

Mixed-Signal Circuits
Thomas Noulis and Mani Soma

Magnetic Sensors: Technologies and Applications
Simone Gambini and Kirill Poletkin

MRI: Physics, Image Reconstruction, and Analysis
Angshul Majumdar and Rabab Ward

Multisensor Data Fusion: From Algorithm and Architecture Design to Applications
Hassen Fourati

Nanoelectronics: Devices, Circuits, and Systems
Nikos Konofaos

Nanomaterials: A Guide to Fabrication and Applications
Gordon Harling, Krzysztof Iniewski, and Sivashankar Krishnamoorthy

Physical Design for 3D Integrated Circuits
Aida Todri-Sanial and Chuan Seng Tan

Power Management Integrated Circuits and Technologies
Mona M. Hella and Patrick Mercier

Radiation Detectors for Medical Imaging
Jan S. Iwanczyk

Radio Frequency Integrated Circuit Design
Sebastian Magierowski

Reconfigurable Logic: Architecture, Tools, and Applications
Pierre-Emmanuel Gaillardon

Silicon on Insulator System Design
Bastien Giraud

FORTHCOMING TITLES:

Structural Health Monitoring of Composite Structures Using Fiber Optic Methods
Ginu Rajan and Gangadhara Prusty

Terahertz Sensing and Imaging: Technology and Devices
Daryoosh Saeedkia and Wojciech Knap

Tunable RF Components and Circuits: Applications in Mobile Handsets
Jeffrey L. Hilbert

Wireless Medical Systems and Algorithms: Design and Applications
Pietro Salvo and Miguel Hernandez-Silveira

Radiation Detectors for Medical Imaging

EDITED BY
JAN S. IWANCZYK
DxRay, Inc., Northridge, CA, United States

KRZYSZTOF INIEWSKI MANAGING EDITOR
CMOS Emerging Technologies Research Inc.
Vancouver, British Columbia, Canada

CRC Press is an imprint of the
Taylor & Francis Group, an **informa** business

CRC Press
Taylor & Francis Group
6000 Broken Sound Parkway NW, Suite 300
Boca Raton, FL 33487-2742

© 2016 by Taylor & Francis Group, LLC
CRC Press is an imprint of Taylor & Francis Group, an Informa business

No claim to original U.S. Government works

Printed on acid-free paper
Version Date: 20150729

International Standard Book Number-13: 978-1-4987-0435-9 (Hardback)

This book contains information obtained from authentic and highly regarded sources. Reasonable efforts have been made to publish reliable data and information, but the author and publisher cannot assume responsibility for the validity of all materials or the consequences of their use. The authors and publishers have attempted to trace the copyright holders of all material reproduced in this publication and apologize to copyright holders if permission to publish in this form has not been obtained. If any copyright material has not been acknowledged please write and let us know so we may rectify in any future reprint.

Except as permitted under U.S. Copyright Law, no part of this book may be reprinted, reproduced, transmitted, or utilized in any form by any electronic, mechanical, or other means, now known or hereafter invented, including photocopying, microfilming, and recording, or in any information storage or retrieval system, without written permission from the publishers.

For permission to photocopy or use material electronically from this work, please access www.copyright.com (http://www.copyright.com/) or contact the Copyright Clearance Center, Inc. (CCC), 222 Rosewood Drive, Danvers, MA 01923, 978-750-8400. CCC is a not-for-profit organization that provides licenses and registration for a variety of users. For organizations that have been granted a photocopy license by the CCC, a separate system of payment has been arranged.

Trademark Notice: Product or corporate names may be trademarks or registered trademarks, and are used only for identification and explanation without intent to infringe.

Visit the Taylor & Francis Web site at
http://www.taylorandfrancis.com

and the CRC Press Web site at
http://www.crcpress.com

Contents

Preface xi
Editors xiii
Contributors xv

Chapter 1 CdZnTe and CdTe Crystals for Medical Applications 1
Csaba Szeles

Chapter 2 Monte Carlo Modeling of X-Ray Detectors for Medical Imaging 27
Yuan Fang, Karim S. Karim, and Aldo Badano

Chapter 3 Medical X-Ray and CT Imaging with Photon-Counting Detectors ... 47
Polad M. Shikhaliev

Chapter 4 Pixelated Semiconductor and Parallel ASIC Design for Spectral
Clinical Radiology 81
William C. Barber, Einar Nygard, and Jan S. Iwanczyk

Chapter 5 Silicon Photomultipliers in Detectors for Nuclear Medicine 119
Martyna Grodzicka, Marek Moszyński, and Tomasz Szczęśniak

Chapter 6 Imaging Technologies and Potential Clinical Applications
of Photon-Counting X-Ray Computed Tomography 149
Katsuyuki Taguchi

Chapter 7 Photon-Counting Detectors and Clinical Applications
in Medical CT Imaging 169
Ira Blevis and Reuven Levinson

Chapter 8 Radiation Detection in SPECT and PET 193
Chin-Tu Chen and Chien-Min Kao

Chapter 9 Review of Detectors Available for Full-Field Digital
Mammography 233
Nico Lanconelli and Stefano Rivetti

Chapter 10 Grating-Based Phase-Contrast X-Ray Imaging Technique 255

Salim Reza

Chapter 11 Emerging Concept in Nuclear Medicine: The Voxel Imaging
PET (VIP) Project 269

Mokhtar Chmeissani, Gianluca De Lorenzo, and Machiel Kolstein

Index 327

Preface

This book aims to provide the current status and prospective of key technologies and applications of photon-counting detectors emerging in medical imaging. Photon-counting detectors have been commonly used in nuclear medicine equipment for many years. However, the use of photon counting in x-ray imaging is very new and only possible due to recent progress in material/detector technologies combined with the availability of application-specific integrated circuits that provide very compact low-noise amplification and processing of the signal from individual detector pixels in imaging arrays. Also, novel methods for photon counting that can replace conventional photon-counting detectors utilizing photomultiplier tubes are starting to impact the design of nuclear medicine equipment.

Since their discovery in 1895 by Wilhelm C. Röntgen, x-rays have played a very important role in medical imaging, helping physicians to detect and characterize disease processes. Detected transmitted x-ray beams can generate a snapshot projection image, a series of projection images, or cross-sectional tomographic images. X-ray radiography (XR) provides 2D projection images of the transmitted x-ray intensities. The first computed tomography (CT) system was built by Godfrey Hounsfield in 1971. Since that time, CT has seen rapid development to become a workhorse in many clinical settings throughout the world. According to the market research firm the IMV Medical Information Division, the CT procedure volume peaked at 85.3 million studies performed in the United States in 2011. Multislice x-ray CT (or MDCT for multi-detector-row CT) scanners provide 3D images of the distribution of the linear attenuation coefficients within a patient by reconstructing 2D projection images acquired from many angles and can accurately delineate organs and tissues. However, there are several major limitations to current XR and CT technologies: (1) The contrast between different tissues is not sufficient, (2) images are not tissue-type specific, (3) CT scanning is a relatively high-dose procedure, and (4) the grayscale pixel values of CT images, which should be the linear attenuation coefficients, are not quantitative but qualitative. These limitations result from the energy-integrating detectors used in CT scanners and XR systems.

Energy-integrating detectors measure the intensity of x-rays, that is, integrate the area under the curve of the energy-weighted transmitted x-ray spectrum, losing all energy-dependent information. Energy-integrating detectors not only add electric noise and Swank noise but also apply a weight in the signal proportional to the energy of the individual x-rays. Energy proportional weighting in the energy-integrated signal minimizes the contribution of the lower-energy photons, which carry larger contrast between tissues, resulting in increased noise and decreased contrast.

In general, dual-energy CT imaging can provide tissue-specific images. However, neither of the current conventional detector systems utilizing rapid switching of the voltage at x-ray tubes (dual kVp) in dual-source techniques or dual-layer detectors can provide optimal results due to cross-talk between the high- and low-energy images and because of the limited number of resolvable basis functions (only two) for material decomposition. The presence of contrast media containing elements with high atomic numbers can be identified with a third basis function. Thus, it is desirable to measure the transmitted x-ray photons with more than two energy windows.

Recently, photon-counting detectors with energy discrimination capabilities based on pulse height discrimination have been developed for medical x-ray imaging as described in this book. X-ray spectral imaging detectors are mostly based on cadmium telluride, cadmium zinc telluride, or silicon semiconductor materials. These detectors count the number of photons of the transmitted x-ray spectrum using two to six energy windows. CT and XR systems based on photon-counting detectors with multiple energy windows have the potential to improve many of the major limitations listed earlier. Electronic and Swank noise affect the measured energy but do not change the output signal intensity (i.e., the counts), and the energy overlap in the spectral measurements is superior to (i.e., smaller than) that from any of the current dual-energy techniques using energy-integrating detectors. In addition, more than one contrast media could be imaged simultaneously and would be distinguishable if the detectors had four or more energy windows. Photon-counting detectors may therefore lead to novel clinical CT and XR applications as discussed in several of the following chapters. One of the most exciting possibilities is the future use of CT scanners not only as an anatomical modality but also as a functional modality entering into areas reserved for nuclear medicine techniques but with much better spatial resolution and a significantly reduced time for examination.

The performance of photon-counting detectors is not flawless, however, especially at the large count rates in current clinical CT. Due to the stochastic nature of the x-ray signal and a limited pulse resolving time, quasicoincident photons (overlapping pulses) can be recorded as a single count with a higher or lower energy. This phenomenon is called pulse pileup and results both in a loss of counts, referred to as dead time losses, and a distortion of the recorded spectrum. It is thus critical to develop detector systems that can handle very high x-ray flux with minimized loss of counts and spectral distortions. Also, software schemes to compensate for these effects are very important. Other phenomena may also degrade the spectral response of photon-counting detectors. These include incomplete charge collection that is generated by x-rays due to charge sharing and charge trapping effects. These phenomena are reviewed in depth in several chapters in this book.

In addition to the exciting development of photon-counting detectors for x-ray imaging, some of the chapters discuss compound semiconductors used as direct converting gamma ray detectors, and silicon photomultipliers used for reading the light from scintillators are starting to make a big impact on the design concepts for new nuclear medicine equipment for the gamma cameras used in single-photon emission computed tomography (SPECT) and positron emission tomography (PET) applications. These new designs allow for the construction of more compact imagers with better performance that are not sensitive to magnetic fields, as are designs utilizing conventional photomultiplier tubes, allowing for the construction of SPECT and PET systems combined with magnetic resonance scanners in multimodality systems.

Jan S. Iwanczyk
Krzysztof Iniewski

Editors

Dr. Jan S. Iwanczyk has served as a president and CEO of DxRay, Inc., Northridge, California, since 2005. He was previously affiliated with several start-up private and publicly traded companies and centered on bringing novel scientific and medical technologies to the market. From 1979 to 1989, Dr. Iwanczyk was an associate professor at the University of Southern California, School of Medicine, Los Angeles, California. He holds an MS in electronics and a PhD in physics. His multifaceted experience combines operations, organizational development with strong scientific research, and technical project management qualifications. Dr. Iwanczyk's technical expertise is in the field of x-ray and gamma ray imaging detectors and systems. In recent years, he has pioneered the development of photon-counting, energy-dispersive x-ray detectors for medical applications. He is the author of over 200 scientific papers, book chapters, and 20 patents. He also lectures at major symposia worldwide as an invited speaker and has received numerous honors and awards, including the 2002 Merit Award, IEEE Nuclear and Plasma Sciences Society. He can be reached at jan.iwanczyk@dxray.com.

Krzysztof (Kris) Iniewski is managing R&D at Redlen Technologies, Inc., a start-up company in Vancouver, Canada. Redlen's revolutionary production process for advanced semiconductor materials enables a new generation of more accurate, all-digital, radiation-based imaging solutions. Kris is also a president of CMOS Emerging Technologies Research, Inc. (www.cmosetr..com), an organization of high-tech events covering communications, microsystems, optoelectronics, and sensors. In his career, Dr. Iniewski held numerous faculty and management positions at the University of Toronto, University of Alberta, SFU, and PMC-Sierra, Inc. He has published over 100 research papers in international journals and conferences. He holds 18 international patents granted in the USA, Canada, France, Germany, and Japan. He is a frequent invited speaker and has consulted for multiple organizations internationally. He has written and edited several books for CRC Press, Cambridge University Press, IEEE Press, Wiley, McGraw-Hill, Artech House, and Springer. His personal goal is to contribute to healthy living and sustainability through innovative engineering solutions. In his leisurely time, Kris can be found hiking, sailing, skiing, or biking in beautiful British Columbia. He can be reached at kris.iniewski@gmail.com.

Contributors

Aldo Badano
Division of Imaging, Diagnostics, and Software Reliability
Office of Science and Engineering Laboratories
Center for Devices and Radiological Health
Food and Drug Administration
Silver Spring, Maryland

William C. Barber
DxRay, Inc.
Northridge, California

and

Interon AS
Asker, Norway

Ira Blevis
Philips Healthcare
Global Research and Advanced Development
Philips, Israel

Chin-Tu Chen
The University of Chicago
Chicago, Illinois

Mokhtar Chmeissani
Institut de Física d'Altes Energies
Universitat Autònoma de Barcelona
Barcelona, Spain

Gianluca De Lorenzo
Institut de Física d'Altes Energies
Universitat Autònoma de Barcelona
Barcelona, Spain

Yuan Fang
Center for Devices and Radiological Health
Food and Drug Administration
Silver Spring, Maryland

and

Department of Electrical and Computer Engineering
University of Waterloo
Waterloo, Ontario, Canada

Martyna Grodzicka
Department of Nuclear Techniques and Equipment
Radiation Detectors Division (TJ3)
National Centre for Nuclear Research
Otwock-Świerk, Poland

Jan S. Iwanczyk
DxRay, Inc.
Northridge, California

Chien-Min Kao
The University of Chicago
Chicago, Illinois

Karim S. Karim
Department of Electrical and Computer Engineering
University of Waterloo
Waterloo, Ontario, Canada

Machiel Kolstein
Institut de Física d'Altes Energies
Universitat Autònoma de Barcelona
Barcelona, Spain

Nico Lanconelli
Alma Mater Studiorum
University of Bologna
Bologna, Italy

Reuven Levinson
Philips Healthcare
Global Research and Advanced
  Development
Haifa, Israel

Marek Moszyński
Department of Nuclear Techniques and
  Equipment
Radiation Detectors Division (TJ3)
National Centre for Nuclear Research
Otwock-Świerk, Poland

Einar Nygard
Interon AS
Asker, Norway

Salim Reza
Electronics Department
Mid Sweden University
Sundsvall, Sweden

and

The Photon Science Detector Group
  (FS-DS)
Deutsches Elektronen-Synchrotron
  (DESY)
Hamburg, Germany

Stefano Rivetti
Ospedale di Sassuolo S.p.A
Modena, Italy

Polad M. Shikhaliev
Department of Nuclear Engineering
Hacettepe University
Ankara, Turkey

Tomasz Szczęśniak
Department of Nuclear Techniques and
  Equipment
Radiation Detectors Division (TJ3)
National Centre for Nuclear Research
Otwock-Świerk, Poland

Csaba Szeles
Nious Technologies, Inc.
Pittsburgh, Pennsylvania

Katsuyuki Taguchi
Division of Medical Imaging Physics
The Russell H. Morgan Department of
  Radiology & Radiological Science
School of Medicine
Johns Hopkins University
Baltimore, Maryland

1 CdZnTe and CdTe Crystals for Medical Applications

Csaba Szeles

CONTENTS

1.1 CdTe and CdZnTe for Medical Applications 1
1.2 CdTe and CdZnTe Materials 2
1.3 Materials Technology 5
    1.3.1 Defect Structure of High-Purity CdTe 6
    1.3.2 Electrical Compensation 8
    1.3.3 Carrier Transport 10
        1.3.3.1 Recombination 12
        1.3.3.2 Uniform Trapping 12
        1.3.3.3 Nonuniform Trapping 15
        1.3.3.4 Carrier Transport under High Photon Flux 15
1.4 Crystal Growth Technology 18
    1.4.1 Parasitic Nucleation 19
    1.4.2 Physical Defect Generation 21
    1.4.3 Defect Interactions 22
    1.4.4 Annealing 24
    1.4.5 Status of Crystal Growth 24
1.5 Summary 25
References 25

1.1 CdTe AND CdZnTe FOR MEDICAL APPLICATIONS

CdTe and CdZnTe semiconductor detectors are solid-state devices that provide direct conversion of the absorbed gamma-ray energy into an electronic signal. Many of the advantages of these detectors for medical applications stem from this inherent energy discrimination and photon-counting capability. In addition, the high radiation stopping power and resulting high detection efficiency, low leakage current at room temperature, good charge transport of the photon-generated carriers, and the favorable chemical and mechanical properties that allow the fabrication of pixelated detectors enable the manufacture of sophisticated X-ray and gamma-ray imaging devices that are compact and can be operated at room temperature and at low voltage.

Nuclear medicine gamma cameras built for cardiac single-photon emission computerized tomography (SPECT) [1] and scintimammography, often referred to as molecular breast imaging (MBI) [2], are taking advantage of the superior energy resolution (2%–5% FWHM at 140 keV) of CdZnTe detectors to improve scatter rejection and optimized pixel dimensions to achieve high intrinsic spatial resolution that is independent of the photon energy and what allows the use of wide-angle collimators to achieve higher sensitivity. Combined with advance image reconstruction techniques, this detector technology provides improved image contrast and resolution.

The high detection efficiency, energy sensitivity, and good spatial resolution of pixelated CdZnTe detectors are exploited in dual-energy X-ray absorptiometry (DEXA) for high-performance bone mineral densitometry [3].

Photon-counting detector technology offer benefits also to digital radiography where both the image signal-to-noise ratio and contrast resolution can be improved if the radiation energy information is used alongside the radiation intensity. The direct conversion detector technology enables very sharp line spread function (LSF) limited only by the pixel size. A sharp LSF together with the high absorption efficiency of CdTe and CdZnTe and low-noise readout circuitry yields high detector quantum efficiency, which ultimately determines the performance of an imaging system. CdTe detectors with fine pixelation coupled to low-noise CMOS readout chips have been developed and successfully deployed in panoramic dental imaging applications [4].

Multienergy computed tomography (CT) is the ultimate challenge for any solid-state detector technology including CdTe and CdZnTe detectors [5]. Just as in the previously mentioned applications, the energy discrimination capability is the key advantage of these photon-counting detectors. The fast data acquisition and high photon flux used in state-of-the-art CT imaging systems require very-fast-response detector technology that can operate under intense photon radiation conditions. CT applications represent a huge challenge for CdTe and CdZnTe detector technology and are the subject of intense research today.

The compact size, capability for fine pixelation, low voltage requirements, and room-temperature operation enable the deployment of CdTe and CdZnTe detectors in compact gamma cameras for prostate imaging [6] and miniature ingestible imaging capsules for colorectal cancer detection [7].

In this chapter, we review the state of the art of CdTe and CdZnTe materials and crystal growth technologies including what these technologies face for deployment in X-ray and gamma-ray detectors and the opportunities these technologies provide for both mainstream and novel medical applications.

1.2 CdTe AND CdZnTe MATERIALS

The binary semiconducting compound cadmium telluride (CdTe) and its ternary cousin cadmium zinc telluride (CdZnTe or CZT) possess material properties that make them uniquely befitting for room-temperature solid-state radiation detectors. The high average atomic numbers ($Z_{CT} = 50$ and $Z_{CZT} = 48.2$) and densities

(ρ_{CT} = 5.85 g/cm^3 and ρ_{CZT} = 5.78 g/cm^3) provide high stopping power for X-rays and gamma rays enabling high sensitivity and high detection efficiency of the detectors.

The band gaps of CdTe and CdZnTe are E_g = 1.5 eV and E_g = 1.572 eV, respectively, at room temperature, making the materials ideal for room-temperature radiation detectors. In electrically compensated semi-insulating CdTe and CdZnTe crystals, the low free-carrier concentration enables achieving large depletion depths ranging from a few mm to a few cm and low leakage currents in the few pA to few nA range.

The moderately high mobility and lifetime of charge carriers in CdTe and CdZnTe allow good charge transport across detector devices depleted to several mm or even cm thickness. State-of-the-art crystal growth technology is regularly producing semi-insulating CdTe crystals with electron and hole mobility-lifetime products in the $\mu_e\tau_e$ = 10^{-3} cm^2/V and $\mu_h\tau_h$ = 10^{-4} cm^2/V range, respectively, and CdZnTe crystals with electron and hole mobility-lifetime products in the $\mu_e\tau_e$ = 10^{-2} cm^2/V and $\mu_h\tau_h$ = 10^{-5} cm^2/V range, respectively, today.

The duality of CdTe and CdZnTe radiation detectors has existed over 30 years in the industry. Acrorad Ltd. in Japan pioneered the semi-insulating CdTe detector technology, while eV Products Inc. in the United States, Redlen Technologies in Canada, and Imarad Imaging Systems Ltd. (now part of GE Healthcare) in Israel have been commercializing CdZnTe detectors since the early 1990s.

Fundamentally, there are no major differences between the two compounds. CdZnTe is an alloy of CdTe and ZnTe and typically about 10% Zn is alloyed in the ternary compound for detector applications. The alloying with Zn causes a few changes to the properties of CdTe.

First, by adding Zn to CdTe the band gap is increased. The wider band gap enables a higher maximum resistivity of the ternary compound. For CdZnTe with 10% Zn, the band gap increases from 1.5 to 1.572 eV and the maximum achievable resistivity increases by a factor of three, typically from 2 × 10^{10} Ω cm for CdTe to 5 × 10^{10} Ω cm for CdZnTe. It is to be noted, however, that these are only the maximum resistivity limits allowed by the band gap of the material that are not always achieved in practice because of incomplete electrical compensation.

Various vendors employ different doping approaches for electrical compensation to achieve the high resistivity. Acrorad Ltd. uses Cl doping of CdTe crystals to achieve *p-type* conductivity and resistivity in the 10^8 to 10^9 Ω cm range. This resistivity is well under the maximum allowed for CdTe. The disadvantage of this approach is the relatively small depletion depth of the detectors typically in the few *mm* range. The *p-type* conductivity of the Acrorad CdTe detectors on the other hand enables the use of In and Al contacts to manufacture high-barrier Schottky devices to achieve very-low-leakage-current detectors even at very high applied bias. The high bias voltage ensures fast charge collection and fast response of the detectors that is a significant advantage in high-flux applications.

CdZnTe vendors typically use In, Ga, or Al doping to achieve *n-type* conductivity and nearly complete electrical compensation with resistivity in the (2–3) × 10^{10} Ω cm range. Because of the low free-carrier concentration in semi-insulating CdZnTe prepared with this compensation approach, the detector devices can be depleted to a few *cm* in depth enabling large active volume detectors. The detectors are typically fabricated with Pt or Au Schottky barrier contacts. However, because the Schottky barrier height of Pt and Au contacts on *n-type* CdZnTe is lower than the barrier height of Al or In contacts on *p-type* CdTe, the leakage current of commercial CdZnTe detectors is typically an order of magnitude higher than that of commercial CdTe detectors. It is to be emphasized that the different bulk resistivity of CdTe and CdZnTe crystals and the different leakage current of the detectors stem from the technology choices made by the vendors and are not the result of the different band gaps of these compounds.

The second difference comes from the different chemical properties of CdTe and ZnTe. ZnTe has a lower iconicity and a higher binding energy than CdTe and the bond length is shorter in ZnTe. The CdTe lattice is strengthened by the incorporation of Zn leading to the increase of the sheer modulus and solution hardening of the ternary compound, a well-known effect often used in metallurgy [8]. The solution hardening of CdZnTe reduces the propensity for plastic deformation and the formation of dislocations; however, it also reduces dislocation motion and makes the ternary compound more brittle than CdTe.

The higher binding energy and shorter bond length between Zn and Te atoms induce a local strain into the host CdTe lattice and relaxation of the Te atoms around the Zn atom. The local lattice distortions increase the migration barrier of interstitial atoms in the proximity of Zn atoms. As a result, diffusion and ionic migration rates are reduced in CdZnTe compared to CdTe. The effect is however minor and both high-purity CdTe and CdZnTe crystals demonstrate excellent long-term stability and are inert against ionic migration when operated under high bias voltages for a prolonged period of time as radiation detectors. This is a critically important property of these compounds that ensures the long-term stable operation of CdTe and CdZnTe detectors under high bias unlike many of the more ionic compounds that suffer from physical polarization because of the migration of the constituent atoms under the applied bias like in TlBr.

The third difference stems from the difference in the chemical potentials in the two material systems. The addition of Zn increases the maximum deviation from stoichiometry in CdZnTe on the Te-rich side of the phase diagram. The maximum Te nonstoichiometry or solid solubility is about 4×10^{18} cm^{-3} in CdTe and about 1.2×10^{19} cm^{-3} in CdZnTe and is reached at about 880°C [9]. Because in thermal equilibrium Cd and Zn vacancies are the dominant native defects in CdTe and CdZnTe, the excess Te is primarily accommodated by the increase of the number of vacancies in the lattice. The higher maximum Te solubility in CdZnTe therefore indicates that the formation energy of Cd (and Zn) vacancy is reduced in CdZnTe relative to CdTe. Although CdTe and CdZnTe show retrograde Te solid solubility (the maximum nonstoichiometry decreases with decreasing temperature) and much of the excess Te segregates into Te precipitates by the time the crystal is cooled to room temperature, there is a significant difference in the residual Cd and Zn vacancy concentration. Because Cd (and Zn) vacancies introduce acceptor levels in the lower half of the band gap [10], the higher vacancy concentration causes increased hole trapping and a lower hole mobility-lifetime product in CdZnTe than in CdTe often observed experimentally.

The fourth difference stems from the different solubility of Zn in solid and liquid CdTe. The ratio of the solubility called segregation coefficient controls the partition of the alloying element between the solid and liquid during crystal growth. The segregation coefficient of Zn is about $k = 1.3$ in CdTe, meaning that the Zn segregates preferentially into the solid rather than the liquid at the solid–liquid interface. This leads to an axial Zn concentration distribution in CdZnTe grown by melt-growth techniques. For a 10% Zn-doped CdZnTe, the actual Zn concentration starts at about 13%–14% at the first to freeze section of the ingot and decreases to about 6% at the last to freeze section of the ingot. With appropriate design of the solvent and feed material Zn compositions, constant Zn concentration can be achieved with solution growth techniques such as the traveling heater method (THM).

The varying Zn concentration causes a varying lattice constant and the development of a built-in constitutional stress and strain in the ingot [8]. This stress can be excessive and can lead to the deformation of the crystals and a higher dislocation density in CdZnTe than in CdTe despite the alloy hardening of the lattice.

1.3 MATERIALS TECHNOLOGY

Semiconductor radiation detectors require high-quality semi-insulating single crystals for satisfactory performance. High resistivity is required in order to attain sufficient carrier depletion of the semiconductor crystal and realize active detector thickness in the few mm to few cm range and to maintain a high electric field across the detector with low leakage current. Too low depletion limits the active depth of the device and the energy range of the detector, while excessive leakage current produces electronic noise that deteriorates the energy resolution of the detector.

Detector applications require high-quality single crystals because defects in the crystals cause carrier trapping, recombination, and distortion of the internal electric field and the deterioration of detector performance. Large-angle grain boundaries are typically very strong carrier traps in CdTe and CdZnTe. In order to avoid the detrimental effects of grain boundaries on detector performance, single crystals are mined from the typically polycrystalline ingots. The single crystal yield from this mining process is usually the largest cost driver of CdTe and CdZnTe detectors. Growing as perfect single crystal ingots as possible or achieving the highest single crystal yield from polycrystalline ingots is the primary goal and challenge of CdTe and CdZnTe crystal growth technologies. The extracted single crystals themselves have to be as perfect as possible and must have a low concentration of point defects and their clusters and low density of extended defects such as dislocations, low-angle grain boundaries (subgrain boundaries), twins and second-phase precipitates, and inclusions in order to minimize carrier trapping.

In the next few paragraphs, we review the challenges and methods to achieve *high electrical resistivity and good carrier transport* in CdTe and CdZnTe crystals *simultaneously*. We start with a review of the point-defect structure of uncompensated high-purity CdTe.

1.3.1 Defect Structure of High-Purity CdTe

The formation energy of native defects (vacancies, interstitial atoms, and antisites) and impurities in semiconductor crystals has three components:

$$\Delta H = (E_{\text{defect}} - E_{\text{host}}) - \sum_i n_i \left(\mu_i + \mu_i^{\text{ref}}\right) + q(\varepsilon_{\text{VBM}} + \varepsilon_F) \quad (1.1)$$

E_{defect} and E_{host} in the first term are the total energies of the defect-containing and the defect-free host crystals. The second term represents the energy contribution from the chemical potential of the species forming the defects. n_i is the difference in the number of atoms for the ith atomic species between the defect-containing and defect-free crystals. μ_i is a relative chemical potential for the ith atomic species, referenced to μ_i^{ref}. For Cd and Te, $\mu_{\text{Cd}}^{\text{ref}}$ and $\mu_{\text{Te}}^{\text{ref}}$ are the chemical potentials in bulk Cd and bulk Te, respectively. This term captures the change in the formation energy of the defect as a function of deviation from stoichiometry and concentrations of impurities. The third term represents the change in energy due to exchange of electrons or holes with the respective carrier reservoirs. ε_{VBM} is the energy of the valence band maximum (VBM) in the host system and ε_F is the Fermi energy relative to the VBM. This term captures the energy contribution from the charges residing in the defects. Once the formation energy of the defects is calculated, the defect concentration at thermal equilibrium can be evaluated using

$$N = N_{\text{site}} \exp\left(-\frac{\Delta H}{kT}\right) \quad (1.2)$$

where
  N_{site} is the number of available sites for the defect in the crystal
  ΔH is the defect formation energy
  k is Boltzmann's constant
  T is the absolute temperature

With steady advancement of first-principles calculations based on the density-functional theory in the past decade, computational materials science techniques today provide invaluable insights to the structure and properties of defects in semiconductor crystals. The theoretical models provide adequate estimates of the formation energies of native defects and impurities and their complexes as well as their ionization energies in the band gap in many semiconductors including CdTe.

Figure 1.1 shows the calculated formation energy of native defects in Te-rich CdTe as the function of the Fermi energy such as the native donors Cd interstitial (Cd_{int}), Te vacancy (V_{Te}), Te antisite (Te_{Cd}, i.e., Te sitting on a Cd site), and Te interstitial (Te_{int}) and the acceptors Cd vacancy (V_{Cd}) and Cd vacancy–Te antisite pair ($V_{\text{Cd}} + Te_{\text{Cd}}$) [11].

What is very important to emphasize is that the defect formation energy of charged defects depends on the Fermi energy. As the Fermi energy increases, the

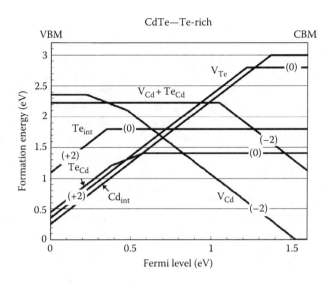

FIGURE 1.1 Calculated formation energy of native defects in CdTe as a function of the Fermi energy.

formation energy of donors increases while the formation energy of acceptors decreases linearly according to Equation 1.1. It is also important to recognize that the specific dependency of donor and acceptor formation energies on the Fermi energy gives a high degree of rigidity to the Fermi level. If we perturb the electronic system and try to move the Fermi level from its equilibrium position dictated by the balance of the concentrations of native defect and impurities to either direction, the Fermi level quickly returns to the equilibrium point. If we would move the Fermi level higher toward the conduction band minimum (CBM), the formation energy of Cd vacancy acceptors decreases and their concentration rapidly rises to pull the Fermi level back to the equilibrium point. Similarly, if we would lower the Fermi level toward the VBM, the formation energy of Cd interstitial donors would decrease and their concentration would increase rapidly to pull the Fermi level up back to the equilibrium point. Because of the exponential dependence of the defect concentrations on the formation energy in Equation 1.2, this negative feedback is very strong giving the Fermi level significant stability. This rigidity of the Fermi level is a critical property of the system that enables implementation of practical doping schemes in CdTe and CdZnTe.

It is important to point out that the formation energy of the defects also depends on the concentration of the constituent Cd and Te atoms in the system (chemical potentials) and any deviation from perfect stoichiometry causes a change in the formation energies of the native defects. Similarly, the formation energy of impurities and doping elements depends on their concentrations.

It is clear from Equation 1.2 that the defect with the lowest formation energy will have the highest concentration in the crystal and be the dominant defect in thermal equilibrium. Figure 1.1 shows that Cd vacancy is the dominant native acceptor and Cd interstitial is the dominant native donor defect in CdTe having the lowest formation energies. When the formation energy of donors and acceptors is equal at a given Fermi energy, their concentrations are equal as well and their free-carrier contributions cancel each other out: *electrical compensation* occurs. In other words, it is energetically more favorable for the electrons liberated from the donor state to occupy an acceptor level than to remain in the conduction band as free electrons. Such a condition occurs when the Fermi energy is at $E_v + 0.7$ eV in Figure 1.1. The formation energies and concentrations of Cd interstitials and Cd vacancies equal and exact electrical compensation occurs between the two dominant defects. Because this point is at the middle of the band gap, the material has very low free-carrier concentration and has high resistivity.

The concentration of the various defects in the crystal and the position of the Fermi level are determined by the minimum of the total Gibbs free energy of the material and can be calculated by taking into account the energies of all defects in the crystal. It is not possible to tell a priori the position of the Fermi level in a semiconductor with a complex defect structure where multiple donors and acceptors are present in the material.

There are however general trends that stem from the Fermi statistics governing the occupancy of defect levels in the band gap. Donor doping moves the Fermi level between the ionization level of the donor and the CBM. Acceptor doping moves the Fermi level between the ionization level of the acceptor and the VBM. When donors and acceptors compete, the Fermi level is stabilized around the middle of the band gap between the acceptor levels and donor levels and electrical compensation occurs. Because the Fermi energy is a representation of the average electron energy in the system, its position represents the statistical average of electron energies on the various acceptor and donor energy levels weighted by the concentration of the defects. If only one donor and one acceptor defect is present in equal concentrations, the Fermi level settles halfway between the donor and acceptor ionization level.

1.3.2 ELECTRICAL COMPENSATION

For the CdTe and CdZnTe crystals to be useful for room-temperature radiation detection, spectroscopy, and imaging applications, the chosen crystal growth technology has to achieve *high electrical resistivity* and *good carrier transport simultaneously*. This is not an easy task because these two requirements are often counteracting each other. Good carrier transport requires the growth of high-purity crystals to minimize carrier trapping at impurity defects. High-purity CdTe and CdZnTe crystals are, however, typically low-resistivity *p*-type with electrical resistivity in the 10^3–10^6 Ω cm range because of the significant concentration of residual Cd (and Zn) vacancies that are the dominant native acceptors in these compounds. As we have discussed in the previous section, there is a significant concentration of Cd interstitials in CdTe in thermal equilibrium. At high temperature, close to the solidification point of CdTe and CdZnTe, the Cd vacancies are largely compensated by Cd interstitials. However, as the crystal is cooled to room temperature, the point-defect diffusion slows and eventually stops and the defects *freeze in* the crystal. Because the diffusivity of Cd interstitials is much higher than the diffusivity of

Cd vacancies, they freeze in at a much lower temperature than vacancies. Because the *residual* concentration of defects corresponds to their equilibrium concentration at their freeze-in temperature, the residual concentration of Cd interstitials is much lower than the residual concentration of Cd vacancies in CdTe and CdZnTe at room temperature.

In order to compensate the doping effect of Cd (and Zn) vacancies and reduce the free-carrier concentration to the 10^5 cm^3 range, the CdTe and CdZnTe crystals are doped with shallow donors such as Al, Ga, In, or Cl. The donor doping induces several effects. First, it supplies free electrons that fill the Cd (and Zn) vacancy acceptor levels and elevate the Fermi level close to the middle of the band gap. Second, the donors combine with the vacancies to form vacancy–donor pairs (such as V_{Cd}–In_{Cd}) called A-centers. This pairing reduces the concentration of Cd vacancies by converting them to A-*centers* with a lower acceptor ionization energy. Because Cd (and Zn) vacancies have two ionization energy levels and contribute two holes to the valence band while the A-centers have a single ionization level, the pair formation reduces the doping effect of Cd vacancies.

Figure 1.2 shows the formation energy of the In_{Cd} donor and the In A-center (V_{Cd} + In_{Cd}) as a function of the Fermi energy for three different In concentrations (low, medium, high). As more In is added to the crystal, the chemical potential of In increases and the formation energies of both the In donor and the A-center acceptor are decreasing. The circles in Figure 1.2 indicate the primary compensation points, namely, the Fermi energy where the lowest-energy acceptor and donor formation energies are equal. For low In doping concentrations, the primary compensation occurs between the Cd vacancy and the Cd interstitial. At medium In concentration, the primary compensation occurs between the Cd vacancy and the In donor.

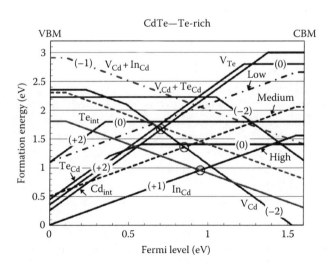

FIGURE 1.2 Calculated formation energy of native defects, In donor, and In A-centers in CdTe as a function of the Fermi energy (After Biswas, K. and Du, M.H., *New J. Phys.*, 14, 063020, 2012). The In donor and A-center formation energies are shown for three In concentration levels: high (—), medium (---), and low (– · –).

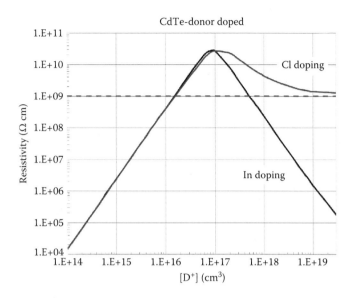

FIGURE 1.3 Calculated electrical resistivity as a function of donor concentration for CdTe doped with In or Cl. (From Biswas, K. and Du, M.H., *New J. Phys.*, 14, 063020, 2012.)

At high In concentration, the A-center formation energy becomes smaller than the Cd vacancy formation energy and the primary compensation occurs between the In donor and the A-center.

The latest theoretical analysis of electrical compensation in CdTe shows that practical electrical compensation is achievable using shallow-level donors such as In, Al, Ga, or Cl [11]. The strong dependence of the defect formation energies on the Fermi energy provides a broad range of donor concentrations where high resistivity is achievable. Figure 1.3 shows that electrical resistivity higher than 10^9 Ω cm is achievable in the 2×10^{16} to 5×10^{17} cm^{-3} donor concentration range with both In and Cl doping [11]. The theoretical analysis indicates that CdTe:In electrical compensation primarily takes place between In_{Cd} donor and V_{Cd} acceptor. In CdTe:Cl, the primary compensation is between Cl_{Te} and V_{Cd} in the lower Cl concentration range (10^{16}–10^{17} cm^{-3}) and between Cl_{Te} and the A-center in the high concentration range (10^{19} cm^{-3}). The different behavior of the In-doped and Cl-doped CdTe stems from the higher binding energy of the $Cl_{Cd} + V_{Cd}$ pair than the $In_{Cd} + V_{Cd}$ pair [11].

1.3.3 Carrier Transport

Gamma- and X-ray detectors require very good carrier transport though the active volume of the device to minimize carrier loss from the charge cloud generated by the photons and preserve the proportionality of the detector signal amplitude to the energy of the photons. Crystal defects cause charge trapping and recombination that reduces the amount of collected charge and cause a low-energy tailing of the photopeaks deteriorating the energy resolution of the detector. It is therefore imperative that the crystal growth technology produces single crystals with as low defect density as possible.

Point defects, point-defect clusters, and extended defects all can cause detrimental trapping. Point defects and small clusters (doublets, triplets) cause uniform trapping if these defects are randomly distributed in the crystal lattice. Extended defects such as dislocations, subgrain boundaries, second-phase precipitates and inclusion, and twin boundaries form spatially correlated structures and typically cause nonuniform trapping. Second-phase precipitates, inclusions, and impurities often decorate grain boundaries, subgrain boundaries, twin planes, and dislocations further enhancing nonuniform carrier trapping. Nonuniform trapping is difficult to correct by electronic or software methods and is particularly harmful for spectroscopic and imaging applications.

Although the physics of carrier trapping and detrapping is exactly the same for both shallow-level defects and deep-level defects, they induce detector performance degradation in a somewhat different way. Although there is no clear delineation between shallow-level and deep-level defects, one can distinguish these defects based on the residence time of trapped charge at the defects relative to the transit time of the carriers through the detector device. If the residence time is shorter than the transit time of the charge carriers, the defect is a shallow-level trap. If the residence time of the trapped carriers is longer than the transit time, the trap can be considered deep-level defect. The typical electron transit times in CdTe and CdZnTe detectors are in the 50–500 ns range depending on the device thickness and applied bias voltage (Figure 1.4). The defects with ionization energies less than 0.35 eV can be considered shallow-level defects, while defects with ionization energies larger than this are considered deep-level defects [12].

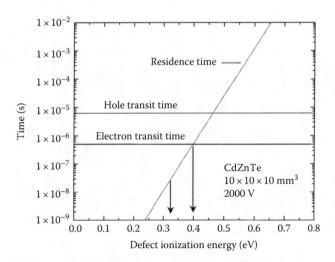

FIGURE 1.4 Residence time versus transit time for electrons and holes in a CdZnTe detector. (From Szeles, C., *IEEE Trans. Nucl. Sci.*, 51(3), 1242, 2004.)

Deep-level defects cause permanent charge loss from the charge cloud generated by the radiation during the collection or transit time of the charge cloud.

1.3.3.1 Recombination

If the defect level is close to the middle of the band gap, the probability of electron and hole trapping are of similar magnitude and the trap acts as a recombination center. At sufficiently high deep-level defect concentration, the recombination causes an immediate charge loss from the charge cloud generated by the radiation even before the electron and hole clouds are separated by the electric field. Group IV elements such as Ge, Sn, and Pb and transition metals like Ti, V, Fe, and Ni are known impurities to introduce deep levels to CdTe. When the deep level of the impurity is in the lower half of the band gap below the Fermi level in semi-insulating CdTe or CdZnTe, it acts as a hole trap. For example, isolated Fe impurity with donor level 0.6 eV above the VBM is a hole trap. If the Fe concentration is high, the trapping of holes and the subsequent recombination of the trapped hole by trapping an electron reduce the size of the charge cloud and as a result deteriorate the proportionality of the signal amplitude to the photon energy. Despite this recombination effect and worsened detector performance, the electron mobility-lifetime product remains high (the Fe donor is not an electron trap) in Fe-doped semi-insulating CdTe and CdZnTe crystals.

1.3.3.2 Uniform Trapping

1.3.3.2.1 Trapping at Deep-Level Defects

Trapping at deep levels causes the trapped charge to be removed from the charge cloud for duration of the transit time of the carriers. As a result, the signal amplitude is reduced and the photopeak suffers a low-energy tailing. Ionized defects with a net charge are stronger traps due to the Coulomb attraction between the localized charge on the defect and the free carriers of the opposite sign. The capture cross section of neutral defects is about an order of magnitude lower than that of charged defects. The trapped state is the metastable state and the trapped carrier either recombines with free carriers of the opposite sign or the carrier escapes the trap by thermal excitation. The residence time of the trapped carriers τ_r is given by

$$\tau_r^{-1} = \nu \exp\left(-\frac{E_t}{kT}\right) \quad (1.3)$$

where
ν is the attempt to escape frequency of the charge, typically in the 10^{13} s^{-1} range
E_t is the ionization energy of the defect level measured from the VBM or CBM
k is Boltzmann's constant
T is the absolute temperature

Carriers liberated from deep traps make a small erroneous contribution to detector signals corresponding to photons that arrived later than the photon generating the trapped charge.

CdZnTe and CdTe Crystals for Medical Applications

Randomly distributed deep-level defects provide a constant trapping rate at thermal equilibrium that can be described by a well-defined carrier lifetime:

$$\tau^{-1} = v_{th} N_{active}^{t} \sigma \tag{1.4}$$

where
v_{th} is the thermal velocity of the carriers
N_{active}^{t} is the density of electrically active traps
σ is the capture cross section of the defect

It is to be pointed out that the active fraction of defects depends on the position of the Fermi level. Because the capture cross section of charged and neutral defects is very different, this is particularly important for deep-level defects close to the middle of the band gap where relatively minor differences in the compensation condition and the resulting shift in the equilibrium Fermi level position cause large change in the trapping probabilities and therefore lifetime of the free carriers.

We can consider the case of tin in CdTe to illustrate the behavior of deep-level defects. Sn incorporates at Cd sites to the CdTe lattice and introduces two deep donor levels. The doubly ionized donor state $Sn^{+1/+2}$ was measured to be at $E_c - 0.85$ eV by photo-EPR at 4.2 K [13]. Considering the shift of the band edges and the reduction of the band gap to 1.5 eV, the $Sn^{+1/+2}$ donor level is expected to be between 0.85 and 0.8 eV from the CBM at room temperature. Theoretical calculations suggest that the singly ionized level $Sn^{0/+1}$ is a few hundredths of an eV above the doubly ionized level $Sn^{+1/+2}$ [14]. For the purpose of this analysis, we assume that the singly ionization level $Sn^{0/+1}$ is at $E_c - 0.8$ eV and the ionization level $Sn^{+1/+2}$ is at $E_c - 0.9$ eV as shown in Figure 1.5.

It is to be emphasized that the occupation of the defect levels is governed by the Fermi distribution and portions of the defects are in different charge states. Figure 1.5 shows the predominant charge state of each defect level of Sn_{Cd} for three different

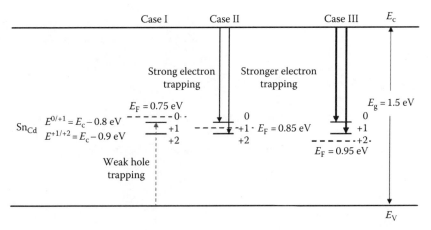

FIGURE 1.5 Charge state of Sn_{Cd} in CdTe at room temperature for three different positions of the Fermi level. Bold arrows indicate strong trapping, while dashed arrow indicates weak trapping.

positions of the Fermi level. For perfectly compensated CdTe, the Fermi level is at midgap around $E_c - 0.75$ eV and both levels are occupied by electrons and most of the Sn atoms are neutral (Case I). In this state, the defect has no effect on electron transport because the defect is a donor and cannot trap an electron when neutral. It can however trap holes (i.e., release an electron to the valence band) to become positively charge. Because the defect is neutral, the trapping cross section is low and the defect is a weak hole trap. If the Fermi level is lowered to $E_c - 0.85$ eV, the defect levels become predominantly singly ionized (+1) with a localized positive charge (Case II). The defect becomes a strong electron trap. By further decrease of the Fermi level to $E_c - 0.95$ eV, the defect becomes doubly ionized (+2) and the fraction of the defects in the ionized state significantly increases. This causes even stronger electron trapping and further decrease of the electron lifetime (Case III). The positively charged Sn_{Cd} defects are repulsive to holes so they do not influence hole transport.

The effect of the Fermi level position can be quite dramatic on carrier trapping at deep-level defects because of the exponential dependence of the Fermi distribution on the Fermi energy. For a donor with defect level E_D below the CBM, the concentration of ionized donors N_d^+ is given by

$$N_D^+ = N_D \frac{1}{1 + g_D \exp\left(-(E_D - E_F)/kT\right)} \quad (1.5)$$

where

E_D is the total concentration of donors
g_D is the degeneracy factor of the donor level
k is Boltzmann's constant
T is the absolute temperature

Combining this with Equation 1.4, it is easy to see that the carrier lifetimes are sensitive functions of the Fermi level position. This dependency is one of the reasons why it is so challenging to reproduce CdTe and CdZnTe crystals with predictable carrier transport properties.

1.3.3.2.2 Trapping at Shallow-Level Defects

Trapping at shallow-level defects, that is, those within 0.35 eV of the CBM and VBM, can also cause significant distortion in the detector signal amplitude if present in significant concentration. When the residence time of trapped carries and the lifetime of the carriers are of the same order of magnitude, the carrier effective velocity can be described by

$$v_{eff} = v \frac{\tau}{\tau + \tau_r} \quad (1.6)$$

where

v is the carrier velocity (either thermal or drift velocity)
τ is the carrier lifetime
τ_r is the residence time of the trapped carrier [15]

The carriers undergo frequent trapping and detrapping cycles that reduce their effective speed and cause a reduction of the signal rise time in the detector. When the signal rise time approaches or exceeds the peaking time of the amplifier, the signal amplitude is truncated and ballistic deficit occurs.

1.3.3.3 Nonuniform Trapping

The situation is significantly more complex when the carrier traps are spatially correlated rather than randomly dispersed in the lattice. This a common situation when trapping occurs at extended defects such as dislocations, subgrain boundaries, grain boundaries, twins, second-phase precipitates, and inclusions and when the point defects and small point-defect clusters form correlated clusters or are associated with extended defects [16].

It is relatively simple to correct for uniform trapping and resulting signal amplitude degradation in CdTe and CdZnTe detectors by electronic and software techniques. It is however significantly more difficult to correct for nonuniform trapping for spatially nonhomogeneous three-dimensional (3D) defect distributions. Techniques are being developed to use fine detector pixelation to perform a 3D signal correction and obtain a superior energy resolution from CdZnTe detectors [17]. Apart from requiring advanced custom electronics and significant software overhead, the challenge remains that these techniques cannot correct for nonuniform charge trapping within the pixel voxel.

An elegant analysis of nonuniform trapping caused by Te inclusions in CdZnTe crystals was developed based on the homogenization theory that incorporates fluctuations in the induced charge, that is, charge collection nonuniformities introduced by the random nature of the Te inclusion population [18]. The results from this general model clearly demonstrate the intricate distortions to pulse-height spectra induced by nonuniform trapping due to nonhomogeneous spatial distribution of the defects and due to the broad distribution of capture cross sections of the various defects.

1.3.3.4 Carrier Transport under High Photon Flux

State-of-the-art CT uses high X-ray flux and fast-response detectors to accumulate images at very high speed in order to surpass physiological limits such as the rate of the beating hearth. In these applications, the semiconductor detectors have to be able to operate under high-intensity radiation and have to respond proportionally to changes in the photon flux nearly instantaneously.

High-flux applications represent a significant challenge for CdTe and CdZnTe detectors. The challenge stems from the charge transport properties of the crystals and most importantly their drift mobility. The drift mobility controls the speed of the photon-generated electrons and holes through the detector:

$$v_{\text{drift}} = \mu_{\text{drift}} E \tag{1.7}$$

where

μ_{drift} is the drift mobility
E is the electric field in the device

If the electric field is constant across the detector, the drift velocity is also constant and the transit time of the carriers is given by

$$t_{tr} = \frac{L}{\mu_{drift} E} = \frac{L^2}{\mu_{drift} V} \qquad (1.8)$$

where
   L is the detector thickness
   V is the applied voltage

If the photon flux is high and multiple photons generate charge carriers during the transit time, the charge-generation rate is higher than the removal rate and an excess charge builds up in the detector. The excess charge is proportional to the photon flux and its magnitude changes with the radiation intensity. The detector is not in thermal equilibrium anymore but in a *dynamic equilibrium with the photon field*.

The buildup of the excess charge is not instantaneous and the electronic system in the semiconductor undergoes a *transient* when the radiation is turned on or when the radiation intensity changes before steady state is achieved. The temporal evolution of the transient is controlled by the carrier dynamics and the resulting changes in the internal electric field distribution in the detector.

In the absence of trapping (ideal detector), the excess charge is in the form of free carriers and easily swept out when the radiation source is removed. In real crystals, however, there are always defects acting as carrier traps and a significant portion of the radiation-induced charge carriers is trapped forming a *space charge*. The excess carriers and particularly the space charge have significant detrimental effect on charge transport in the crystal and the performance of the detector under intense radiation.

First, the space charge reduces the electric field in the bulk of the detector device. Figure 1.6 shows a cartoon illustrating the charge distribution in a semiconductor detector and the calculated steady-state electric field distribution in a CdZnTe detector irradiated with high flux of 60 keV monoenergetic X-rays at the cathode side [15]. Because the X-rays have an exponential absorption profile, most of the space charge is formed by trapped holes in the proximity of the cathode. However, once a steady state is reached between the radiation and the charge transport, a low-field *pinch* point develops in the nonuniform electric field. Electrons generated in the high-field region under the cathode move fast for a short distance until the pinch point. Electrons generated in the middle of the device beyond the pinch point move slowly in the very weak electric field. Electrons generated deep in the detector close to the anode move a short distance in a field region where the field strength is closest to the value of the initial constant electric field.

The collapse of the internal electric field in the middle of the device and the much reduced drift speed of the carriers enhance carrier trapping dramatically. As a result, the signal amplitude of the detector is dramatically reduced and the pulse-height spectra show a striking shift toward low energies: detector *polarization* occurs [15]. This is a catastrophic loss of performance and the detector is not a useful device in this state.

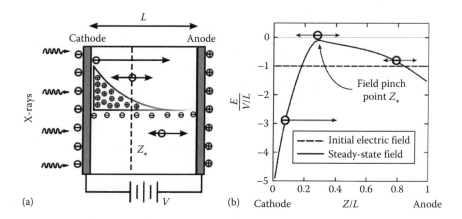

FIGURE 1.6 Space-charge distribution in a CdZnTe detector irradiated with high-intensity X-rays from the cathode side. (a) Steady-state electric field distribution. (b) Z_* indicates the location of the electric field pinch point. The circles illustrate the charge clouds generated by the X-ray photons at three different depths in the device. The arrows indicate the direction of the motion of electrons and holes. (From Bale, D.S. and Szeles, C., *Phys. Rev. B*, 77, 035205, 2008.)

It is important to understand that electrical polarization is not a permanent change in the detector performance and there is no physical or chemical change in the detector material and device. Because the detector is in a dynamic equilibrium with the photons, a large part of its performance recovers when the radiation source is removed. However, the recovery is not instantaneous because it takes finite time until the excess charge carriers are swept out of the device and the electric field distribution restored. In the presence of deep-level defects, the recovery time could be many orders of magnitude longer than the transit time corresponding to the constant electric field. The thermal emission of carriers from deep-level traps generates signals much later than the moment when the radiation is removed. These delayed signals cause afterglow in imaging devices. If the deep defect levels are close to the middle of the band gap, the residence time is hundreds of milliseconds (Figure 1.4). Coupled with the longer transit times caused by the reduced electric field, the recovery of the detector may take several seconds or even minutes. The recovery may be sped up by temporarily removing the bias voltage, applying heat or irradiating the detector with light. Numerous such techniques have been proposed and implemented in the industry to deal primarily with the catastrophic loss of performance due to polarization. It is however far more challenging to improve the response speed, stability, and uniformity of the detectors and to minimize the detrimental effect of signal delays caused by the carrier dynamics in the semiconductor crystals.

In order to improve the maximum X-ray and gamma-ray flux that CdTe or CdZnTe detectors can tolerate, the removal rate of the charge carriers has to be increased, or in other words, the transit time of the carriers has to be reduced. Because the drift mobility is a material property controlled by the carrier scattering mechanism in the crystal, it is not possible to modify. The transit time can be shortened by applying a higher electric field and using thinner detectors. While higher bias voltage always comes at the price of increased leakage current, and the resulting detector noise reducing the thickness of the detector is a particularly effective method of reducing the transit time because of the quadratic dependence shown in Equation 1.8. In order to preserve the detection efficiency, thin detectors are typically side irradiated and a number of such detector configurations have been proposed and under development for CdTe and CdZnTe detectors for CT and SPECT applications [19].

Controlling the detector response transients in CdTe and CdZnTe detectors is significantly more difficult. These transients are the function of carrier dynamics occurring as the detector attains a new steady state once the photon flux is changed. When the intensity of radiation is changed, the generation rate of electron–hole pairs is changed and the electronic system has to reach a new dynamic equilibrium through the drift, trapping, and detrapping of carriers. This process is controlled by the defect structure of crystals and the position of the Fermi level set by the electrical compensation applied during crystal growth. The carrier dynamics and the temporal evolution of the concentrations of free and trapped carriers and the internal electric field are fairly complex for CdTe and CdZnTe crystals containing multiple types of native and impurity defects. Spatially nonhomogeneous defect distribution in the crystals adds further complications and not just that it causes pixel-to-pixel response nonuniformity in the steady state (what is correctable in imaging applications) but it also causes pixel-to-pixel temporal response variation that is nearly impossible to correct for by electronic or software means.

Although application of CdTe and CdZnTe detectors to high-flux applications such as multienergy CT is a subject of very active research and development, it faces sizeable challenges today. Because of the complexities of carrier dynamics under high X-ray or gamma-ray fluxes, substantial improvements are needed in the perfection of the CdTe and CdZnTe crystals. It is safe to say that few orders of magnitude reduction of the residual defect density and great improvement of the spatial uniformity of defect distribution will be needed before this detector technology can be deployed in mainstream CT.

1.4 CRYSTAL GROWTH TECHNOLOGY

CdTe and CdZnTe crystals are typically grown by directional solidification from their melts or from solution. In melt-growth techniques, the CdTe or CdZnTe is melted few degrees above their melting points and very slowly solidified in a temperature gradient. The solution growth process is very similar except the melt is enriched in one of the constituents to form a solution with a lower melt temperature and crystallize the material at a lower temperature. The most often used techniques for the growth of CdTe and CdZnTe single crystals are Bridgman, gradient freeze, and electrodynamic gradient freeze growth that are melt-growth techniques and THM that is a solution growth technique.

Crystal growth is aiming at achieving atomic level perfection of the crystal lattice by controlling macroscopic parameters. This is fundamentally an impossible problem: the thermodynamics of the material system dictates the *formation of point defects* during crystallization. The unavoidable formation of defects during solidification and the ensuing *defect interactions* during cooling lead to a rich and complex defect structure in the crystals. In addition to point defects, the formation of extended structural defects occurs because the physical system lowers its total energy in the fields and forces imposed during crystal growth. These outside fields and forces are the driving forces of the extended-defect generation. Gaining increasing control over the atomic perfection of the crystals and the suppression of the proliferation of crystal defects is the continuing principal challenge of CdTe and CdZnTe crystal growth technologies.

The magnitude of the crystallization challenge cannot be sufficiently emphasized: we are aiming at controlling the *multitude of microscopic interfacial phenomena and parameters* with few far-field *macroscopic parameters* to *minimize the formation of crystal defects* at the growth interface and in the solidified crystal.

The principal crystal growth challenges of CdTe and CdZnTe can be categorized into three main areas: *parasitic nucleation*, *physical defect generation*, and *defect interactions*.

1.4.1 Parasitic Nucleation

Parasitic nucleation is the process where a microscopic cluster of atoms with a different orientation than the surrounding crystal is formed at the growth interface. Depending on the relative growth rates of the parasitic grain and the surrounding matrix, the cluster either grows into a macroscopic crystallite of a different orientation or is overcome by the surrounding crystal and retained as a microscopic defect. Solid-state recrystallization of the surrounding matrix may also occur completely eliminating the misoriented crystallite. Parasitic nucleation is caused by a few fundamental processes.

If the imposed solidification rate (i.e., rate of temperature gradient movement, pull rate) exceeds the *natural crystallization rate*, parasitic nucleation is guaranteed because the newly attached atoms have no sufficient time to diffuse along the growth interface and find their ideal position. The resulting local microscopic nonstoichiometry and subsequent atomic relaxation lead to the formation of parasitic crystallites.

A poor choice of the *crystal orientation relative to the direction of the temperature gradient* is another common source of parasitic nucleation. If the parasitic grains have higher natural growth rate than the surrounding crystal, they grow faster into large grains. Unfortunately for CdTe and CdZnTe, most crystal orientations have similar growth rates, and there is no orientation with a significantly faster growth rate that would help to stabilize the growth of a primary single crystal. In addition, the growth interface is almost never flat but typically has a constant or more commonly a varying curvature. The interface curvature is controlled by the 3D spatial distribution of the temperature gradient (the growth interface is perpendicular to the local temperature gradient). Because of this curved interface, the relative orientation of the crystal to the direction of the temperature gradient changes point by point along the growth interface. For a crystal growth process employing significant interface curvature, it is unavoidable to have segments on the growth interface where the orientation of the crystal relative to the temperature gradient is unfavorable and this section of the interface is prone to parasitic nucleation.

High interfacial *thermomechanical stress* is another cause of parasitic nucleation. High interfacial stress causes strain in the growing crystal at the interface. This strain is reduced if the crystal is deformed and dislocations are generated. Dislocations interact with each other to reduce the total energy of the system and form dislocation arrays equivalent to small-angle grain boundaries. If the stress is sufficiently high and persistent for extended periods of time, the dislocation density in the vicinity of the interface may be sufficiently high for the arrays to rearrange into large-angle grain boundaries and thereby segment the growth interface into several parasitic crystallites.

Finally, *instability and drift of the growth parameters* are another cause of parasitic nucleation. CdTe and CdZnTe melts exhibit a significant undercooling before solidification is initiated. Constitutional undercooling caused by the enrichment of the melt in Te in front of the growth interface is a particularly common phenomenon. This enrichment is caused by the incongruent solidification of CdTe and CdZnTe, which means that the composition of the solid is different from the composition of the liquid. If the temperature or pressure control system of the crystal growth equipment is not sufficiently stable, short few degrees drops in the heater temperature initiate crystal nucleation in the undercooled melt in front of the interface. The microscopic crystal nuclei formed in the melt are obviously of random orientation and, if they get attached to the growth interface, may grow into parasitic grains of different orientation.

Parasitic nucleation is the primary reason behind the polycrystalline nature of most CdTe and CdZnTe ingots. Due to the multitude of processes that can cause parasitic nucleation, it is difficult to avoid particularly in large-diameter (50–200 mm) ingots. Figure 1.7 shows a cross section of a typical 50 mm diameter CdZnTe ingot along with the corresponding orientation map obtained by electron backscattering diffraction (EBSD). Each color represents a different orientation. Thick black lines show large-angle grain boundaries. Thin lines show twin boundaries. This particular ingot consists of a large primary crystal albeit with significant twinning and a few parasitic grains at the periphery. There is a proliferation of densely packed grain boundaries in the top-right quadrant of the orientation map. This area probably contains a high density of low-angle grain boundaries and microcrystallites that are not resolved with the chosen spatial resolution of the EBSD mapping. Twins are errors in the stacking order of a crystal plane. They are typically not electrically

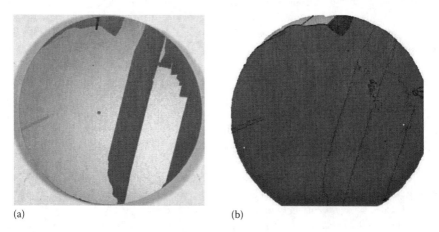

FIGURE 1.7 (**See color insert.**) (a) Cross section of a typical 50 mm diameter CdZnTe ingot and (b) the matching orientation map measured by EBSD.

active defects unless they are decorated with other defects such as impurities and second-phase precipitates. This ingot has a high single crystal mining yield if the twins are not electrically active or moderate-to-low single crystal yield if the twins are electrically active.

The initial and early stages of solidification pose the biggest challenge for single crystal growth. Without a seed crystal, the nucleation of the first crystallites is random and occurs either on the wall of the growth crucible or in case of significant *undercooling* in the molten material. The latter can be particularly severe in liquid CdTe and CdZnTe that are known to undergo as much as 20°C undercooling before crystallization initiates. The initial rapid solidification of the undercooled liquid leads to the formation of many small crystallites with random orientation. Even if the undercooling of the CdTe and CdZnTe melt is minimized, the numerous crystallites are nucleated at random locations at the crucible wall with random orientation. Crystallites with the highest growth rates survive and grow into dominant larger grains as crystallization proceeds.

Seeding is a time-tested technique to achieve controlled crystallization from the beginning of solidification and is used for the growth of many semiconductor crystals. Seeding helps to minimize the detrimental effects from undercooling and provides a single perfect crystal over the entire growth interface and a distinct favorable fast-growth orientation for stable crystallization from the very beginning of the solidification process. The one caveat is that any defects present in the seed crystal are replicated in the grown crystals. Seed crystal selection, characterization, and preparation are critically important for successful seeded growth. Poor choice of seed orientation, presence of defects and strain, and poor seed preparation can easily diminish all the benefits of seeding and may produce ingots with high defect concentration and even with polycrystalline structure.

1.4.2 Physical Defect Generation

It is impossible to produce crystals free of defects in practical bulk crystal growth processes employing directional solidification from a melt or solution at high temperature. Equilibrium point defects are always present in substantial concentration at the growth temperature as dictated by the thermodynamics of the material. A fraction of these defects are frozen in the crystals during cooling because at lower temperatures, the defects have insufficient diffusivity to attain thermal equilibrium. In addition, extended structural defects are nearly always generated because of the physical and thermal conditions imposed during crystal growth.

Instabilities in the thermodynamic parameters such as temperature, pressure, and constituent concentrations may cause the formation of second-phase particles such as Te inclusions.

Thermomechanical and constitutional stresses induced by the imposed temperature gradient and by the compositional variation in the ternary compound CdZnTe due to Zn distribution induce considerable strain to the growing crystal. If the stress exceeds the yield stress of the crystal, dislocations are generated to reduce the strain energy. The barrier to nucleation of dislocations is low in CdTe and CdZnTe due to the low stacking fault energy and low yield stress of the materials. Zn alloying increases the yield stress of the material but it is not sufficient to completely suppress dislocation generation and it comes at the price of the additional constitutional stress. The density of the dislocations depends on the various formation, multiplication, and annihilation mechanisms, and their interactions lead to dislocation networks and small-angle subgrain boundaries. As the density of dislocations in subgrain boundaries grows, they may rearrange to large-angle grain boundaries releasing some of the strain energy.

The presence of insoluble contaminant particles (such as particles from crucible and ampoule materials: quartz, graphite, alumina, etc.) in the CdTe and CdZnTe liquid may lead to complex dislocation structures if the particles are attached to the growth interface and grown into the crystal. Cd vacancy supersaturation may lead to the precipitation of vacancies and the formation of stacking faults and dislocation loops upon cooling.

Numerous growth parameters have conflicting effects on crystal growth: beneficial for some aspect of the process but detrimental for another aspect of the growth process. Temperature gradient is a pertinent example. High temperature gradient is usually beneficial to the stability of the crystallization process and helps to suppress parasitic nucleation and the formation of Te inclusions. On the other hand, the high gradient induces large thermomechanical stress and causes the generation of dislocations and subgrain boundaries.

1.4.3 Defect Interactions

Defects formed at the crystallization interface or within the body of the crystal during cooling migrate under the influence of thermodynamic parameters and imposed outside fields and forces. As the defects move across the crystal, they intersect each other's paths and interact with each other. Often the defect interactions lower the energy of the system by forming a new combined defect. The stress field of grain boundaries, subgrain boundaries, and dislocations attracts excess alloy constituents, dopants, and impurities to form boundary decorations.

For example, Te inclusions that are extensively studied extended defects in CdTe and CdZnTe are typically pinned to other defects such as grain boundaries, subgrain boundaries, and twin planes forming correlated clusters of inclusions and causing nonuniform trapping and distortion of charge transport. Figure 1.8 shows correlated clusters of Te inclusions in CdZnTe trapped at dislocations,

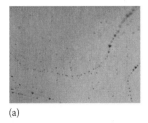
(a)
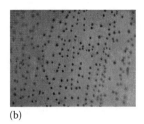
(b)
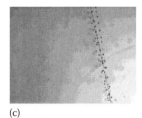
(c)

FIGURE 1.8 (a) Correlated clusters of Te inclusions in CdZnTe trapped along dislocations, (b) grain boundaries, (c) and twin planes. The size of the Te inclusions in these infrared microscopy images is in the 10–70 μm range.

CdZnTe and CdTe Crystals for Medical Applications

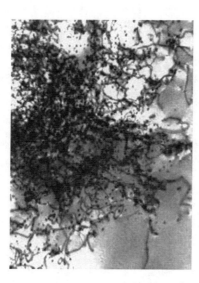

FIGURE 1.9 Dislocation network surrounding a Te inclusion in CdZnTe shown by TEM. The dislocations are decorated with Te precipitates seen as dark spots. The area of the image is about 3 μm × 5 μm.

grain boundaries, and twin planes. The inclusions seen in the infrared microscopy images are dark triangular or hexagonal shape features and their size is in the 10–70 μm range.

Figure 1.9 shows a transmission electron microscopy image of a dislocation network surrounding a Te inclusion in CdZnTe (the Te inclusion is outside of the visible area in the direction of the top-left corner of the image). The dense network of dislocations is probably formed when the Te inclusion froze out during cooling of the crystal. Because the melting temperature of Te is low (450°C), the Te inclusions are liquid droplets at higher temperatures. Because the thermal expansion coefficient of the Te is larger than the host CdZnTe matrix, the large stress induced to the crystal during Te solidification gave way to dislocation generation. Notice that many of the dislocations are terminated by dark spots. These are Te precipitates of about 50 nm size. Unlike to their much larger cousins Te inclusions that are generated at the crystallization interface, Te precipitates are formed during cooling. The excess Te atoms dissolved in the matrix at high temperature are rejected by the crystal as the temperature lowered: CdTe and CdZnTe have a retrograde solubility of Te. The excess Te atoms form the small precipitates. If there are other defects in the crystals, the Te precipitates preferentially nucleate along those defects. As illustrated in Figure 1.9, the end result is a correlated cluster of dislocations and Te precipitates.

There is a high equilibrium concentration of native point defects at the growth temperature. As the CdTe and CdZnTe ingots are cooled from the solidification temperature, the crystals get supersaturated by the native point defects. The diffusion rate of the defects also rapidly decreases with temperature. At certain temperature, the defects cannot diffuse sufficient distances anymore to reach the ingot surface and maintain the equilibrium defect concentration. Because the average defect-to-defect distances are much shorter, the point defects form pairs and triplets and precipitate to small clusters to maintain their equilibrium concentration in the host matrix. Similarly they are attracted to extended structural defects where they get trapped. By the time the crystals are cooled to room temperature, there is an excess of point-defect pairs and small clusters of native defects above their equilibrium concentrations.

Impurities undergo a similar process and one typically finds that a significant fraction of the total impurities are trapped at extended defects such as inclusions, precipitates, grain boundaries, and dislocations.

Cooling the crystals from the growth temperature to room temperature is an important element of the crystal growth process. It is easy to see that too rapid cooling freezes in a higher concentration of defects than slow cooling. Typically slow cooling or annealing at low temperature (100°C–300°C) is required to provide sufficient time for the defects to reach the crystal surface or extended defects and to remove them from the pool of electrically active defects. This defect relaxation is a critical phase of the crystal growth process for achieving good charge transport in the crystals.

1.4.4 Annealing

Annealing of the crystals following the growth is often proposed to reduce the concentration of defects and improve the charge transport properties of CdTe and CdZnTe crystals. Annealing can be implemented as part of the cooling procedure (in situ ingot annealing) or performed in a separate process after the ingots are sliced to wafers (wafer-level annealing). While annealing in controlled vapor pressures of the constituents is an effective way to dissolve and eliminate Te inclusions and precipitates and adjust the residual concentration of native defects, it is ineffective in eliminating most extended defects such as dislocation networks, subgrain boundaries, twins, and grain boundaries. It has to be pointed out also that annealing may also cause deterioration of the charge transport properties of the crystals because impurities trapped at Te inclusions, precipitates, and other extended defects that are liberated and released to the host crystal may perturb electrical compensation and enhance carrier trapping.

In summary, CdTe and CdZnTe crystals typically have a rich and complex defect structure stemming from the numerous defect generation mechanisms and defect interactions operating during crystallization and cooling of the crystals. Many of these defects are electrically active and trap electrons and holes hampering the transport of carriers. It is also more typical than not that the defects are forming correlated clusters resulting in nonuniform trapping, electric field, and carrier transport what ultimately leads to nonuniform detector response.

1.4.5 Status of Crystal Growth

CdTe and CdZnTe crystal growth made a steady progress in the past three decades. The growth of 100 mm diameter CdTe ingots that are nearly entirely single crystal has been demonstrated by the THM technique [20]. The growth of large CdZnTe single crystals proved to be considerably more difficult due to the complexities induced by Zn. The highest perfection CdZnTe single crystals are grown by the vertical gradient freeze (VGF) technique. CdZnTe ingots as large as 125 mm diameter with high single crystal yield and low defect density have been demonstrated with the VGF technique mostly for infrared detector substrate applications [21]. Interestingly, despite numerous development efforts, the VGF technique did not yet gain a foothold in radiation detector crystal manufacturing. Much of the CdZnTe crystals for X-ray and gamma-ray radiation detectors are produced by the THM technique although the high-pressure and horizontal Bridgman techniques are also used by various vendors [22].

1.5 SUMMARY

CdTe and CdZnTe materials and crystal growth technology demonstrated a remarkable progress in the past few decades and enabled the development and deployment of room-temperature photon-counting X-ray and gamma-ray detectors and imaging arrays in a number of medical applications. Utilization in low- and moderate-flux applications such as cardiac SPECT, MBI, and DEXA is expected to grow as the performance and availability of these detectors improve and their cost decreases. The performance requirements in numerous medical imaging applications particularly in applications where high photon fluxes are used and high-speed detectors are needed are still challenging for this detector technology.

As with other semiconductor detector technologies, the performance of CdTe and CdZnTe detectors is controlled by the electrically active point defects and extended structural defects in the crystals. Our understanding of the nature of these defects, their formation, diffusion, and electrical properties as well as their effects on the charge transport properties of the crystals significantly expanded in the past decade. Similarly impressive progress is occurring in understanding the crystallization phenomena, defect formation processes, and interactions during solidification and cooling of the crystals to room temperature. The maturing scientific understanding of the crystallization and defect phenomena fuels further advancement of materials and crystal growth technologies and facilitates the development of exciting new CdTe and CdZnTe detectors with enhanced performance.

REFERENCES

1. K. Erlandsson, K. Kacperski, D. van Gramberg, and B. F. Hutton, Performance evaluation of D-SPECT: A novel SPECT system for nuclear cardiology, *Physics in Medicine and Biology*, 54, 2635–2649, 2009.
2. C. B. Hruska and M. K. O'Connor, CZT detectors: How important is energy resolution for nuclear breast imaging? *Medical Physics*, 21(Suppl. 1), 72–75, 2006.
3. J. Wear, M. Buchholz, R. K. Payne, D. Gorsuch, J. Bisek, D. L. Ergun, J. Grosholz, and R. Falk, CZT detector for dual-energy x-ray absorptiometry (DEXA), in *Proceedings of the SPIE 4142, Penetrating Radiation Systems and Applications II*, San Diego, 2000.
4. K. Spartiotis, J. Havulinna, A. Leppanen, T. Pantsar, K. Puhakka, J. Pyyhtia, and T. Schulman, A CdTe real time imaging sensor and system, *Nuclear Instruments and Methods A*, 527, 478–486, 2004.

5. J. P. Schlomka, E. Roessl, R. Dorscheid, S. Dill, G. Martens, T. Istel, C. Baumer et al., Experimental feasibility of multi-energy photon-counting K-edge imaging in pre-clinical computed tomography, *Physics in Medicine and Biology*, 53, 4031–4047, 2008.
6. Y. G. Cui, T. Lall, B. Tsui, J. H. Yu, G. Mahler, A. Bolotnikov, P. Vaska et al., Compact CdZnTe-based gamma camera for prostate cancer imaging, in *Proceedings of the SPIE 8192, International Symposium on Photoelectronic Detection and Imaging: Laser Sensing and Imaging; and Biological and Medical Applications of Photonics Sensing and Imaging*, Beijing, China, 2011.
7. H. Chatrath and D. K. Rex, Potential screening benefit of a colorectal imaging capsule that does not require bowel preparation, *Journal of Clinical Gastroenterology*, 48(1), 52–54, 2014.
8. R. Triboulet, Fundamentals of the CdTe and CdZnTe bulk growth, *Physica Status Solidi C*, 2(5), 1556–1565, 2005.
9. J. H. Greenberg, V. N. Guskov, M. Fiederle and K.-W. Benz, Experimental study of non-stoichiometry in $Cd_{1-x}Zn_xTe_{1-\delta}$, *Journal of Electronic Materials*, 33(6), 719–723, 2004.
10. S. H. Wei and S. B. Wang, Chemical trends of defect formation and doping limit in II-VI semiconductors: The case of CdTe, *Physical Review B*, 66, 155211–155221, 2002.
11. K. Biswas and M. H. Du, What causes high resistivity in CdTe, *New Journal of Physics*, 14, 063020, 2012.
12. C. Szeles, Advances in the crystal growth and device fabrication technology of CdZnTe room temperature radiation detectors, *IEEE Transactions on Nuclear Science*, 51(3), 1242–1249, 2004.
13. W. Jantsch and G. Hendorfer, Characterization of deep levels in CdTe by photo-EPR and related techniques, *Journal of Crystal Growth*, 101, 404–413, 1990.
14. J. E. Jaffe, Computational study of Ge and Sn doping of CdTe, *Journal of Applied Physics*, 99, 33704–33708, 2006.
15. D. S. Bale and C. Szeles, Nature of polarization in wide-bandgap semiconductor detectors under high-flux irradiation: Application to semi-insulating $Cd_{1-x}Zn_xTe$, *Physical Review B*, 77, 035205–0352021, 2008.
16. A. E. Bolotnikov, S. O. Babalola, G. S. Camarda, H. Chen, S. Awadalla, Y. Cui, S. U. Egarievwe et al., Extended defects in CdZnTe radiation detectors, *IEEE Transactions on Nuclear Science*, 56(4), 1775–1783, 2009.
17. F. Zhang, Z. He, and C. E. Seifert, A prototype three-dimensional position sensitive CdZnTe detector array, *IEEE Transactions on Nuclear Science*, 54(4), 843–848, 2007.
18. D. S. Bale, Fluctuations in induced charge introduced by Te inclusions within CdZnTe radiation detectors, *Journal of Applied Physics*, 108, 24504–24512, 2010.
19. I. Takahashi, T. Ishitsu, H. Kawauchi, J. Yu, T. Seino, I. Fukasaku, Y. Sunaga, S. Inoue, and N. Yamada, Development of edge-on type CdTe detector module for gamma camera, *IEEE Nuclear Science Symposium Conference Record (NSS/MIC)*, pp. 2000–2003, 2010.
20. H. Shiraki, M. Funaki, Y. Ando, A. Tachibana, S. Kominami, and R. Ohno, THM growth and characterization of 100 mm diameter CdTe single crystals, *IEEE Transactions on Nuclear Science*, 54(4), 1717–1723, 2009.
21. R. Hirano and H. Kurita, Bulk growth of CdZnTe/CdTe single crystals, in *Bulk Crystal Growth of Electronic, Optical and Optoelectronic Materials*, P. Capper, Ed., John Wiley & Sons, Chichester, U.K., 2005, pp. 241–268.
22. H. Chen, S. A. Awadalla, J. Mackenzie, R. Redden, G. Bindley, A. E. Bolotnikov, G. S. Camarda, G. Carini, and R. B. James, Characterization of traveling heater method (THM) grown Cd0.9Zn0.1Te crystals, *IEEE Transactions on Nuclear Science*, 54(4), 811–816, 2007.

2 Monte Carlo Modeling of X-Ray Detectors for Medical Imaging

Yuan Fang, Karim S. Karim, and Aldo Badano

CONTENTS

2.1 Introduction 27
2.2 X-Ray Detector Technologies 28
    2.2.1 Scintillator-Based Detectors 28
    2.2.2 Semiconductor-Based Detectors 29
2.3 Modeling Approaches 29
    2.3.1 Photon–Electron Interactions 30
    2.3.2 Electron–Hole Pair Transport 31
    2.3.3 Coupled Monte Carlo Simulation 32
    2.3.4 Analytical Methods 32
2.4 Monte Carlo Simulation of Radiation Transport 32
    2.4.1 Theory 32
        2.4.1.1 Charge Generation 32
        2.4.1.2 Recombination and Trapping 34
    2.4.2 Implementation 35
    2.4.3 Indirect Detectors 38
2.5 Applications 41
    2.5.1 Pulse-Height Spectroscopy 41
    2.5.2 Information Factor and Detective Quantum Efficiency 41
2.6 Summary 43
Acknowledgments 43
References 44

2.1 INTRODUCTION

Semiconductor X-ray detectors are important components of medical imaging systems and are used in a wide range of modalities including general radiography, full-field digital mammography [1], and computed tomography [2]. Characterization of semiconductor X-ray detectors provides insight into performance limitations and guides the development and optimization of imaging systems. By employing semiconductive materials to convert X-rays directly into electric signals, these detectors allow for good energy resolution, high efficiency, and high carrier yield [3].

Monte Carlo (MC) simulation is a statistical numerical technique that relies on random numbers to sample models that describe stochastic physical processes in radiation transport. This technique provides the ability to manage very complex models using well-known and simpler atomic interaction models. Using MC methods, the mean value of the outcome of a stochastic process can be estimated by integrating the results of many random trials. Due to these inherent advantages, MC methods can be used for the study of charge generation and transport in semiconductor materials, for design validation and optimization of imaging systems.

In this chapter, direct and indirect detection methods using semiconductor and scintillator materials are presented to give background on X-ray detection materials used in medical imaging applications (Section 2.2). A summary of modeling approaches using MC methods are covered (Section 2.3). The theory and implementation of a detailed MC model for a direct X-ray detector is then described, with parallels to indirect detectors for scintillator materials (Section 2.4). Practical applications of the model and simulation results are presented (Section 2.5).

2.2 X-RAY DETECTOR TECHNOLOGIES

2.2.1 Scintillator-Based Detectors

In scintillator-based detectors, a phosphor scintillator converts X-ray photons into multiple optical light photons detected as electric signals in a photodiode or photomultiplier tube, hence an indirect conversion process [4]. Figure 2.1a shows the structure of a scintillator detector.

One of the advantages of the indirect detection method is the high absorption efficiency of the scintillator material (e.g., CsI) capable of absorbing a high percentage of incident X-ray photons. However, one disadvantage of the indirect method is the loss of resolution due to isotropic generation of optical photons, shown in Figure 2.1a. A thinner scintillator material can be used to limit the effect of spreading, at a cost of reduced absorption efficiency of incident X-rays. Scintillators with columnar structures are also used to confine and reduce the spreading of optical photons.

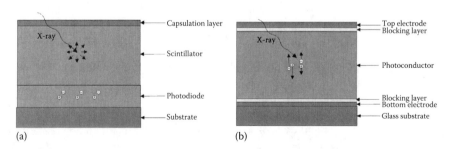

FIGURE 2.1 (a) Scintillator-based indirect digital X-ray imaging detector. (b) Semiconductor-based direct digital X-ray imaging detector.

FIGURE 2.2 (**See color insert.**) Commercial a-Se semiconductor direct X-ray detector for full-field digital mammography. (From http://www.anrad.com/products-direct-xray-detectors.htm, Analogic, Direct Conversion X-ray Detectors, accessed August 12, 2013.)

2.2.2 Semiconductor-Based Detectors

In semiconductor-based detectors, X-ray photons are absorbed in the photoconductor and converted directly into charge carriers called electron–hole pairs (EHPs). Figure 2.1b is an illustration of the direct conversion method. As X-ray photons are absorbed in the photoconductive material, many EHPs are generated near the region of interaction and are eventually collected at the electrodes.

Due to the direct conversion process, the resolution of semiconductor X-ray detectors depends only on the spreading of incident X-ray photons, secondary high-energy electrons, and EHPs. In general, photoconductive materials used in semiconductor detectors have a lower atomic number compared to scintillators and require a thicker detector to absorb the same amount of incident X-rays. Often, a biasing voltage is used to collect the EHPs, and blocking or (electron/hole) transport layers are used to prevent leakage. Figure 2.2 is a commercially available amorphous selenium (a-Se) direct X-ray detector developed by ANRAD Corporation.

2.3 MODELING APPROACHES

The complete signal formation process in semiconductor X-ray detectors from incident X-rays to electric signal can be divided into four subprocesses: incident X-ray interactions, secondary electron interactions, EHP generation, and charge transport, illustrated in Figure 2.3. Incident X-ray photons can interact within the semiconductor material through Rayleigh scattering, Compton scattering, and photoelectric effect. High-energy electrons can be generated from photoelectric absorption and Compton scattering events, where the electron's kinetic energy is deposited in the semiconductor via inelastic scattering events. Models for the generation of EHPs include sampling algorithms and spatial distribution of EHPs through calculation of burst and thermalization distances. The charge transport model should include recombination

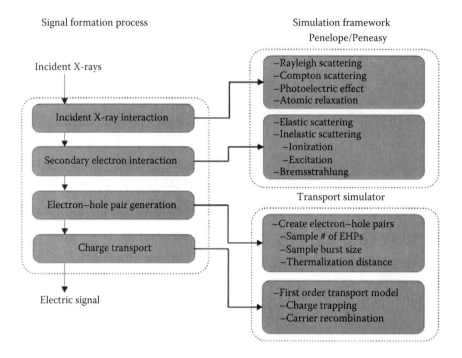

FIGURE 2.3 Block diagram of the signal formation process in semiconductor X-ray detectors including incident X-ray interaction, secondary electron interaction, electron–hole pair generation and charge transport. (Reprinted from Fang Y. et al., Monte Carlo simulation of amorphous selenium imaging detectors, *Proc of SPIE*, 7622, 762214, 2010. With permission.)

and trapping effects. Photon–electron and EHP interactions can be simulated separately or coupled together in MC models. This section briefly describes some existing methods available for modeling direct X-ray imaging detectors.

2.3.1 Photon–Electron Interactions

Incident X-ray photon and secondary electron interactions can be modeled with a number of available MC simulators: Penetration and Energy Loss of Positrons and Electrons (PENELOPE), Electron Gamma Shower National Research Council (EGSnrc), Geometry and Tracking (Geant4), Monte Carlo N-Particle, electronic transport, internal transcribed spacer 3, and Fluktuierende Kaskade [6–12]. Figures 2.4a and b show the photon and electron interaction cross sections in selenium, generated with PENELOPE 2006 [8]. The interactions are a function of particle energy and material properties. For the X-ray energy range of medical applications, the main interaction mechanisms are Rayleigh scattering, Compton scattering, and photoelectric absorption. For electrons, the main mechanisms are elastic scattering, inelastic scattering, and Bremsstrahlung. The total interaction cross section can be computed as the sum of the cross sections for all possible interaction mechanisms and is used to compute the mean free path required for sampling the location of scattering events in the material.

Existing MC models offer advanced geometric packages that allow for simulation of complex detector geometries and experimental setups. The interaction models

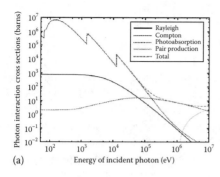 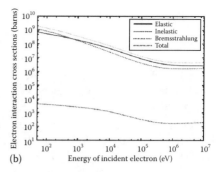

FIGURE 2.4 (**See color insert.**) (a) PENELOPE photon interaction cross sections in selenium from 100 eV to 10 MeV. (b) PENELOPE electron interaction cross sections in selenium from 100 eV to 10 MeV. *Note*: 1 barn = 10^{-24} cm^2. (Reprinted from Fang Y. et al., Monte Carlo simulation of amorphous selenium imaging detectors, *Proc of SPIE*, 7622, 762214, 2010. With permission.)

and cross sections have already been benchmarked and validated with established databases and offer an accurate model for the simulation of various interactions. MC models require simulation of many histories in order to reach high accuracy and low variance. Some MC codes allow for different modes of simulation such as condensed, detailed, and mixed. In condensed simulation of electrons, many soft interactions that do not change significantly the direction and energy of the particle are reproduced into a single interaction using multiple scattering theories to reduce simulation time. In the case where interaction locations and small energy changes are needed, the detailed simulation mode can be used to track all interactions by the electron. A mixed simulation mode can be used to optimize simulation detail and runtime. One limitation of most available MC simulators for modeling of semiconductor detectors is the lack of simulation libraries and models for generation and transport of EHPs.

2.3.2 Electron–Hole Pair Transport

Some custom MC simulators have been developed to focus on modeling EHP interactions [13,14]. The effect of trapping and recombination of EHPs on sensitivity reduction and ghosting [15] and time-of-flight (TOF) simulations of EHP to determine the density of state in a-Se [14] has been previously studied. The sensitivity reduction in a-Se detectors was found to depend on different detector operating conditions such as applied electric field, X-ray spectrum, and photoconductor thickness in turn affecting recombination and trapping inside the detector. The TOF simulation takes into account carrier drift and trapping.

Custom EHP simulators allow for focused studies of carrier transport in semiconductor X-ray detectors and offer significant flexibility for the implementation of complex recombination and trapping models. Compared to detailed photon–electron simulations, in some cases, exponential attenuation models for X-ray photons are used assuming complete absorption of incident energy for carrier generation and 1D model for EHP transport. This ignores the charge spreading due to high-energy photoelectric and Compton secondary electrons, the lateral spreading of EHPs due to diffusion, and noise in the detector response from Compton scattering and fluorescent X-rays.

2.3.3 COUPLED MONTE CARLO SIMULATION

The coupled simulation method takes advantage of the previous two methods, by combining the simulation of X-ray radiation with EHP transport. For example, energy deposition events in the semiconductor or scintillator material can be simulated with an available MC simulator in combination with a custom simulator for EHPs. The Monte cArlo X-ray, electroN Transport Imaging Simulation (MANTIS) package is an efficient and flexible simulation tool for the research and development of scintillator-based indirect radiation systems. The package consists of PENELOPE for photon–electron interactions including X-ray scattering, and DETECTII routines for the simulation of optical photons allow for detailed studies of 3D optical blur with realistic columnar model.

Combined simulation can be used to simulate the complete signal formation process in X-ray detectors, by taking advantage of existing validated MC simulators for photon–electron interactions and allowing for significant customization of EHP transport models. However, in-depth knowledge and modifications to the existing simulation packages are often required to efficiently interface the codes. A high number of simulation histories are required to achieve low variance for studies such as the point response function needed for modulation transfer function (MTF) and detective quantum efficiency (DQE) calculations. These limitations drive the need to further improve simulation efficiency, including parallelization implementations with computer clusters and utilization of graphical processing units for further speedups [16].

2.3.4 ANALYTICAL METHODS

It is important to note that even though this chapter focuses on MC methods, analytical methods have also been widely used for modeling X-ray detectors. For example, the small pixel effect for minimization of trapping of slow carriers on the electric signal has been previously studied by Barrett et al. [17]. This work assumes a homogeneous slab, where the current induced on each pixel electrode is calculated via the Shockley–Ramo theorem, and reductions in low-energy tails of the pulse-height spectra (PHS) are validated with experimental results. Compared to MC methods, analytical methods do not require a long simulation time and are efficient at solving radiation transport problems with simple electric field distributions that can be mathematically represented. However, analytical methods can be limited in modeling 3D charge distributions of secondary carriers inside the detector material and taking into account the stochastic events that affect radiation transport, such as trapping and recombination inside a nonuniform electric field.

2.4 MONTE CARLO SIMULATION OF RADIATION TRANSPORT

2.4.1 THEORY

2.4.1.1 Charge Generation

For optical photon detection, only one EHP is generated, and the carriers lose their initial kinetic energy and separate by a finite distance r_0 in a thermalization. This distance can be estimated by the photon energy, hv, and applied electric field,

E_{app}, using the Knight–Davis equation [18], where D is the diffusion constant, E_{gap} is the bandgap of the semiconductor material, ε is the dielectric constant, and e is the elementary charge:

$$\frac{r_o^2}{D} = \frac{(hv - E_{gap}) + e^2/4\pi\varepsilon r_o + eE_{app}r_o}{hv_p^2}. \quad (2.1)$$

Compared to optical detectors, the charge generation models for radiation detectors is more complex due to the generation of many EHPs by a single incident photon. Photoelectric absorption is the dominant X-ray interaction mechanism in the energy range of interest and creates a secondary photoelectron with most of the energy of the initial X-ray capable of ionizing the material and producing many EHPs in the detector. X-ray photons that are Compton scattered can also produce energetic electrons capable of creating many EHPs; however, the particle's kinetic energy is lower compared to the photoelectron. As the high-energy ionizing electron travels through the detector material, it gradually loses energy through inelastic scattering, and the energy lost, E_d, is deposited in the semiconductor material leading to the generation of many EHPs. The mean number of EHP generated, $\bar{N}_{EHP}$, can be estimated via Poisson sampling from the energy deposited and the material ionization energy, W_0:

$$\bar{N}_{EHP} = \frac{E_d}{W_o}. \quad (2.2)$$

The ionization energy for semiconductors was originally developed by Klein [19], with the following sampling equation:

$$W_o \approx K * E_{gap} + rhv_p, \quad (2.3)$$

where
  hv_p is the phonon energy
  r is a uniform random number between 0 and 1, representing the ionization and photon emission components

The constant K is found to be 2.8 for crystalline materials in the semiempirical formula and to be 2.2 for amorphous materials [20].

Several models have been developed to model the carrier generation process in silicon. Some models assume all the EHPs generated in a sphere following either Gaussian or uniform distribution [21–23], while others use MC simulations of a large number of electron tracks to estimate the center of gravity and uniformly distribute portions of the photon energy into a bubble and a line [24]. In silicon, W_0 is not field dependent, and the dominant effect of charge sharing is diffusion of carriers. However, in a-Se, carrier drift also plays a major role due to the field dependence of carrier generation and transport. The concept of EHP bursts is proposed for modeling carrier generation in a-Se. A burst is defined as the cloud (spatiotemporal distribution) of electrons and holes generated after a local deposition of energy [25]. Energy deposited in electron inelastic collisions with outer-shell electrons can lead to excitation of plasma waves and generate multiple EHPs [26]. These pairs constitute a burst, and the burst size is dependent on the energy of the incident particle and the material plasma frequency. According to Bohr's adiabatic criterion [27], the burst size, r_b, can be approximated using the following expression:

$$r_b = \frac{\upsilon}{\omega_{pe}}, \quad (2.4)$$

where

υ is the velocity of the incident particle
ω_{pe} is the plasma frequency, dependent on the material electron mass and density

The concept of a burst is introduced in conjunction to the thermalization of carriers, in order to provide a 3D distribution model for EHP generation.

2.4.1.2 Recombination and Trapping

There are two models to study recombination of carriers in a-Se: geminate and columnar recombination. Geminate recombination is used by Onsager to model EHP recombination due to optical photons and assumes carriers can only recombine with their original geminate pair. Columnar recombination occurs when the high-energy electron produces EHPs continuously in a column surrounding its track, and carriers from different interactions recombine in a columnar fashion. Our model takes into account both processes, by considering both geminate and columnar recombination in bursts. Recombination can occur between any electron and hole traveling toward each other, and trapping can occur when an electron or hole reaches a lower-energy state due to material impurities. The drift component takes into account both the applied electric field, $\vec{E}_{app}$, and the Coulomb field due to other charge carriers. For the ith charge carrier, the resulting electric field acting on it is given by

$$\vec{E}_i = \vec{E}_{app} + \sum_{j \neq i} \frac{1}{4\pi\varepsilon} \frac{e_j}{r_{ij}^2} \hat{r}_{ij}, \quad (2.5)$$

where

r_{ij} is the separation distance between charge carrier i and j
$\hat{r}_{ij}$ the field direction vector

In turn, the displacement in x, y, and z direction due to drift can be found:

$$\Delta x_{drift} = \mu E_x \Delta t, \quad (2.6)$$

where

μ is the carrier mobility
E_x is the x component of the electric field
Δt is the simulation time step

The components of the y and z directions can be found similarly. The diffusion component can be found by sampling the polar and azimuth angles from a uniform distribution, where the diffusion distance is given by $\sqrt{(6D\Delta t)}$ [28]. And the total displacement in each direction is a sum of the drift and diffusion components:

$$\Delta x = \Delta x_{\text{drift}} + \Delta x_{\text{diffusion}}. \qquad (2.7)$$

During transport, both drift and diffusion of carriers are calculated at each time step. The drift component depends on the carrier mobility, the electric field acting on the carrier, and the simulation time step. The diffusion component depends on the diffusion coefficient and the time step as shown in Equations 2.5 through 2.7. Depending on the material properties, the carrier mobilities may differ and the drift and diffusion components are affected, causing the carrier to travel faster/slower in the semiconductor material.

Many trapping effects have been modeled previously in 1D (thickness) for a-Se detectors [15]. These include deep trap, shallow trap, trap releasing, trap filling, and trap center generation due to incident X-rays. Deep and shallow trapping differ in the trapping time of carriers. Deep traps have long trapping times on the order of seconds to minutes, while shallow traps may release carriers in fractions of a microsecond or less. For simulation purposes, when a carrier is trapped in a deep trap, it is considered lost. However, when a carrier is trapped in a shallow trap, the release of this trapped carrier (perhaps in subsequent exposures) can also contribute to the detected signal. As EHPs start to move in the material and get trapped, the number of available traps decreases as a function of time, X-ray exposure, and carrier concentration. At the same time, a competing process of trap center creation is occurring due to X-ray bombardment of the semiconductor material. The current implementation of trapping uses a simple model that only considers deep trapping. The probability of trapping, P_{trapping}, can be calculated as [15]

$$P_{\text{trapping}} = 1 - e^{-\Delta t / \tau_{\text{trapping}}}, \qquad (2.8)$$

where τ_{trapping} is the trapping time. Constant trapping times are used for electrons and holes, to give an estimate of the average carrier lifetime and the effect of applied electric field on carrier trapping probabilities in the semiconductor material. The probability of carrier trapping depends on the total carrier transit time, and an applied electric field can be used to collect carriers from the interaction site to the appropriate electrodes.

2.4.2 Implementation

An MC transport code, ARTEMIS (pArticle transport, Recombination, and Trapping in sEMiconductor Imaging Simulation), was developed for the purpose of simulation of the signal formation process in direct X-ray detectors [29]. Various functions are

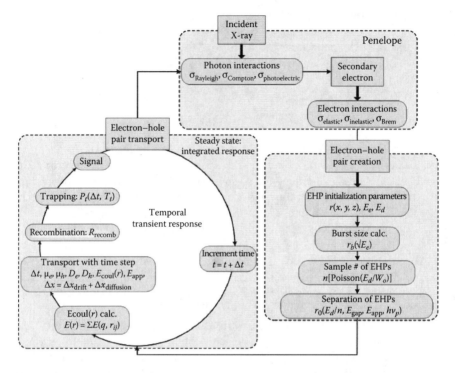

FIGURE 2.5 Flow chart for the simulation of the signal formation process in semiconductor X-ray detectors. Simulation of photon and secondary electron with PENELOPE is coupled with transport for detailed spatiotemporal simulation of electron–hole pairs. (Reprinted from Fang Y. et al., *Med. Phys.*, 39, 308, 2012. With permission.)

implemented to model the physics outlined in the theory section. The flow diagram for the implemented simulation framework is shown in Figure 2.5.

X-ray photon and secondary electron interactions in the presence of an external electric field are modeled by PENELOPE [8], and the locations of inelastic electron interactions with energy deposition are coupled with the transport routines for EHP simulations. Figure 2.6 shows the photon and electron particle tracks of 100 keV monoenergetic X-rays. Figure 2.6a depicts the absorption of a pencil beam of X-ray photons perpendicularly incident on the a-Se detector (in green). Most photons are absorbed in the center of the detectors, and the off-center photons are due to Compton scattering and fluorescence. Figure 2.6b is a close-up showing the secondary electrons move in random walk and deposit energy at random locations in the photoconductor (in red).

To further show the energy deposition events as high-energy electron lose kinetic energy in the semiconductor material, Figure 2.7a and b depict the single high-energy electron track produced by a 40 and 140 keV X-ray photon. The bubble size is largest at the beginning of the track where the high-energy electron is created and gradually decreases as energy are deposited in the semiconductor material. PENELOPE has been modified to take into account the effect of electric field for high-energy electron interactions.

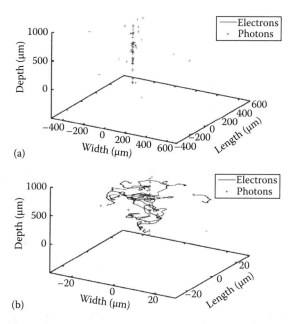

FIGURE 2.6 (See color insert.) (a) Particle track of 100 keV incident photons (100 histories) in selenium. (b) Close-up of (a). (Reprinted from Fang Y. et al., Monte Carlo simulation of amorphous selenium imaging detectors, *Proc of SPIE*, 7622, 762214, 2010. With permission.)

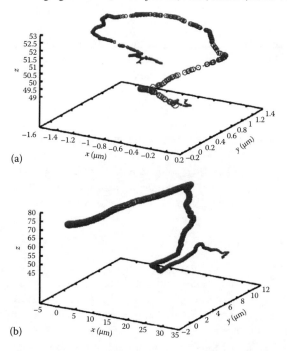

FIGURE 2.7 (a) 3D bubble plot of energy deposition events by secondary electrons in PENELOPE by a 40 keV photon. (b) 140 keV photon.

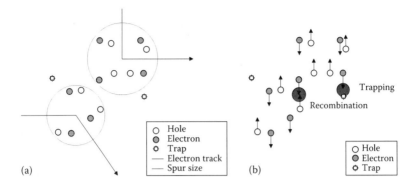

FIGURE 2.8 (a) Generation of electron–hole pairs from inelastic electron interactions, with varying burst size and thermalization distance. (b) Transport of electron–hole pairs, charged carriers can be lost due to recombination and trapping. (Reprinted from Fang Y. et al., Monte Carlo simulation of amorphous selenium imaging detectors, *Proc of SPIE*, 7622, 762214, 2010. With permission.)

Figure 2.8a illustrates the generation of two burst of EHPs from sites of energy deposition. Once the EHPs are generated, the applied electric field pulls the holes and electrons to the opposing electrodes. However, these charge carriers could be lost as they travel within the photoconductor shown in Figure 2.8b by two processes: recombination and trapping [5]. Currently, due to the large number of EHPs, each burst is simulated separately for the transport including recombination and trapping considerations.

The recombination of carriers is checked at each simulation step. Recombination occurs when an electron and a hole are sufficiently close together, making the Coulomb attraction so strong that they cannot escape each other. As carriers approach each other due to Coulomb attraction, their drift component increases inversely proportional to the separation distance squared. Thus, as the separation distance is reduced, the simulation time step also should be reduced in order to accurately capture the movement of the carriers as they come close to each other. However, this comes at the expense of simulation time. To solve this problem, a recombination distance was used by Bartczak et al. [30] in their study of ion recombination in irradiated nonpolar liquids. Figure 2.9a and b shows the sample transport tracks of three EHPs in electric field taking into account drift alone and drift and diffusion.

2.4.3 Indirect Detectors

Several models exist for modeling indirect detectors taking into account both ionizing radiation and optical particle transport. In this section, we review and compare recent efforts in modeling indirect detectors. Moisan et al. at Tri-University Meson Facility [31,32] developed an MC simulator for positron emission tomography imaging detectors utilizing a gamma-ray interaction tracking (GRIT) for high-energy photons and DETECT [33] for optical photon transport simulations. The GRIT program uses a simple gamma-ray model taking into account photon interactions. The locations of interaction are saved as input to the DETECT routine allowing for only

Monte Carlo Modeling of X-Ray Detectors for Medical Imaging

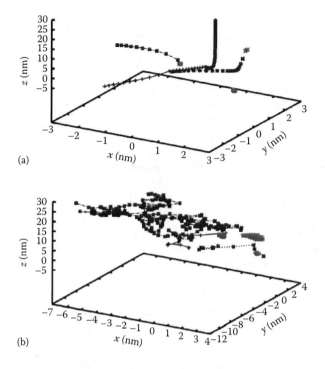

FIGURE 2.9 (**See color insert.**) (a) Sample transport simulation track of three electron–hole pairs in electric field taking into account drift. (b) Sample transport simulation track of three electron–hole pairs in electric field taking into account drift and diffusion. (Blue and red dots represent hole and electron tracks, respectively.)

simple detector geometries. The DETECT routines are used for optical photon transport simulation allowing for realistic transport in scintillator detectors. More sophisticated gamma- and X-ray simulations can be achieved by replacing GRIT with more complex programs such as Geant4 [11], EGSnrc [9], or PENELOPE [8].

Blakesley et al. performed work in the area of modeling organic X-ray imagers utilizing EGSnrc [34]. The DOSxyznrc code was used to generate the photon absorption probability distribution function, and the optical photon transport was modeled by DETECTII routines [35–37]. For simplicity, this model assumes incident photons only interact and deposit energy in a single location of the detector.

Recently, Blake et al. utilized Geant4 and its standard electromagnetic and optic physics modules for investigating optical transport in electronic portal imagers [38]. This model has the advantage of being able to simulate X-ray and optical photon interactions all in a single simulator. Another approach based on Geant4 through the use of the inherent UserSteppingAction and UserEventAction classes [39] has enabled transport of optical photons analogous to any other secondary particle for DQE simulations.

More recently, Poludniowski and Evans modeled the light transport in powdered-phosphor scintillator screens based on Boltzmann transport equations (BTE) [40,41]. The 3D dose distribution was calculated using DOSRZnrc, and the optical photon transport was performed using the BTE model and input parameters calculated with geometric optic and diffraction models. The BTE approach has the advantage of faster calculation times compared to ray-tracing methods but have limitations in modeling the absorbed fraction and MTF for higher binder-to-phosphor relative refractive indices and screen thickness.

Extensive work has been published by Badano et al. on modeling indirect X-ray detectors by interfacing PENELOPE with DETECTII routines [42,43]. This model, known as MANTIS, takes into account not only X-ray photon interactions but also high-energy secondary electron interactions and the associated spreading that may further degrade detector performance. The generation of optical photons inside the scintillator material is coupled from energy deposition events caused by incident X-ray photons. In addition, the optical model takes into account the gain variance due to conversion with a Poisson random variable, analogous to direct X-ray detectors described in Equation 2.2. The models also support a realistic columnar model for anisotropic blur. The disadvantage of long simulation time has been addressed with the development of hybridMANTIS for parallelizing simulations utilizing graphics processing units (GPUs) [16]. In this novel approach, PENELOPE simulations are ran in the CPU, while the slow optical DETECTII transport is massively parallelized in the GPU to achieve faster simulation speeds. Table 2.1 summarizes these models and approaches.

TABLE 2.1
Comparison of Indirect Models

Group/Author	Application	Availability	Model Features	Limitations
Moisan et al.	Positron emission tomography	Available by contacting author	GRIT and DETECT optical models	Not consider electron interactions
Blakesley et al.	Conceptual flat-panel X-ray imager based on organic semiconductors and plastic substrate	Not available	EGSnrc and DETECTII optical models	Not consider electron interactions
Blake et al.	Electronic portal imaging device	Not yet available	Geant4 with electromagnetic and optic physics modules	Unknown
Poludniowski and Evans	Powdered-phosphor screens for X-ray medical imaging	Available on the medical physics website [40,41]	DOSRZnrc with phosphor program for optical models	Limitations for higher binder-to-phosphor relative refractive indices and screen thickness
Badano et al.	Flat-panel X-ray imager based on microcolumnar scintillators	Available on Google code: mantismc and hybridmantis	PENELOPE and DETECTII optical models	Long simulation time for large amount of histories

Monte Carlo Modeling of X-Ray Detectors for Medical Imaging

2.5 APPLICATIONS

2.5.1 PULSE-HEIGHT SPECTROSCOPY

When an X-ray beam with constant energy is absorbed, ideally the detector response is constant in terms of electric signal. However, in reality the detector response typically consists of a distribution of pulses with varying heights. Signal level variations are due to a range of effects, such as generation, reabsorption, and escape of fluorescent X-rays above the material k-edge, Compton scattering at high energies, depth-dependent absorption of incident X-rays, and variations in EHP generation and transport under bias. Figure 2.10 shows the simulated PHS using detailed spatiotemporal MC simulation for 12 and 100 keV monoenergetic X-ray photons, with applied bias of 4 and 30 V/μm. During simulation, many bursts of EHPs are generated, initialized, and transported. The transport takes into account carrier diffusion due to Brownian motion and drift due to external applied electric field and Coulomb attraction/repulsion due to neighboring carriers. Both recombination and trapping of EHPs are taken into account for carriers moving in the detector. The x-axis shows the number of EHPs per keV (normalized by the incident photon energy). In Figure 2.10a, the incident photon energy is 12 keV. There are two distinct peaks corresponding to the two different applied voltages. The PHS consists of a single spectral peak because the incident photon energy is below the k-edge of a-Se. At 30 V/μm, many more EHPs are detected compared to 4 V/μm because a high electric field allows for more EHPs to escape recombination and also reduces the number of carriers trapped. In Figure 2.10b, above the k-edge energy, secondary peaks are observed due to the generation and escape of fluorescent photons and low-energy electron creation from Compton scattering.

2.5.2 INFORMATION FACTOR AND DETECTIVE QUANTUM EFFICIENCY

An important performance metric to capture the statistical variation of detector response to primary quanta is the information factor (or Swank factor) [44]. It is closely related to the PHS, where if $p(x)$ is the probability distribution of fluctuations

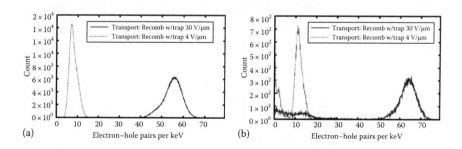

FIGURE 2.10 (See color insert.) (a) Results of the detailed spatiotemporal Monte Carlo simulations with ARTEMIS. Plots of the pulse-height spectra, for transport with 4 and 30 V/μm applied electric field with recombination only and with recombination and trapping for 12 keV monoenergetic incident photon energies. (b) For 100 keV.

in the number of detected EHPs, x, and M_n is the nth moment of the detected EHP PHS distribution,

$$M_n = \sum_x p(x) x^n, \qquad (2.9)$$

then the information factor can be defined as the following [45,46]:

$$I = \frac{M_1^2}{M_0 M_2}. \qquad (2.10)$$

When the detector responses consist of only a single photo peak such as in Figure 2.10a, the information factor can be calculated using the mean and standard deviation of the distribution:

$$I = \frac{m^2}{m^2 + \sigma^2}. \qquad (2.11)$$

Figure 2.11a shows the information factor calculated from the simulated PHS as a function of incident photon energy calculated using Equation 2.11. At 30 V/μm, the information factor is close to one below the k-edge of a-Se and drops sharply at the energies slightly higher than the k-edge. This is due to the generation and reabsorption of fluorescent X-rays, which gives variation in the detector response. The information factor slowly recovers as the photon energy increases, up to approximately 40 keV. Afterward, the information factor starts to fall again due to Compton scattering of incident X-rays. The DQE at zero spatial frequency can be calculated from the information factor and absorption efficiency, η:

$$DQE(0) = I * \eta, \qquad (2.12)$$

where DQE is a measure of the combined effects of the signal and noise performance of the imaging system. The simulated DQE at zero spatial frequency

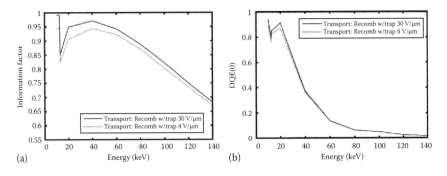

FIGURE 2.11 (See color insert.) (a) Simulated information factor as a function of incident photon energy. (b) Simulated DQE at zero spatial frequency as a function of incident photon energy. (Reprinted from Fang Y. et al., *Med. Phys.*, 39, 308, 2012. With permission.)

results takes into account the detailed transport of EHPs, which results in lower information factor and DQE as shown in Figure 2.11b. At lower energies, the DQE follows the information factor trends closely. Since the absorption efficiency of a-Se is low for high-energy X-rays, the DQE drops sharply for higher incident photon energies.

2.6 SUMMARY

Semiconductor X-ray detectors are important components of medical imaging systems and can be used in a wide range of modalities and applications. MC methods can be used for modeling both direct and indirect detectors and provide insight into the fundamental physics and theoretical performance limitations of imaging detectors. There are still many areas that need improvement in modeling the complete signal formation process in semiconductor detectors for X-ray imaging applications. As the high-energy photoelectron or Compton electron deposits energy in the detection material, each burst of EHPs is simulated separately due to the large number of carriers to be considered at one time. When the electron energy is high, the mean free path is larger and deposition events occur far and apart. However, as the electron energy is reduced, energy deposition events occur more locally, and a need to consider multiple bursts may arise to more realistically model the charge generation and recombination processes.

Detector thickness and carrier mobility can affect greatly the transport properties and hence detector performance. As the detector thickness increases, the carriers require more time to travel to the electrode thus increasing the probability of recombination and trapping.

Experimental validation of MC models is always a challenge. Pulse-height spectroscopy measurements of a-Se detectors can give much detail regarding detector response to each incident photon. However, low signal levels in a-Se, combined with noise introduced by the high-voltage supply and long pulse shaping times, make direct experimental observations for validation difficult.

One disadvantage of MC models is the large number of X-ray histories required, thus causing long simulation times to achieve low statistical variance in the estimates. Many efforts have been proposed to improve this area, such as the hybridMANTIS [16] package designed with runtime bottlenecks in mind, allowing for parallelization of detailed carrier transport code in the GPU for additional performance.

ACKNOWLEDGMENTS

The authors would like to thank Dr. Andreu Badal and Dr. Nicholas Allec for long discussions on the development and implementations of the MC models for direct detectors. The mention of commercial products herein is not to be construed as either an actual or implied endorsement of such products by the Department of Health and Human Services. Y.F. acknowledges funding by an appointment to the Research Participation Program at the Center for Device and Radiological Health administered by the Oak Ridge Institute for Science and Education through an interagency agreement between the U.S. Department of Energy and U.S.

Food and Drug Administration. This work was also financially supported in part by the Natural Sciences and Engineering Research Council of Canada.

REFERENCES

1. R. Schulz-Wendtland, K. Hermann, E. Wenkel, B. Adamietz, M. Lell, K. Anders, and M. Uder, First experiments for the detection of simulated mammographic lesions: Digital full field mammography with new detector with double plate of pure selenium, *Radiologe*, 51, 130–134 (2004).
2. A.L. Goertzen, V. Nagarkar, R.A. Street, M.J. Paulus, J.M. Boone, and S.R. Cherry, A comparison of X-ray detectors for mouse CT imaging, *Phys. Med. Biol.*, 49, 5251–5265 (2004).
3. G. Knoll, *Radiation Detection and Measurement*, Wiley Interscience/John Wiley & Sons, Inc., Hoboken, NJ, pp. 265–366 (2010).
4. S.O. Kasap and J.A. Rowlands, Direct-conversion flat-panel X-ray image detectors, *IEE Proc. CDS*, 149, 85 (2002).
5. Y. Fang, A. Badal, N. Allec, K.S. Karim, and A. Badano, Monte Carlo simulation of amorphous selenium imaging detectors, *Proc. SPIE*, 7622, 762214 (2010).
6. M.J. Berger and S.M. Seltzer, *Monte Carlo Transport of Electrons and Photons*, T.M. Jenkins, W.R. Nelson, and A. Rindi (eds.). Plenum Press, New York, p. 153, (1988).
7. J.A. Halbleib, P.P. Kensek, G.D. Valdez, S.M. Seltzer, and M.J. Berger, ITS: The Integrated TIGER Series of electron/photon transport codes-version 3.0. Nuclear Science, IEEE Transactions on, 39(4), 1025–1030 (1992).
8. F. Salvat, J.M. Fernandez-Varea, and J. Sempau, PENELOPE-2006: A code system for Monte Carlo simulation of electron and photon transport, OECD/NEA Data Bank, Issy-les-Moulineaux, France (2006).
9. I. Kawrakow and D.W.O. Rogers, The EGSnrc system, a status report. In Advanced Monte Carlo for Radiation Physics, Particle Transport Simulation and Applications, Springer Berlin Heidelberg. pp. 135–140 (2001).
10. J.F. Briesmeister, X-5 Monte Carlo Team, MCNP—A general Monte Carlo N-particle transport code, Version 5, Report No. LA-UR-03-1987 (2003).
11. S. Agostinelli et al., Geant4—A simulation toolkit, *Nucl. Instrum. Methods A*, 506, (2003).
12. A. Ferrari, P.R. Sala, A. Fassµo, and J. Ranft, FLUKA: A multi-particle transport code (Program version 2005), CERN-2005-10, France, INFN/TC-05/11, SLAC-R-773 (2005).
13. E. Fourkal, M. Lachaine, and B.G. Fallone, Signal formation in amorphous-Se-based X-ray detectors, *Phys. Rev. B*, 63, 195204 (2001).
14. M. Yunus, M.Z. Kabir, and S.O. Kasap, Sensitivity reduction mechanisms in amorphous selenium photoconductive X-ray image detectors, *Appl. Phys. Lett.*, 85, 6430–6432 (2004).
15. K. Koughia, Z. Shakoor, S.O. Kasap, and J.M. Marshall, Density of localized electronic states in a-Se from electron time-of-flight photocurrent measurement, *J. Appl. Phys.*, 97, 033706-1–033706-11 (2005).
16. D. Sharma, A. Badal, and A. Badano, hybridMANTIS: A CPU-GPU Monte Carlo method for modeling indirect X-ray detector with columnar scintillators, *Phys. Med. Biol.*, 57, 2357 (2012).
17. H.H. Barrett, J.D. Eskin, and H.B. Barber, Charge transport in arrays of semiconductor gamma-ray detectors, *Phys. Rev. Lett.*, 75, 156–159 (1995).
18. J. Knight and E. Davis, Photogeneration of charge carrier in amorphous selenium, *J. Phys. Chem. Solids*, 35, 543–554 (1975).
19. C.A. Klein, Bandgap dependence and related features of radiation ionization energies in semiconductors, *J. Appl. Phys.*, 39, 2029 (1968).

20. W. Que and J. Rowlands, X-ray photogeneration in amorphous selenium: Geminate versus columnar recombination, *Phys. Rev. B*, 51, 10500–10507 (1995).
21. J.R. Janesik, *Scientific Charge-Coupled Devices*, SPIE Press, Bellingham, WA (2001).
22. L.K. Townsley, P.S. Broos, G. Chartas, E. Moskalenko, J.A. Nousek, and G.G. Pavlov, Simulating CCDs for the Chandra advanced CCD imaging spectrometer, *Nucl. Instrum. Methods A*, 486, 716 (2002).
23. O. Godet, P. Sizun, D. Barret, P. Mandrou, B. Cordier, S. Schanne, and N. Remoue, Monte-Carlo simulations of the background of the coded-mask camera for X-and Gamma-rays on-board the Chinese-French GRB mission SVOM, *Nucl. Instrum. Methods A*, 494, 775 (2009).
24. C. Xu, M. Danielsson, and H. Bornefalk, Validity of spherical approximation of initial charge cloud shape in silicon detectors, *Nucl. Instrum. Methods A*, 648, S190 (2011).
25. M. Lachaine et al., Calculation of inelastic cross-sections for the interaction of electrons with amorphous selenium, *J. Phys. D: Appl. Phys.*, 33, 551–555 (2000).
26. D. Sharma, Y. Fang, F. Zafar, K.S. Karim, and A. Badano, Recombination models for spatio-temporal Monte Carlo transport of interacting carriers in semiconductors, *Appl. Phys. Lett.*, 98, 242111 (2011).
27. N. Bohr, The penetration of atomic particles through matter, *K. Dan. Vidensk. Selsk. Mat. Fys. Medd.*, 18, 8 (1948).
28. H.H. Barrett and K. Myers, *Foundation of Imaging Science*, Wiley Interscience/John Wiley & Sons, Inc., Hoboken, NJ, pp. 748–763 (2004).
29. Y. Fang, A. Badal, N. Allec, K.S. Karim, and A. Badano, Spatiotemporal Monte Carlo transport methods in X-ray semiconductor detectors: Application to pulse-height spectroscopy in a-Se, *Med. Phys.*, 39, 308–319, 2012.
30. W.M. Bartczak, M.P. DeHaas, and A. Hummel, Computer simulation of the recombination of the ions in tracks of high-energy electrons in nonpolar liquids, *Radiat. Phys. Chem.*, 37, 401–406 (1991).
31. C. Moisan, D. Vooza, and M. Loope, Simulating the performance of an LSO based position encoding detector for PET, *IEEE Trans. Nucl. Sci.*, 44, 2450–2458 (1997).
32. D. Vooza, C. Moisan, and S. Pasquet, An improved model for the energy resolution of multicrystal encoding detector for PET, *IEEE Trans. Nucl. Sci.*, 44, 179–183 (1997).
33. G.F. Knoll, T.F. Knoll, and T.M. Henderson, Light collection in scintillation detector composites for neutron detection, *IEEE Trans. Nucl. Sci.*, 35, 872–875 (1988).
34. J.C. Blakesley and R. Speller, Modeling the imaging performance of prototype organic X-ray imagers, *Med. Phys.*, 35, 225–239 (2008).
35. T. Radcliffe, G. Barnea, B. Wowk, R. Rajapakshe, and S. Shalev, Monte Carlo optimization of metal/phosphor screens at megavoltage energies, *Med. Phys.*, 20, 1161–1169 (1993).
36. R. Fasbender, H. Li, and A. Winnacker, Monte Carlo modeling of storage phosphor plate readout, *Nucl. Instrum. Methods Phys. Res. A*, 512, 610–618 (2003).
37. C. Kausch, B. Schreiber, F. Kreuder, R. Schmidt, and O. Dossel, Monte Carlo simulation of the imaging performance of metal/plate phosphor screens used in radiotherapy, *Med. Phys.*, 26, 2113–2124 (1999).
38. S. Blake, P. Vial, L. Holloway, P. Greer, and Z. Kuncic, An investigation into optical photon transport effects on electronic portal imaging performance using Geant4, *AAPM/COMP Meeting*, Vancouver, Canada, SU-F-BRA-02 (2011).
39. E. Abel, M. Sun, D. Constantin, R. Fahrig, and J. Star-Lack, User-friendly, ultra-fast simulation of detector DQE(f), *Proc. SPIE*, 8668, 86683O-1 (2013).
40. G.G. Poludniowski and P.M. Evans, Optical photon transport in powdered-phosphor scintillators. Part 1. Multiple-scattering and validity of the Boltzmann transport equation, *Med. Phys.*, 40, 041904-1–141904-11, (2013).

41. G.G. Poludniowski and P.M. Evans, Optical photon transport in powdered-phosphor scintillators. Part II. Calculation of single-scattering transport parameters, *Med. Phys.*, 40, 041905-1–141905-9 (2013).
42. A. Badano, Optical blur and collection efficiency in columnar phosphor for X-ray imaging, *Nucl. Instrum. Methods Phys. Res. A*, 508, 467–479 (2003).
43. A. Badano and J. Sempau, MANTIS: Combined X-ray, electron and optical Monte Carlo simulation of indirect radiation imaging systems, *Phys. Med. Biol.*, 51, 15454 (2006).
44. Y. Fang, K.S. Karim and A. Badano, Effect of burst and recombination models for Monte Carlo transport of interacting carriers in a-Se X-ray detectors on Swank noise, *Med. Phys.*, 41, 011904-1–011904-5 (2014).
45. A. Ginzburg and C. Dick, Image information transfer properties of X-ray intensifying screens in the energy range from 17 to 320 keV, *Med. Phys.*, 20, 1013–1021 (1993).
46. R.K. Swank, Absorption and noise in X-ray phosphors, *J. Appl. Phys.*, 44, 4199–4203 (1973).

3 Medical X-Ray and CT Imaging with Photon-Counting Detectors

Polad M. Shikhaliev

CONTENTS

- 3.1 Introduction 48
- 3.2 Historical Overview of Photon-Counting X-Ray and CT Systems 48
- 3.3 Advantages of PCXCT 49
  - 3.3.1 Electronic Noise Rejection 49
  - 3.3.2 SNR Improvement with Photon Energy Weighting 50
    - 3.3.2.1 Generalized Weighting Approach 50
    - 3.3.2.2 Energy Weighting in Projection X-Ray Imaging 51
    - 3.3.2.3 Energy Weighting in CT 52
  - 3.3.3 Material Decomposition 53
- 3.4 Design Concepts of PCXCT 55
  - 3.4.1 High Demands to Clinical Imaging Systems 55
  - 3.4.2 Practical PCXCT Detector Configuration 55
  - 3.4.3 PCXCT Detector Electronics 56
  - 3.4.4 Material Selection for PCXCT Detectors 57
  - 3.4.5 Imaging Configurations 58
- 3.5 PCXCT Detector Technologies 60
  - 3.5.1 PCXCT with Si Strip Detectors 60
  - 3.5.2 PCXCT with CZT and CdTe Detectors 61
  - 3.5.3 Medipix Detectors 64
- 3.6 Problems with PCXCT and Future Developments 65
  - 3.6.1 Count Rate Limitations 65
  - 3.6.2 Low-Energy Spectral Tailing 66
  - 3.6.3 Intensity-Dependent Line Artifacts 67
  - 3.6.4 Charge Sharing 70
  - 3.6.5 Suboptimal Energy Resolution 72
- 3.7 Feasible Clinical Application: Photon-Counting Spectral Breast CT 74
- 3.8 Conclusion 75
- References 75

3.1 INTRODUCTION

Projection radiography and computed tomography (CT) have been gold standards for screening, diagnostic, and therapeutic medical imaging in the last several decades [1–4]. Radiography and CT systems employ polyenergetic X-ray beams with broad energy spectra generated by conventional X-ray tubes. Until now, these systems use radiation detectors based on the energy integration principle, that is, X-ray photons (usually several thousand photons per pixel) are absorbed and energies of all photons summed up to provide a single analog signal. On the other hand, X-ray photons are inherently discrete, and it would be straightforward to detect (count) each photon separately during image acquisition. Such a photon-counting mode would allow for rejecting electronics noise by setting an appropriate energy threshold for the detector electronics. In addition, the photon-counting detector would allow for measuring the energy of each X-ray photon. Energy-selective data acquisition, in turn, would enable material-selective imaging for major clinical applications.

The advantages of photon-counting radiography and CT were well known for a long time. However, practical applications of photon-counting radiography and CT were hampered by serious problems. The major problems for using photon-counting detectors are associated with very high demands of medical imaging. Thousands of individual detector pixels with small sizes should be developed and packed in a small area; each pixel should employ an individual photon-counting readout circuit running at multimillion count per second rates. A large amount of digital data should be recorded and saved in a few seconds needed for image acquisition. The detectors and readout electronics, as well as necessary computational power and memory, were not available for a long time.

In the last decade or so, substantial improvements in semiconductor detector technologies were made. Simultaneously, chip-level application-specific integrated circuit (ASIC) electronics were developed. This enabled large numbers of small readout circuits to match the detector pixel arrays. Additionally, spectacular improvements have been made on computer memory and speed. All these factors were integrated together and allowed for moving the photon-counting detector technologies to a level where it now meets many of the demands for clinical applications. Consecutively, a number of start-up companies are developing prototype photon-counting detectors for potential clinical applications. Thus, photon-counting radiography and CT are now a "hot" topic.

3.2 HISTORICAL OVERVIEW OF PHOTON-COUNTING X-RAY AND CT SYSTEMS

Photon-counting radiation detectors were first applied in isotope emission imaging in gamma camera–based systems with Anger logic, introduced in 1953 [1,3]. The detectors used in these applications were scintillation detectors that at this time were well developed, particularly the detectors based on NaI scintillators coupled to photomultiplier tubes.

Further advancement in isotope imaging led to a different type of experiment, namely, transmission imaging, which was considered as adjunct to emission imaging [3]. The purpose was to combine transmission and isotope imaging to be able to observe simultaneously functional and anatomical information. For transmission imaging, a narrow gamma ray beam was used. The gamma ray was generated by collimation of the isotope source. A photon-counting scintillation detector and isotope source were installed on opposite sides of a patient, and transmission scan across the patient was performed on a point-by-point basis. This kind of transmission imaging had a major limitation; however, low activities of the isotope sources resulted in prohibitively long scan times. Also, large size of the sources decreased spatial resolution. Thus, transmission imaging with isotope sources and photon-counting scintillation detectors were considered suboptimal and did not find wide clinical application.

The aforementioned limitations of transmission imaging with isotope sources were later addressed when X-ray tubes were used as radiation sources. The X-ray tubes provided sufficiently high photon outputs and small focal spot sizes. Using high-flux X-ray tubes, however, created another problem: the photon-counting scintillation detectors could not handle the high count rates needed to detect the photons from the X-ray tubes. Thus, the count rate problem was prohibitive for further applications of photon-counting scintillation detectors in transmission imaging with X-ray tubes.

To address the count rate problem, a method was proposed that was fatal for future applications of photon-counting detectors: the scintillation detectors were used in the energy integration mode instead of the photon-counting mode. The output signal pulses of the scintillation detectors were integrated, amplified, digitized, and used to estimate the photon flux. Further developments of the transmission scans, including CT applications, were based on energy-integrating rather than photon-counting detectors. Thus, although the first clinical CT system reported by Hounsfield used a scintillation detector with a NaI scintillator coupled to a photomultiplier tube [5], this detector was not a photon-counting detector; it was operated in the energy-integrating mode [6]. Later on, new generations of energy-integrating detectors, such as pixilated Xe ionization chambers and pixilated scintillators coupled to photodiode arrays, were introduced [1,3]. The new types of ceramic scintillators coupled to photodiode arrays are now the state of the art in CT technologies [2,3].

The photon-counting X-ray/CT (PCXCT) imaging remained dormant until the early 1990s, when different groups started research and developments in this area. The pixilated photon-counting detector technologies based on semiconductors such as high-purity Ge [7], CdTe [8], CZT [9], and Si [10–12] have been investigated. Some other types of photon-counting detector technologies such as microchannel plates [13–15] and gas-filled detectors [16,17] have also been investigated. Some of the aforementioned technologies have shown promising perspectives for clinically applicable PCXCT systems and are being investigated.

3.3 ADVANTAGES OF PCXCT

3.3.1 Electronic Noise Rejection

One of the major advantages of photon-counting detectors is electronics noise rejection. A well-designed photon-counting detector allows for setting electronics threshold low enough to reject noise pulses while counting useful signals.

Therefore, a quantum-limited operation of the photon-counting detector is provided as image noise is determined by only statistical variations of X-ray photons. On the other hand, energy-integrating detectors suffer from electronics noise, which is mixed with useful photon signals, and separating it from statistical noise is not possible. Electronics noise rejection is important because its magnitude for currently used digital X-ray detectors (flat-panel detectors) is not negligible [18]. As high signal-to-noise ratio (SNR) is required to visualize low-contrast features of human anatomy, electronics noise must be suppressed. This requires additional X-ray exposure to the patient, which is not desirable due to the increased risk factor of radiation-induced cancer.

3.3.2 SNR Improvement with Photon Energy Weighting

3.3.2.1 Generalized Weighting Approach

The energy-integrating detectors convert energies of X-ray photons to charge and each X-ray photon contributes to the signal proportionally to its energy. In other words, the X-ray photons are weighted proportionally to their energies. However, such weighting is not optimal to achieve the highest SNR in the image. Tapiovaara and Wagner first showed that the highest SNR can be achieved if the lower-energy X-ray photons are weighted higher than the higher-energy photons [19]. Such weighting might be possible if each X-ray photon is detected separately and its energy is measured using a photon-counting detector. The photon energy weighting in PCXCT has been investigated and quantified later in a number of studies [20–25].

In 2010, Shikhaliev has shown that the energy weighting in X-ray/CT imaging is a particular case of a general image weighting (GIW) method [26]. The GIW method states the following: if an object is imaged in a series of image acquisitions (not necessarily with X-ray) with different acquisition parameters, then the resulting subimages can be combined in a unique way that provides the highest SNR in the combined image. The subimages should be combined with unique weighting factors that are determined from magnitudes of the signals and noises in each subimage. Assume that there are n subimages I_i acquired at n different settings of a particular acquisition parameter. Assume further that each subimage includes a region of interest with signal S_i and noise σ_i (Figure 3.1).

 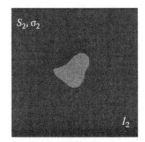 

FIGURE 3.1 Illustration of the generalized image weighting method: the subimages are acquired at n different parameters, weighted according to their signal and noise content, and combined to provide the highest signal-to-noise ratio.

The final image can be composed from these subimages after weighting each subimage with an optimal weight factor w_i^{opt} uniquely determined as $w_i^{opt} = S_i/\sigma_i^2$. The maximum SNR in the composed image is then achieved, which is determined as

$$SNR_{max}^2 = \sum_{i=1}^{i=n}\left(\frac{S_i}{\sigma_i}\right)^2 = \sum_{i=1}^{i=n} SNR_i^2,$$

where $SNR_i^2 = S_i^2/\sigma_i^2$ is the SNR in each subimage I_i [26]. Notice that weighting by itself does not change SNR in each subimage, but the SNR in the composed image is maximized.

The GIW method can be applied to any image data regardless of the contrast mechanism used for generating this image. These may include projection X-ray image, reconstructed CT image, or other types of image created with different contrast mechanisms. For example, the GIW method can be used as a *temporal weighting* method when contrast kinetics imaging is performed in radiography, CT, or MRI [26]. In this case, a series of images are acquired after contrast agent is injected into the patient, and contrast uptake and washout are analyzed with respect to time. Each of the images in the chronological series, therefore, has different magnitudes of the signal and noise, depending on the time of the measurement. Therefore, the subimages can be optimally weighted and composed to provide a final image with the highest SNR.

3.3.2.2 Energy Weighting in Projection X-Ray Imaging

The GIW method can be applied directly in PCXCT imaging when subimages are acquired with an energy-resolving detector simultaneously at different energies over the X-ray energy spectrum. Assume that a contrast element with thickness d and linear attenuation coefficient μ_{ci} is located within the background material with thickness L and a linear attenuation coefficient of μ_{bi}, where i indicates the measurement at energy E_i. Then the optimal weighting factors for subimages corresponding to different energies are determined as

$$w_i^{opt} = \frac{1-e^{-\Delta\mu_i d}}{1+e^{-\Delta\mu_i d}},$$

where $\Delta\mu_i = \mu_{ci} - \mu_{bi}$. If N_{0i} X-ray photons with energy E_i are further assumed, then according to the GIW method, the maximum SNR with energy weighting is determined as

$$SNR_{weight}^2 = \sum_{i=1}^{i=n} N_{0i} e^{-\mu_{bi} L} \frac{(1-e^{-\Delta\mu_i d})^2}{1+e^{-\Delta\mu_i d}}.$$

Notice that in practical imaging situations, the SNR enhancement with energy weighting is most important when the signal magnitude is small, which may occur if the contrast element has a small thickness and/or its linear attenuation coefficient is similar to the background. In this case $\Delta\mu_i d \ll 1$, and the aforementioned expression for an optimal weight factor is reduced to $w_i^{opt} = (1/2)\Delta\mu_i d$. The SNR for small signal is now determined as

$$SNR_{\text{weight}}^2 = \frac{1}{2}d^2 \sum_{i=1}^{i=n} N_{0i} e^{-\mu_{bi} L} \Delta\mu_i^2.$$

Therefore, once the X-ray spectrum and parameters of the object are known, both weighting factors and expected maximum SNR can be calculated theoretically.

Notice that special care should be taken due to the fact that the aforementioned calculations assume an ideal PCXCT detector. Current energy-resolving PCXCT detectors are far from ideal in many aspects, including suboptimal energy resolution and inaccurate response function. Some of the limitations of PCXCT detectors are of fundamental nature and cannot be addressed in principle. For this reason, theoretically predicted weighting factors may not be optimal in real imaging situations. In this case, one needs to return back to the GIW method described earlier and find optimal weighting factors directly from measured signal and noise in experimental subimages [26]. This is appropriate because signal and noise in experimental subimages already include all possible image deteriorations due to suboptimal detector performance. On the other hand, these deteriorations could not be accounted when theoretical weighting factors are calculated.

3.3.2.3 Energy Weighting in CT

Energy weighting can be applied in CT in different ways. Energy-resolved (monoenergetic) CT projections can be acquired with a photon-counting detector. Similarly to the projection imaging, the monoenergetic CT projections acquired at different energies can be weighted with appropriate weighting factors optimized for a particular contrast element and then combined to yield the final set of CT projections with the highest SNR. Once the CT image is reconstructed from these energy-weighted projections, improved SNR in CT projections transfers to the reconstructed CT image and improved contrast-to-noise ratio (CNR) is achieved in CT images. This energy weighting method is called projection-based energy weighting in CT. Although projection-based weighting allows for substantial CNR improvement in CT [23], depending on the particular type of contrast element, such as calcification, iodine contrast, and adipose tissue, it increases beam-hardening artifacts in reconstructed CT images [21]. The reason for the increased beam hardening is that the energy weighting scores low-energy photons with higher weight than the higher energy photons, but the low-energy part of the X-ray spectrum is primarily responsible for beam hardening [21].

The energy weighting in CT can also be performed using monoenergetic CT images reconstructed from CT projections acquired at corresponding energies [26]. The reconstructed monoenergetic CT images can be weighted according to the GIW method and composed to yield the highest CNR for a particular type of contrast element. Because reconstructed monoenergetic CT images do not suffer from beam-hardening artifacts, the composed CT image also does not exhibit beam-hardening artifacts. This weighting method is called image-based energy weighting in CT [26].

Comprehensive investigation and comparison of projection-based and image-based energy weighting methods have been performed in [26]. It has been shown that projection-based weighting in CT always provides a slightly higher CNR improvement as compared to image-based weighting, and differences between the two methods on CNR improvements vary depending on the types and sizes of contrast elements, as well as on the imaged object.

It should be noticed that once the energy-selective CT projections are acquired in multiple energy bins and saved, a variety of options for CT reconstruction are available including different energy weighting options. Modern computers with high speeds and large memories will allow fast CT reconstructions and displays using energy-selective CT data.

3.3.3 Material Decomposition

Material decomposition has important applications in many areas of medical X-ray and CT imaging. Some of these applications include decomposition of iodine and other contrast material for quantitative imaging, bone–tissue decomposition for bone densitometry, soft tissue–fat decomposition for breast density and body fat measurements, calcium-tissue decomposition for diagnostics of vascular diseases, and many other applications.

The material decomposition with photon-counting detectors can be performed using dual-, triple-, or multiple-energy subtraction methods. In general, if an object is composed of n materials and an X-ray/CT image is acquired at n energies, then these n materials can be decomposed one from another. Assume further that the n materials have effective thicknesses of $t_1, t_2, \ldots, t_n$, and image acquisition is performed at energies $E_1, E_2, \ldots, E_n$. Once the type of materials and photon energies are known, the linear attenuation coefficients can be determined. Assume that the linear attenuation coefficient of the material i at energy j is μ_j^i. Further, assume that the numbers of the X-ray photons with energy E_i before and after passing the object are N_{0i} and N_i, respectively. Then following n equations with n unknown, t_i can be written as

$$\begin{cases} \mu_1^1 t_1 + \mu_1^2 t_2 + \cdots + \mu_1^n t_n = \ln \dfrac{N_{01}}{N_1}, \\[6pt] \mu_2^1 t_1 + \mu_2^2 t_2 + \cdots + \mu_2^n t_n = \ln \dfrac{N_{02}}{N_2}, \\[6pt] \cdots \\[6pt] \mu_n^1 t_1 + \mu_n^2 t_2 + \cdots + \mu_n^n t_n = \ln \dfrac{N_{0n}}{N_n} \end{cases}$$

The unknown thicknesses t_i can be easily found by solving these equations. In the case of two materials, it is sufficient to have the data acquired at two energies, and

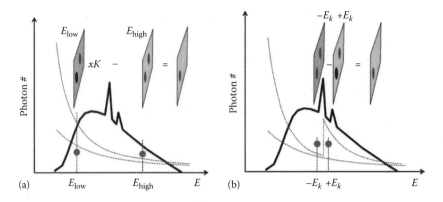

FIGURE 3.2 (See color insert.) Schematics of the material decomposition using a photon-counting X-ray/CT system: material decomposition by dual-energy subtraction (a) and by K-edge subtraction (b).

the aforementioned equation is reduced to a simple system of two equations with a straightforward solution. Figure 3.2 shows a schematic of the two-material decomposition with dual-energy subtraction.

Some clarifications should be made with respect to the dual-/multiple-energy subtraction method described earlier using photon-counting detectors. (1) The materials of interest in medical imaging are primarily compound materials. Each of these materials is determined by its effective atomic number and density. However, the same effective atomic number can be achieved using a variety of combinations of simple elements and other subcompounds [27]. Therefore, material decomposition with PCXCT requires *a priori* knowledge of the type of materials included in the object. (2) In a real PCXCT system, the narrow energy bins centered at energies E_i are used. The numbers of the X-ray photons in each energy bin are small compared to the total numbers of the X-ray photons in the spectrum. Therefore, the photon statistics in each energy bin is decreased and statistical noise is increased. In addition, the noise in decomposed images is further increased when dual and multiple subtractions of noisy data are performed. Often, a high noise level in decomposed images is a primary limiting factor for dual- and multiple-energy subtraction. (3) Generally, two-material decomposition is of primary interest, and in this case, two energy bins are used for dual-energy subtraction as shown in Figure 3.2. The position and width of the energy bins in this case should be optimized to achieve the highest SNR in a decomposed image.

The second method for material decomposition with PCXCT is the K-edge subtraction method. This method is applied when a contrast material with high atomic number (Z) such as iodine (I), gadolinium (Gd), and gold (Au) is used. The K-edge energies of these materials are 33.17, 50.24, and 80.73 keV, respectively, and at these energies, the linear attenuation coefficients of the materials increase stepwise. Two energies immediately before and after the material K-edge are used for K-edge subtraction (Figure 3.2). If both energies are sufficiently close to the K-edge, the linear attenuation coefficient of background material will change slightly between these two energies. Therefore, when two images are subtracted, the signal from the contrast material remains, while the signal from the background is cancelled out. Although the K-edge subtraction method appears to be attractive, the difficulties exist in that the small fraction of the X-ray photons with energies immediately before and after the K-edge is used and statistical noise in subtracted images can be high.

3.4 DESIGN CONCEPTS OF PCXCT

3.4.1 High Demands to Clinical Imaging Systems

The PCXCT imaging requires counting each X-ray photon and simultaneously measuring its energy similarly to gamma camera systems for isotope emission imaging. However, gamma camera systems use Anger logic with single-channel readout electronics that cannot be applied in PCXCT systems due to very high count rate requirements. The general-purpose X-ray imaging detectors use 0.2–0.3 mm size pixels and have a detector field of view of 43 × 43 cm, which amounts to several million detector pixels. In CT, the detector pixel size is approximately 1 mm and 16–256 detector rows are used with approximately 800 detector pixels in each row; this amounts to >100,000 detector pixels. Furthermore, clinical X-ray/CT imaging systems should provide fast data acquisition to avoid possible artifacts due to the patient motion during the image acquisition. This requires, in turn, a fast scan and running X-ray tubes at 200–600 mA tube currents, which is equivalent to the photon fluxes of up to 10^9 photon/mm²/s at the detector surface. Taking into account detector pixel size and image acquisition time, the detector count rates should be 10^7–10^9 count/pixel/s. It is a challenging task to run so many detector pixels at so high count rates, in addition, at high energy resolution.

3.4.2 Practical PCXCT Detector Configuration

In order to move PCXCT technology forward for the clinical applications, some sort of simplifications and trade-offs were necessary. These simplifications were applied as shown in schematics of the pixilated photon-counting detector and its readout electronics (Figure 3.3). Two major simplifications were made in the detector architecture, which, however, does not substantially deteriorate performance of a

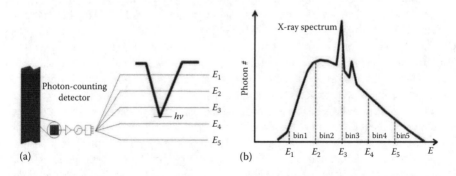

FIGURE 3.3 Architecture of the energy-selective photon-counting detector (a) and energy bins where separate X-ray data are acquired (b).

clinical imaging system. First, in order to decrease the total numbers of the detector pixels, multiple rows of 1D pixel arrays are used instead of full-size 2D pixel arrangement. Similar detector configuration is well matched to CT acquisition because current CT systems use multiple rows of 1D pixel detectors. However, for planar X-ray imaging such as radiography and mammography, the image acquisition with 1D detector rows requires scanning the linear detector arrays across the imaged object. Although scanning takes a longer time and potentially some motion artifacts may take place, careful system design can minimize the negative effects and scanning radiography systems can still be applied for routine clinical imaging.

Such scanning systems based on energy-integrating detectors have been developed and introduced into clinics for general-purpose radiography [28] and mammography [29]. Thus, photon-counting detectors with linear arrays of multiple rows and smaller total numbers of pixels can be developed and used for clinical X-ray imaging.

3.4.3 PCXCT Detector Electronics

The second simplification in PCXCT design is related to energy-selective image acquisition. Current detectors used for gamma spectroscopy utilize sophisticated analog to digital convertors (ADCs) that allow digitization with very high accuracy. These ADC, however, serve for single-channel spectroscopic detectors that require sophisticated readout electronics with a large size and high power consumption. For a PCXCT system with thousands of independent detector pixels, such electronics and ADC obviously cannot be used. Fortunately, for clinical applications of the energy-resolved PCXCT, such a high-energy resolution is not necessary [22,23]. The clinical advantages of the energy-resolved PCXCT system can be fully utilized using lower-energy resolution. This, in turn, allows for great simplification of the energy-resolved readout electronics decreasing its size. Figure 3.3 shows a schematic of such readout electronics. Each detector pixel is connected to an independent charge-sensitive amplifier and shaper. The amplifier generates an analog signal with an amplitude proportional to the absorbed X-ray energy. The analog signal from each amplifier is sent to several, say five, independent comparators with digitally adjustable threshold levels. Each comparator is connected to an individual counter, and the pulses with amplitudes higher than the comparator threshold are counted. The data acquisition yields five independent sets of photon counts D_1–D_5, and each set D_i corresponds to a unique energy threshold E_i. The difference of counts $\Delta D = D_i - D_{i+1}$ provides the number of the X-ray photons with energies between E_i and E_{i+1}. The X-ray energy spectrum is therefore split into five regions (energy bins), and five quasi-monochromatic X-ray images are generated in a single image acquisition. The positions and widths of the energy bins can be adjusted by setting appropriate energy thresholds prior to data acquisition.

Once acquired and saved in computer memory, the five energy-selective image data can be used in a variety of ways, according to the requirements of a particular imaging task. For example, the sum of all five energy bins will provide a simple photon-counting image acquired with a lowest energy threshold. The five energy bins can also be appropriately weighted before summing, and this will provide the highest SNR for a particular type of contrast element. Furthermore, two or more energy bins can be used for decomposition of two or more materials. Using appropriately developed software and modern computer power and memory, the previously listed operations with energy-selective data can be performed almost instantaneously, and image data with qualitatively new content can be presented to a clinician's review.

Notice that number 5 of the energy bins was selected in the aforementioned descriptions as an example. On the other hand, the first full-size energy-resolved PCXCT scanner used five energy bins (see Section 3.5.2), and this system has been investigated both experimentally and theoretically. Systems with different numbers of energy bins have also been developed and are being investigated.

3.4.4 Material Selection for PCXCT Detectors

Numerous semiconductor detector materials have been investigated for potential applications in PCXCT imaging. However, the requirements to clinical X-ray and CT imaging are extremely demanding, and majority of these materials have been considered suboptimal and filtered out. Some of the major requirements imposed to the detector materials for PCXCT are (1) high X-ray attenuation with the photoelectric effect being the primary interaction, (2) low carrier creation energy to achieve high energy resolution, (3) room temperature operation, (4) mass production at affordable cost, (5) long-term stability, and (6) negligible lag effect to achieve fast data acquisition at high count rates. Only a few materials, namely, Si, CdTe, and CdZnTe (CZT), are now being considered as the material of choice for PCXCT.

In medical X-ray/CT imaging, the radiation dose to the patient should be minimized. Therefore, to provide the highest dose efficiency, the detector should absorb nearly all X-ray photons passing through the patient and arriving the detector surface. Furthermore, the X-ray photons should interact primarily via the photoelectric effect because other competing interactions (Compton scatter) deteriorate the spectral resolution of the detector. This latter requirement can be fulfilled if the detector material has a sufficiently high atomic number (Z). The linear attenuation coefficients of primary detector materials Si and CZT are shown in Figure 3.4 (data for CdTe are not shown because they are close to CZT). The X-ray energy range of 10–150 keV covers all medical imaging applications including mammography, radiography, and CT. The photoeffect fractions of the linear attenuation coefficients are also shown.

The density and atomic number of Si are 2.33 g/cm^3 and 14, respectively; both are relatively low, which creates certain problems. The carrier creation energy for Si is 3.62 eV per electron–hole pair. As can be seen from Figure 3.4, the photoeffect component of Si is strongly decreased at higher photon energies. The photoeffect fraction is <50% at photon energies >50 keV, which creates a problem for practical applications. The photoeffect fraction is sufficiently high only at lower energies of 10–40 keV used in mammography. The linear attenuation coefficient of the Si is also suboptimal. A typical Si wafer with 0.5 mm thickness attenuates only 39% and 9.5% of photons with 20 and 40 keV energies, respectively, and at higher energies, photon attenuation becomes even smaller. However, this problem is partially addressed when the Si wafer is used in edge-on or tilted angle irradiation geometry (see Section 3.5.1) where X-rays are directed perpendicular to the edge of the Si wafer, or at a small angle to its surface, and the effective attenuating thickness of Si is increased. Nevertheless, low photoelectric attenuation of Si at higher energies still remains a problem.

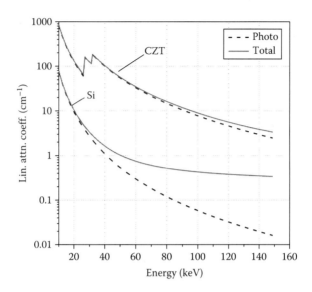

FIGURE 3.4 Linear attenuation coefficients of two major photon-counting detector materials, Si and CZT. Photoeffect components of linear attenuation coefficients are also shown for comparison.

The CZT semiconductor has a sufficiently high density of 5.78 g/cm^3 and effective atomic number of 50, which provides appropriate photon attenuation and photoelectric interaction component. Its carrier creation energy is 4.64 eV per electron–hole pair [30]. The CZT crystals can be fabricated with thicknesses of 1–10 mm and linear sizes of several centimeters. They can be tiled up to provide a full size of the imaging system matched to clinical requirements. A typical CZT crystal with 3 mm thickness provides nearly 100% photon absorption at 60 keV, and its attenuation is still 86% at 120 keV photon energy. The photoelectric fraction is 93% at 60 keV and 85% at 120 keV.

When comparing different materials, it should be noticed that CZT and CdTe semiconductors provide similar photon attenuation properties and carrier creation energies and similar performance characteristics. Although CdTe detectors exhibit lower resistance and higher leakage current, for the applications requiring very high count rates and short pulse shaping times, the negative effect of the leakage current might be negligible. Nevertheless, the energy resolution of the existing CdTe and CZT detectors needs further improvement for their best performance in medical imaging applications. For comparison, high-purity Ge detectors provide superior energy resolution due to the lowest carrier creation energy of 2.95 eV and highest charge collection efficiency, but its best performance is achieved at liquid nitrogen temperatures [31].

3.4.5 Imaging Configurations

The ideal detector for X-ray and CT imaging should be a 2D digital detector that should provide fast image acquisition and, in the case of CT, acquire the complete CT dataset in a single gantry rotation. In the case of spectroscopic PCXCT, given the complexity of the readout electronics, it seems impossible to build a 2D photon-counting detector with clinically feasible 2000 × 2000 pixel arrays, submillimeter pixel sizes, and several independent energy channels per pixel. However, even if such a 2D photon-counting detector would be feasible, it would have a major problem associated with scattered radiation. It is well known that detected scatter exhibits a major problem when 2D digital flat-panel detectors are used for projection radiography and CT [32–34]. Although scatter rejecting grids are used in digital mammography with small flat-panel detectors, they are not used in general radiography with larger flat-panel detectors because overlapped periodic structures of the grid and detector pixel array result in Noire artifacts [33]. Furthermore, rejecting scatter with grids is not efficient because the grid also rejects part of the primary X-rays passing through the patient. While detected scatter decreases SNR, in the case of the spectroscopic PCXCT system, it will also deteriorate spectral information.

To avoid the aforementioned problems, scanning multislit and scanning-slot imaging configurations can be used (Figure 3.5). The scanning multislit image acquisition was originally proposed for dose-efficient scatter rejection in projection radiography [35–37] and mammography [38,39]. Later on, it was also used in photon-counting mammography systems based on the edge-on Si detector [40,41]. Scanning multislit acquisition provides a nearly complete rejection of the scattered radiation while no absorption of the primary X-rays takes place. The scanning multislit acquisition was also proposed for CT imaging, including photon-counting breast CT [23,42]. In multislit CT acquisition, the multiple fan beams shift together with matched 1D detectors, and simultaneously, the gantry rotates around the object while detectors sample CT projections (Figure 3.5). More details of this system are reported in [23,42]. The advantage of multislit data acquisition is that commercially available, full-size 1D energy-selective photon-counting detectors can be used.

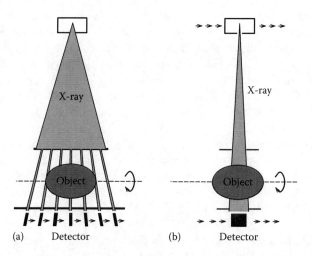

FIGURE 3.5 X-ray and computed tomography image acquisition using multislit scan (a) and slot-scan (b) methods.

The scanning-slot image acquisition uses multiple detector rows tiled next to each other. This detector configuration has been used in commercial CT systems for a long time, and current CT systems use as many as 64–256 detector rows with approximately 800 detector pixels in each row. The scanning-slot acquisition with multiple detector rows was also used in projection X-ray imaging including mammography and radiography [28,29]. Currently, no full-size energy-resolving PCXCT detectors with multirow configuration are available. However, as PCXCT technology improves, similar detectors also meeting clinical requirements might be available. At the present time, only single-row and two-row full-size spectroscopic PCXCT detectors are available. Efforts are being made to increase the numbers of detector rows while providing three to six independent energy channels per detector pixel.

3.5 PCXCT DETECTOR TECHNOLOGIES

3.5.1 PCXCT with Si Strip Detectors

Although Si semiconductor detectors were widely used in high-energy physics, their application in medical X-ray imaging was hampered by low X-ray attenuation due to its small thickness of typically 0.5 mm and low atomic number. It is suggested that the X-ray beam can be directed perpendicular to the edge of the Si wafer, which greatly increases the absorption of the beam [10]. The idea was proposed for detecting inherently planar synchrotron beam for mammography imaging [10]. Later on, different groups have investigated edge-on Si detectors for photon-counting mammography using synchrotron radiation and conventional X-ray sources [11,12,40,43–47]. The edge-on Si detectors were also investigated for general radiography applications [48].

The most successful application of the Si photon-counting detector for X-ray imaging was the MicroDose mammography (MDM) system [47] developed and commercialized by Sectra AB (Linkoping, SE). This system is now used in many hospitals worldwide (except the United States where its clinical application has not been approved yet). The MDM system uses a Si wafer with 0.5 mm thickness, which is irradiated in nearly edge-on geometry; more precisely, the X-ray hits the Si wafer at a tilted angle of approximately 8°, which provides approximately 3.6 mm absorption thickness (Figure 3.6). The bias voltage applied to the detector is approximately 150 V.

The tilted angle irradiation of the Si wafer also allows avoidance of parasitic absorption of the X-rays by an insensitive rim around the Si wafer that could decrease dose efficiency in edge-on irradiation. The Si wafer used in the MDM system is approximately 5 cm length and includes strip-like Al electrodes with a 50 μm pitch arranged at the Si surface. Each Al strip is connected to an independent amplifier, discriminator, and counter. Several Si wafers are tiled to provide a 1D detector with approximately 25 cm length. Several 1D Si detectors 25 cm in length are arranged 2 cm apart parallel to each other in a multislit configuration to provide 2D imaging. During image acquisition, the entire bank of detectors and collimated beams scan across the compressed breast. Each 1D detector scans the 2 cm gap

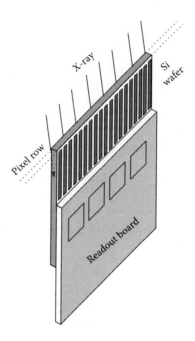

FIGURE 3.6 Schematics of the tilted angle Si strip detector used in the photon-counting mammography system. The X-ray hits the Si surface at a small angle to provide high attenuation and to avoid dead areas of the Si wafer over the edge.

between two neighboring slits, and then the full-size image is constructed using the data from each of the 1D detectors. The MDM system meets all the requirements imposed to clinical mammography units, and at the same time, it applies a lower radiation dose to the patient. Comprehensive evaluation of the MDM system can be found in [47]. Notice that this detector is a simple photon-counting detector in the sense that it does not provide energy information; rather, it counts all photon signals with the amplitudes higher than a predetermined threshold level.

3.5.2 PCXCT with CZT and CdTe Detectors

The first full-size energy-selective photon-counting X-ray imaging detector was developed by NOVA R&D, Inc. (Riverside, CA) [49]. The system called N-Energy X-ray Imaging System (NEXIS) was originally developed for security applications in airport baggage inspection, and later, it was investigated for medical imaging applications [50–53]. The full-size detector includes two rows of CZT detector pixels with 1×1 mm^2 pixel size and 1024 pixels in each row. Thus, the sensitive area of the detector has 2 mm width and 102.4 cm length. The thickness of the CZT crystals is 1.75 or 3 mm depending on the application. Each detector pixel is connected to an individual amplifier and shaper that generates an analog signal with an amplitude proportional to the energy deposited by the X-ray photon. The amplified signal is then sent to five independent discriminators and counters, which provides five energy bins per pixel. The energy-selective data

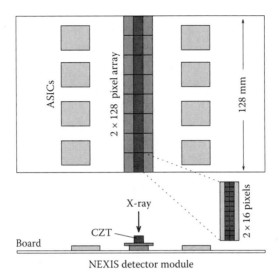

FIGURE 3.7 Schematics of the single detector board of the N-Energy X-ray Imaging System detector. Up to 8 boards can be tiled to provide a 102.4 cm field of view.

are acquired as described in the previous section. The advantages of the NEXIS detector is that it is designed in a modular basis (Figure 3.7).

Each detector module includes eight monolithic CZT crystals with a 2 × 16 pixel arrangement. The substrates with monolithic CZT crystals can easily be plugged in and out if one needs to replace some of them. The module includes 2 × 128 pixels 12.8 cm in length. Such modular detector configuration allows for building imaging CZT detectors with variable sizes ranging from 12.8 to 102.4 cm with 12.8 cm increments, which in turn provides great flexibility for the users. The measured full width at half maximum (FWHM) energy resolution of the detector is 25% and 17% at 60 and 122 keV photon energies, respectively, and the measured count rate linearity extends at least to 1 Mcount/pixel/s [51]. The factors limiting the energy resolution and count rate will be discussed in Section 3.6.5. As mentioned earlier, the NEXIS detector has been investigated for applications in medical X-ray and CT. Particularly, experimental comparison of the photon-counting CT with the NEXIS detector and conventional CT system Siemens Sensation 16 was performed in [52].

Another full-size energy-resolving PCXCT detector was developed by Gamma-Medica Ideas (Northridge, CA). This detector is conceptually similar to the NEXIS detector with a few differences: It uses CdTe instead of CZT, includes a single row of detector pixels, and provides six energy bins. The pixel size is approximately 0.4 mm along the detector row and 1.6 mm in the direction perpendicular to the row. The CdTe thickness is 3 mm. The number of detector pixels in the row is 1024, which provides approximately a 41 cm length of the detector. The energy resolution of this detector is approximately 17% FWHM measured at 60 keV photon energy, and the count rate is linear up to at least 1 Mcount/pixel/s. The detector has been investigated primarily for energy-selective CT imaging of small objects in the magnification mode [54].

The PCXCT detectors described earlier include 1–2 pixel rows that, while appropriate for research purposes, would be suboptimal for clinical radiography and CT imaging. Current CT systems, for example, include multiple detector rows of 16–64 or even more, and each detector row includes approximately 800 detector pixels. Although some of the major problems of CZT and CdTe detectors associated with a high count rate and energy resolution still remain unsolved, work is being performed to extend the numbers of the detector rows in the PCXCT system. Efforts are being made by DxRay, Inc. (Northridge, CA) to adopt a PCXCT detector in a full-size 32-row CT system. The detector is being developed in a modular basis. The approximate schematic of the detector module is shown in Figure 3.8. The module includes two monolithic CZT or CdTe crystals with 3 mm thickness, and each crystal includes a 16 × 16 array of the detector pixels with 1 × 1 mm² pixel size [55,56]. The detector module is being designed such that many modules can be tiled up without a gap and populate gantry of the commercial CT systems including approximately 800 pixels per detector row. Thus, if successfully developed and implemented, this detector can provide a full-size clinically applicable PCXCT system with 32 detector rows. The current version of the detector readout electronics provides only two energy threshold per detector pixel, which allows data acquisition with two energy bins. This is a potential limitation because more than two energy bins are needed to fully realize the advantages of spectral CT imaging. Also, dual-energy CT systems with conventional energy-integrating detectors are already in the market and may successfully compete with PCXCT with two energy bins. The limited number of the energy bins is generally a result of the

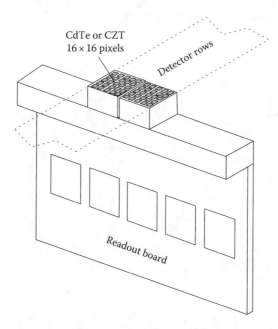

FIGURE 3.8 Photon-counting detector module developed by DxRay that includes 2 CdTe (or CZT) crystals with 16 × 16 pixel arrays with 1 × 1 mm² pixel size. Many modules can be tiled up to extend the detector rows to a clinically applicable level.

larger numbers of the detector rows that require more compact electronic circuits and higher power consumption. Nevertheless, efforts should be continued to increase the numbers of the energy channels per pixel at least to 4–5.

3.5.3 Medipix Detectors

Medipix is a family of 2D pixilated photon-counting detectors that were developed in collaboration with research groups from several universities with center at the CERN [57–59]. These detectors were originally developed for high-energy physics, and later on, developments branched toward medical applications. The first detector of this family was Medipix1, which was introduced in 1997. Since this time, several modifications and advancements have been made, and the Medipix2 and Medipix3 series have been developed. Over 200 scientific papers have been published on Medipix detectors in the last decade (Medipix.web.cern.ch). Because Medipix3 is the most advanced version of the Medipix detectors, we briefly describe it here and refer readers to the web page of the Medipix collaboration outlined earlier for more detailed information. Some of the recent advancements made on Medipix detectors can be found in [57–59].

The Medipix3 detector itself has several modifications, and its basic version includes a 256 × 256 pixel array with a pixel pitch of 55 μm. The sensitive area of the 256 × 256 detector array is approximately 1.4×1.4 cm^2. The detector is designed such that semiconductor wafers such as Si, CdTe, CZT, and GaAs can be bump bounded to the pad array of the detector substrate with corresponding ASIC readout circuits (Figure 3.9). Each pixel is connected to an individual pulse counting circuit that has two independent discriminators and counters. Thus, the Medipix3 detector can provide data acquisition with two energy bins per pixel. Each pixel can run at 100 Kcount/pixel/s rate. This pixel count rate may appear to be not high; however, taking into account that the detector includes approximately 160 pixels/mm^2, the count rate per unit area is 16 Mcount/mm^2/s. Thus, Medipix3 can provide imaging X-rays with highest fluxes as compared to other photon-counting detectors.

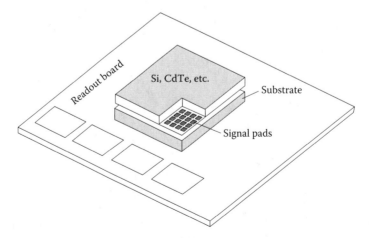

FIGURE 3.9 Schematic of the Medipix detector.

Medical X-Ray and CT Imaging with Photon-Counting Detectors

The Medipix3 detector provides several additional possibilities that are useful for research, optimization, and practical use of the detector. For example, it is possible to electronically combine four neighboring pixels with 55 × 55 μm pitch into one larger pixel with 110 × 110 μm pitch. Another possibility is disabling readout of 3 pixels in a 2 × 2 array and connecting a single 55 × 55 μm pixel to additional three readout channels with two thresholds per channel. Thus, in this case, spectroscopic imaging with eight energy thresholds can be possible. It has also a charge summing option that allows summing up the signals from pixels that helps decrease the negative effects of charge sharing between the pixels. The Medipix3 detector can be butted from three sides to further extend its sensitive area.

A potential limitation of Medipix detectors is their small sensitive area, making it difficult for full-size clinical imaging. Although several detector modules can be tiled side by side to extend the detector-sensitive area in one direction, it will still be limited to 2.8 cm (2 × 1.4 cm) in another direction, and it is not clear how to sample the large areas, for example, 25 × 25 cm^2 needed in mammography. Nevertheless, it seems that Medipix3 can be tiled up for using full-size CT systems similar to the DxRay detector described earlier, but this may require using a larger sensor thickness, larger pixel size, and modification of the readout to sample fast CT projections.

3.6 PROBLEMS WITH PCXCT AND FUTURE DEVELOPMENTS

3.6.1 COUNT RATE LIMITATIONS

The requirement on count rates of the photon-counting detectors can be derived from the X-ray tube currents of 30–800 mA used in clinical CT systems. For example, at a 120 kVp tube voltage and 6.6 mm Al equivalent half value layer of the beam, the photon flux at 1 m from the tube focal spot is 2×10^6 photon/mm^2/mA/s, and the exposure rate is 7.8 mR/mA/s [60,61]. At 800 mA tube current, the photon flux at 1 m would be 1.6×10^9 photon/mm^2/s. For the detector pixel size of approximately 1 mm^2 used in CT, the aforementioned photon flux converts to 1.6×10^9 count/pixel/s count rate, which is unachievable with any existing photon-counting detector.

The problems with high count rates are particularly emphasized for CdTe and CZT detectors that are considered detectors of choice for PCXCT. The two major factors limiting the count rate of these detectors are pulse pileup and polarization of the material due to hole trapping. The drift time of the carriers across the thickness d of the semiconductor is determined as $t = d^2/\mu U$, where μ and U are the carrier mobility and voltage applied to the detector, respectively [31]. For a typical CZT thickness of 3 mm used for PCXCT and 600 V applied voltage, the maximum electron drift time is 150 ns. To provide high energy resolution, a complete charge collection is necessary, and the shaping time of the input circuit should be equal to or larger than 150 ns. Assuming the nonparalyzable model of the detector operation, the pulse pileup fraction α can be linked to the true count rate n as $n = \alpha/((1-\alpha)\tau)$, where τ is the pulse width that can be approximated with the pulse shaping time [31]. If one desires to limit the pileup fraction to 10%, then with 150 ns pulse shaping time, the true count rate (which is the X-ray arrival rate assuming complete absorption) should be limited to 0.74 Mcount/pixel/s, which is much lower than that needed for clinical CT systems.

One potential possibility for decreasing pulse pileup and improving detector count rate is using the so-called small pixel effect. The aforementioned estimated charge collection time was derived assuming that the signal is generated during the entire carrier drift time. However, this is true for the detectors with continuous pixels or with pixilated detectors having pixel sizes comparable to the detector thickness. If the pixel size is much smaller than the detector thickness, then signal generation effectively starts when the charge approaches the pixel, that is, at the distance that is comparable to the pixel size [31,62–65]. When the charge approaches the pixel, the signal is induced primarily at this particular pixel, and induced signal sharply rises during the much shorter drift time. The small pixel effect can allow using pulse shaping times substantially shorter than 150 ns, that is, in the order of 20–50 ns, with a corresponding increase in count rate. However, despite its advantage, the usefulness of the small pixel effect is limited because with small pixels, charge sharing between the pixels is increased.

Another factor limiting the count rate of the CZT (and to some degree CdTe) detectors is the polarization effect due to hole trapping. Hole trapping occurs at defect and impurity sites of semiconductor materials. Generally, both electrons and holes can be trapped. Also, trapping is not permanent, and after some specific time, trapped charges are detrapped. Thus, drifting electrons and holes in CZT/CdTe are accompanied by a series of trappings and detrappings with time constants specific to each type of carrier. The positive space charge of the trapped holes disturbs the electric field within the semiconductor, and the drift and collection of the charges become deteriorated. Thus, polarization negatively affects several key parameters of the detector including not only the count rate but also the energy resolution and spatial uniformity of the pixel response (see Section 3.6.3).

3.6.2 Low-Energy Spectral Tailing

As holes are more likely to be trapped, the total energy (and signal amplitude) depends on the depth of interaction, and the X-ray photons with the same energies can generate signal pulses with different amplitudes [31,62–65]. As a result, the low-energy tailing of the energy spectrum occurs where X-rays absorbed closer to the positively biased electrode.

Low-energy tailing can be substantially decreased if tilted angle irradiation of the detector surface is used. Figure 3.10 shows schematics of the normal irradiation and tilted angle irradiation and corresponding improvement in low-energy tailing. When tilted angle irradiation is used, most X-rays are absorbed closer to the negatively biased electrode so that the electrons drift approximately the same distance and the hole drift length is minimized, which substantially decreases depth dependence of the signal amplitude. The effect of the tilted angle irradiation has been investigated in [22,66,67] and its positive effect has been demonstrated.

Although tilted angle irradiation can decrease low-energy tailing, its application is limited to only a 1D detector as in the case with 1D silicon strip detectors discussed earlier. However, 1D detectors can also be used for certain applications such as in projection X-ray imaging [11,12,40,43–48] and CT [23,42].

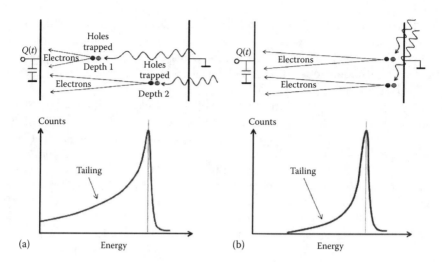

FIGURE 3.10 Low-energy tailing of the energy spectrum with normal irradiation of the CZT detector (a) and decreased tailing effect when tilted angle irradiation is used (b).

Another factor that can partially compensate the low-energy tailing is the small pixel effect discussed earlier. When the pixel size is small enough, the main part of the signal is generated when the electrons drift a shorter distance at close proximity to the pixel; thus, depth dependence of the signal amplitude is decreased. However, the degree of this compensation depends on pixel size, detector thickness, and charge sharing effect. In fact, charge sharing creates low-energy tailing too, and two types of low-energy tailings are mixed together making it difficult to apply correction methods.

3.6.3 Intensity-Dependent Line Artifacts

The pixel response of the radiation imaging detectors generally may vary from pixel to pixel. This effect is observed in conventional energy-integrating detectors including CT detectors and digital flat-panel detectors used for planar X-ray imaging. The reason for nonuniform pixel response can be associated with inherent defects of the detector material and different gain factors of the readout amplifiers. This nonuniform pixel response is corrected by the so-called flat field correction. Notice that although pixel response of the aforementioned detectors may vary from pixel to pixel, the magnitude of this variation is stable over time and does not depend on the intensity of the X-ray.

In photon-counting CZT/CdTe detectors, the mechanisms of nonuniform pixel response described earlier exist too and could be corrected for in a similar way as for other detectors. However, additional mechanisms of pixel nonuniformity appear due to the hole trapping in the detector volume. The positive charge of the trapped holes steers the drifted electrons changing the path in the lateral direction and pushing them from pixel to pixel, which serves as an additional mechanism of nonuniform pixel response [51,68]. The density of the space charge depends on the local X-ray intensity, and the degree of the steering and pixel nonuniformity also depends on the X-ray intensity. For this reason, the pixel response to X-ray intensity is nonlinear, and a nonuniform pixel response cannot be corrected with flat field correction. Notice that intensity-dependent nonuniformity occurs at low and high count rates of the detector.

In practice, for flat field correction, the flat image of the uniform X-ray flux is acquired when no object is placed in the beam, and the average pixel value in the image is normalized to 1. This normalized image includes information about pixel nonuniformities that is used for pixel-by-pixel correction. For flat field correction, the image of the object is simply divided by the normalized flat field image, which corrects each pixel value for nonuniform response. Figure 3.11 shows flat field corrected image of the acrylic slab with 3.6 cm thickness placed on a flat X-ray beam and imaged with a pixilated CZT detector described elsewhere [51–53]. The nonuniformity of the X-ray intensity due to the slab results in line artifacts. The intensity of the artifacts is increased when the slab thickness increases. In clinical practice, projection X-ray and CT imaging deals with much greater changes of the X-ray intensities when the X-ray passes through the patient's body. The above 3.6 cm thick acrylic absorbs the 120 kVp X-ray beam by a factor of 2.4, while a soft tissue with 15 cm thickness would absorb the same beam by a factor of 28, which would result to a larger intensity of the line artifacts.

One potential correction method for the intensity-dependent line artifacts might be direct measurements of the pixel response versus X-ray intensity for each individual pixel in the large range of X-ray intensities [69]. These data could be saved in lookup files and used for correction of the pixel response when the actual image is acquired. However, this method would require generating correction functions for hundreds or thousands of individual pixels. Also, these functions would depend

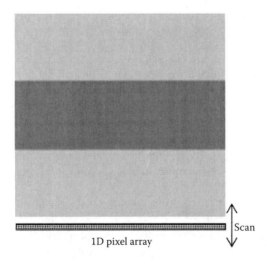

FIGURE 3.11 1D pixel array—demonstration of intensity-dependent line artifacts: An acrylic slab with 3.6 cm thickness was placed in a flat X-ray beam and imaged at 120 kVp tube voltage. Line artifacts remain after flat field correction.

also on the X-ray beam quality. Besides, the correction functions might be unstable because the pixel response is known to change over time. Thus, this correction method does not appear to be reliable.

Another method for addressing the intensity-dependent line artifact problem is using beam flattening filters. A specifically designed filter can be installed between the X-ray tube and the object such that the low-attenuating part of the object receives less radiation and vice versa so that the X-ray flux at the detector surface is more or less uniform. The problem with this method is that the human body has a complex and variable shape and it is difficult to fabricate a filter with the corresponding attenuation profile. In fact, a similar filter called *bow-tie* filter is used in commercial CT systems. However, this filter provides approximate and partial compensation because it has fixed thickness and shape that cannot be matched to a particular patient.

Attempts are also being made to develop a dynamic filter for imaging objects with complex attenuation structures such as the human chest. For example, some works propose using an array of moving wedges made of high-Z material such as Fe or Cu for application in CT [70,71]. This method can potentially provide real-time dynamic intensity compensation. However, a high-Z filter material may result in substantial beam hardening over the least attenuating parts of the object such as the lung in chest CT. The subject contrast may deteriorate over these least attenuation parts, which may not be acceptable. Second, the variation of beam hardening across the field of view may result in spatially variable spectral performance, and quantitative analysis of spectral CT data may become inaccurate.

For some particular type of imaging such as breast CT, a full compensation of intensity variation can be provided using a relatively simple filter [52,72]. In breast CT, the breast is imaged in pendant geometry, so it can be placed in a cylindrical holder. Thus, a filter can be fully adapted to the cylindrically shaped breast with uniform tissue content (Figure 3.12). It is made of tissue-equivalent material (acrylic), which preserves spectral information. The adaptive filter provides many advantages: it prevents intensity-dependent line artifacts, decreases the required detector count rate, and provides uniform beam hardening across the field of view. Thus, using an adaptive filter makes it possible to build a photon-counting spectral breast CT system using commercially available pixilated CZT detectors [52].

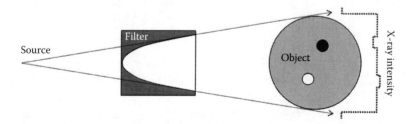

FIGURE 3.12 Flattening X-ray beam with an adaptive filter made of tissue-equivalent material (acrylic). The intensity variation due to the cylindrical shape of the phantom is fully compensated while attenuation profiles of the contrast elements remain.

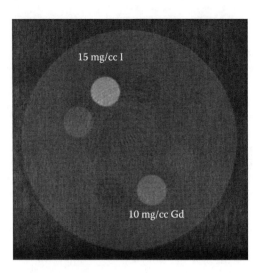

FIGURE 3.13 Computed tomography image of the acrylic phantom with 14 cm diameter acquired with 256-pixel photon-counting CZT detector at 80 kVp.

The CT image of the cylindrical breast phantom with 14 cm diameter acquired with a photon-counting CZT detector is shown in Figure 3.13. The CT image was acquired at 80 kVp tube voltage and the adaptive filter with 14 cm maximum thickness made of acrylic was used. The phantom included contrast agents of iodine and Gd with densities of 15 and 10 mg/cm^3 in water solution. The contrast elements were filled in cylindrical holes with 2 cm diameter made in the phantom. As can be seen from the image, some line artifacts appear over the areas where contrast agents are located. These are typical intensity-dependent line artifacts as they appear in the reconstructed CT image. Although the adaptive filter fully compensates the intensity variation associated with the cylindrical shape of the phantom, it cannot compensate the intensity changes due to the contrast elements.

In this particular case, the iodine and Gd contrasts with 2 cm thicknesses decreased the beam intensity in corresponding CT projections by 28% and 17%, respectively. In practice, the presence of adipose and glandular tissue in the breast is not expected to provide such a large nonuniformity in the beam intensity, and no artifacts are expected. Also, the typical iodine contrast in contrast-enhanced breast imaging is about 2–5 mg/cm^3, which also would not cause line artifacts.

3.6.4 Charge Sharing

Charge sharing occurs when the charge generated by a single X-ray photon is accepted at more than one detector pixel. There are different physical mechanisms that result in charge sharing. These mechanisms could take place separately or simultaneously for the same detected photon. In all cases, the charge created by a single photon is expanded at the time when it arrives at the pixel and can be shared between two or more pixels. Charge sharing deteriorates the spatial resolution, energy resolution, and SNR of the photon-counting detector. The negative effect of charge sharing is mostly emphasized in Medipix detectors due to their small pixel sizes. The corresponding effect for CT detectors with larger pixel sizes is relatively small but still is not negligible.

In the detection process, the X-ray photon is converted to an energetic photo or Compton electron, and the electron travels some distance ionizing the detector material and creating the charge carriers. Thus, the charge cloud at the time of origination has already some expansion due to the electron track. Further, the charge is expanded by electrostatic repulsion during the drift toward the pixel electrodes. A lateral diffusion due to the thermal motion of the charge occurs during the drift, which further expands the charge cloud. If the X-ray interacts via Compton effect (which primarily occurs in Si detectors and is less probable in CZT/CdTe detectors), the scattered photon may be reabsorbed at some distance from the scatter point, and this may result in additional charge sharing. Similarly, if the X-ray interacts via photoeffect, then the characteristic K-X-rays may be reabsorbed at some distance from the original interaction point, which may further expand the charge cloud. Charge sharing due to the aforementioned mechanisms has been investigated in numerous works for different types of semiconductor detectors. However, it is difficult to accurately predict the net effect of charge sharing due to multiple mechanisms occurring at a time, due to the complexity of simulating some of them and, also, due to a wide energy range of the X-ray spectrum [73–84].

For the charge carriers created by X-ray photons, the electrostatic repulsion component of the expansion is small and the diffusion component dominates. The lateral diffusion can be approximated when a point-like charge is drifted by an electric field E to a distance x at temperature T by a Gaussian shape with the σ parameter determined as $\sigma = \sqrt{2kTx/eE}$, where e is the electron charge and k is the Boltzmann constant [31]. Taking into account that at room temperature $kT/e = 0.0253\,V$ and assuming that charge drifts across the entire detector thickness d, a simple expression for the FWHM of the charge cloud $FWHM = 0.374d/\sqrt{U}$ can be established, where U is the potential applied to the detector (in V). For a typical CZT detector with a thickness of 3 mm and applied voltage of 600 V, the magnitude of the charge expansion would be 46 µm. For Si strip detectors with 0.5 mm thickness and 150 V applied voltage, the magnitude of the charge expansion would be 15 µm.

Although charge sharing due to Compton scatter may take place in Si strip detectors, its negative effect is expected to be small due to the small thickness of the Si wafer so that most scattered photons would leave the detector volume.

The K-edge characteristic X-rays of Cd and Te with average energies of 23.4 and 27.5 keV, respectively, are generated in CZT/CdTe materials. The K-X-rays are created when the X-ray energy is higher than the K-edge energy of Cd and Te that are 26.7 and 31.8 keV, respectively. The characteristic X-rays of Cd and Te have mean free path in the material 116 and 64 µm, respectively. Therefore, they can well penetrate to the neighboring pixels and be absorbed there. They may also leave the detector volume from the front and back surface of the crystal. If the characteristic X-ray is detected in neighboring pixel, this results in double counting and also in deterioration of the energy spectrum. If it leaves the detector volume, then the energy spectrum is deteriorated, but double counting does not occur. The effect of the characteristic X-ray escape on the spatial and energy resolution of the pixilated CZT detectors for photon-counting X-ray imaging has been investigated in [84]. It has been shown that 1D strip–pixel detectors suffer from K-X-ray escape less than 2D square-pixel detectors. Additionally, tilted angle irradiation further decreases the negative effect of the K-X-ray escape [84].

One method for addressing the charge sharing problem could be developing a sophisticated readout circuit that could detect the shared charge signals in coincidence and apply summation. This would partially restore the energy information. Also, the position of the detected photon could be determined from the pixel with the highest signal, which could decrease the fraction of the double or multiple counting. Similar circuits have been developed for Medipix detectors that strongly suffer from charge sharing due to its small pixel size [57,59]. To further investigate charge sharing between the neighboring pixels of the Medipix detectors, one pixel from each 2 × 2 or 3 × 3 pixel array was connected to readout circuit, and the effect of the sparse pixel configuration on charge sharing was quantified [58]. Nevertheless, charge sharing still remains a serious problem degrading the spatial and energy resolution and SNR of the spectroscopic PCXCT detectors.

3.6.5 Suboptimal Energy Resolution

As discussed in the previous sections, a very high energy resolution is not necessary for clinical application of the PCXCT. However, the energy response of the PCXCT detectors is deteriorated even more strongly and in a very complicated way so that developing reliable correction methods is difficult. Figure 3.14 demonstrates on how the X-ray spectrum deteriorates as a result of several different distortion mechanisms that act at a time.

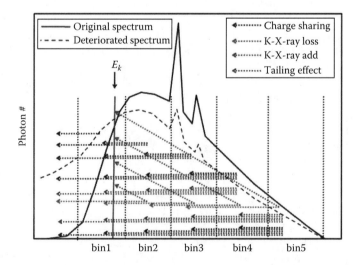

FIGURE 3.14 (See color insert.) Schematics that show moving photon counts from one to another energy bin due to the different mechanisms. The original spectrum of the X-ray is deteriorated in unpredictable manner.

The photon counts shift from higher energy bins to lower energy bins due to several mechanisms including charge diffusion and sharing, K-X-ray loss, K-X-ray add, and low-energy tailing. In addition, the intrinsic energy resolution is decreased due to the electronic noise and leakage current, and energy bins are broadened and partially overlapped (not shown in Figure 3.14). Although the original and deteriorated spectra shown in Figure 3.14 look more or less close to each other, this is a misleading impression because spectral content of the energy bins deteriorate more strongly. For example, in the energy bin2, the count numbers under original and deteriorated spectra look similar. However, a substantial fraction of the original counts of bin2 are shifted to lower bins and the corresponding "vacancy" is "filled" by the counts coming from higher bins. Notice also that the fraction of the counts that fall behind the lowest energy threshold of bin1 and that are not detected is unknown. The minimum threshold level for existing photon-counting systems based on CZT/CdTe detectors is approximately 15–20 keV, below which electronic noise dominates. On the other hand, the fraction of the counts below this threshold can be substantial as can be seen from Figure 3.14.

While there are no optimal ways to correct the deteriorated energy spectra, some methods can still be applied that can decrease negative effects of this deterioration. For example, as discussed earlier, the image data acquired in multiple energy bins can be weighted according to the GIW method using the weighting factors derived from already deteriorated multibin images. In this way, the highest SNR can be achieved in the conditions of the deteriorated energy information [26].

Another problem that can be partially addressed is overlapping of the energy bins. Existing pixilated CZT/CdTe detectors provide an FWHM energy resolution of 15%–25% due to relatively high electronic noise, higher carrier creation energy, and to some degree, leakage current. The limited energy resolution results in blurring energy bin borders and partial overlapping of the energy bins (Figure 3.15).

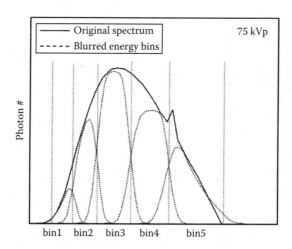

FIGURE 3.15 Blurring of the energy bins due to suboptimal energy resolution was simulated assuming 15% full width at half maximum energy resolution of the detector. Substantial overlapping of the bin borders occurs.

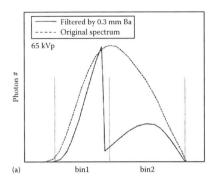 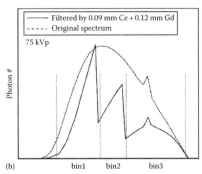

FIGURE 3.16 Demonstration of the K-edge filtration of the X-ray beam to physically separate the energy bins using (a) one and (b) two K-edge filters. Two positive effects of K-edge filtration are decreasing overlapped fraction of the energy bins and providing average bin energy that is more distinct and better separated. The K-edge filter is installed at the X-ray tube output so that it does not increase patient dose.

The overlapped energy bins, in turn, provide deteriorated material decomposition and other spectral performance of the system.

To address the bin separation problem, it has been proposed that the X-ray beam can be prefiltered by high-Z material with appropriate location of the K-edge energies [53]. The optimal filter materials are installed at the X-ray tube output and shape the X-ray energy spectrum such that some gaps between the energy bins are created. It has been demonstrated experimentally that this method substantially improves SNR in material-decomposed images [53]. Although some fraction of the X-ray beam is absorbed by K-edge filters, the method remains dose efficient because beam absorption occurs before the patient is exposed to an X-ray. Figure 3.16 shows single and double K-edge filtered X-ray beams to separate two and three energy bins, respectively, and substantial separation of the energy bins by K-edge filtration is observed. Notice that K-edge filtration also makes the bin energies narrower and bin separation better. Therefore, K-edge filtration method would be useful even for the detectors with high energy resolution when no blurring of the bin borders occurs.

Nevertheless, problems of suboptimal energy response of the existing photon-counting detectors associated with charge sharing, hole trapping, and low energy resolution remain largely unsolved, even though spectroscopic CXCT does not require very high energy resolution.

3.7 FEASIBLE CLINICAL APPLICATION: PHOTON-COUNTING SPECTRAL BREAST CT

Despite the fact that current PCXCT technology does not generally meet the demands of clinical applications, PCXCT can be feasible for some particular applications such as dedicated breast CT. The dedicated breast CT by itself is not a new topic. It has been developed and investigated in the 1970s, but did not find widespread application due to suboptimal spatial resolution and long imaging time of CT systems of that era [85–88]. The dedicated breast CT has been revisited later in the 2000s, and cone beam breast CT systems based on flat-panel detectors have been under investigation by several groups [89–92]. All these systems used conventional energy-integrating detectors.

Photon-counting breast CT has been proposed in 2004 [93] and investigated in [23,42,52]. It has been shown that breast CT with commercially available photon-counting CZT/CdTe detectors is feasible. Furthermore, an experimental photon-counting CT system based on 2 × 256 pixel array of CZT detector has been developed and compared directly to clinical CT system (Siemens Sensation 16) by imaging a breast CT phantom [52]. It has been experimentally demonstrated that the performance parameters of the photon-counting breast CT system is comparable or better than that of Siemens Sensation 16, while photon-counting system provides material decomposition in a single CT scan and at a fixed X-ray tube voltage using energy-selective data.

The clinically applicable photon-counting breast CT is possible mainly due to the fact that the breast is imaged in pendant geometry and can be placed in a cylindrical holder. As discussed earlier, this cylindrical geometry, as well as uniform tissue content of the breast, makes it possible using adaptive filtration, which provides nearly complete flattening of the beam intensity [53,72]. Using adaptive filter decreases the required count rate of the detector to approximately 1–2 Mcount/pixel/s, which is achievable with current CZT detectors [23,52]. The flat X-ray intensity also eliminates intensity-dependent line artifacts and beam hardening.

3.8 CONCLUSION

Extensive research and development efforts taken in the last two decades brought the PCXCT technology closer to the clinical applications. However, some of the key parameters such as high count rates and accurate pixel response still exhibit a problem. Generally, the vector of the advancements of PCXCT technology seems to be in the right direction, and one can hope that some clinical PCXCT systems might be utilized in hospitals in the coming decades. Furthermore, some particular applications such as photon-counting spectral breast CT are already feasible today and can be built based on existing PCXCT detectors.

REFERENCES

1. Webb, S., *The Physics of Medical Imaging*. New York: Taylor & Francis Group, 1988, p. xv, 633pp.
2. Bushberg, J.T. et al., *The Essential Physics of Medical Imaging*, 2nd edn. Baltimore, MD: Williams & Wilkins, 2002, 933pp.
3. Hendee, W.R. and E.R. Ritenour, *Medical Imaging Physics*. New York: John Wiley & Sons, 2002.
4. Khan, F.M., *The Physics of Radiation Therapy*, 3rd edn. Philadelphia, PA: Lippincott Williams & Wilkins, 2003, 560pp.
5. Hounsfield, G.N., Computerized transverse axial scanning (tomography): Part I. Description of system. *Br. J. Radiol.*, 1973;46:1016–1022.
6. McCullough, E.C. and J.T. Payne, X-ray transmission computed tomography. *Med. Phys.*, 1977;4(2):85–98.

7. Hasegawa, B.H. et al., A prototype high-purity germanium detector system with fast photon-counting circuitry for medical imaging. *Med. Phys.*, 1991;18(5):900–909.
8. Tsutsui, H. et al., X-ray energy separation method using a CdTe semiconductor X-ray imaging sensor and photon counting method. *IEEE Trans. Nucl. Sci.*, 1993;40(1):40–44.
9. Kravis, S.D. et al., Readout electronics for nuclear applications (RENA) chip. In *1997 IEEE Nuclear Science Symposium Conference Record*, Albuquerque, NM, 1997, Vol. 1, pp. 700–703.
10. Benini, L. et al., Synchrotron radiation application to digital mammography. A proposal for the Trieste project "Elettra". *Phys. Med.*, 1990;6:293–298.
11. Arfelli, F. et al., Design and evaluation of AC-coupled, FOXFET-biased, "edge-on" silicon strip detectors for X-ray imaging. *Nucl. Instrum. Methods Phys. Res.*, 1997;A385(2):311–320.
12. Danielsson, M. et al., Dose-efficient system for digital mammography. *Proc. SPIE*, 2000;3977:239–249.
13. Shikhaliev, P.M., Detector of ionizing radiation. Patent #2066465, Russian Federation, 1993.
14. Shikhaliev, P.M., A novel gamma-ray imaging concept using "edge-on" microchannel plate detector. *Nucl. Instrum. Methods Phys. Res.*, 2001;A460(2–3):465–468.
15. Shikhaliev, P.M. et al., Scanning-slit photon counting X-ray imaging system using a microchannel plate detector. *Med. Phys.*, 2004;31(5):1061–1071.
16. Babichev, E.A. et al., Photon counting and integrating analog gaseous detectors for digital scanning radiography. *Nucl. Instrum. Methods Phys. Res.*, 1998;A419:290–294.
17. Thunberg, S. et al., Dose reduction in mammography with photon counting imaging. *Proc. SPIE*, 2004;5368:457–465.
18. Beutel, J., H.L. Kundel, and R.L. Van Metter, *Medical Imaging*, Vol. I. Bellingham, WA: SPIE Press, 2000, p. 931.
19. Tapiovaara, M.J. and R.F. Wagner, SNR and DQE analysis of broad spectrum X-ray imaging. *Phys. Med. Biol.*, 1985;30:519–529.
20. Cahn, R.N. et al., Detective quantum efficiency dependence on X-ray energy weighting in mammography. *Med. Phys.*, 1999;26(12):2680–2683.
21. Shikhaliev, P.M., Beam hardening artefacts in computed tomography with photon counting, charge integrating and energy weighting detectors: A simulation study. *Phys. Med. Biol.*, 2005;50:5813–5827.
22. Shikhaliev, P.M., Tilted angle CZT detector for photon counting/energy weighting X-ray and CT imaging. *Phys. Med. Biol.*, 2006;51:4267–4287.
23. Shikhaliev, P.M., Computed tomography with energy resolved detection: A feasibility study. *Phys. Med. Biol.*, 2008;53:1475–1495.
24. Niederlohner, D. et al., The energy weighting technique: Measurements and simulations. *Nucl. Instrum. Methods Phys. Res.*, 2005;A546:37–41.
25. Schmidt, T.G., Optimal "image-based" weighting for energy-resolved CT. *Med. Phys.*, 2009;36(7):3018–3027.
26. Shikhaliev, P.M., The upper limits of the SNR in radiography and CT with polyenergetic X-rays. *Phys. Med. Biol.*, 2010;55:5317–5339.
27. Shikhaliev, P.M., Dedicated phantom materials for spectral radiography and CT. *Phys. Med. Biol.*, 2012;57:1575–1593.
28. Samei, E. et al., Comparative scatter and dose performance of slot-scan and full-field digital chest radiography systems. *Radiology*, 2005;235:940–949.
29. Mainprize, J.G., N.L. Ford, S. Yin, T. Tumer, and M.J. Yaffe, A slot-scanned photodiode-array/CCD hybrid detector for digital mammography. *Med. Phys.*, 2002;29(2):214–225.
30. Fink, J. et al., Characterization of charge collection in CdTe and CZT using the transient current technique. *Nucl. Instrum. Methods Phys. Res. A*, 2006;A560:435–443.

31. Knoll, G.F., *Radiation Detection and Measurement*, 3rd edn. New York: John Wiley & Sons, Inc., 2000, p. 802.
32. Siewerdsen, J.H. and D.A. Jaffray, Cone-beam computed tomography with a flat-panel imager: Magnitude and effects of X-ray scatter. *Med. Phys.*, 2001;28(2):220–231.
33. Siewerdsen, J.H. et al., The influence of antiscatter grids on soft-tissue detectability in cone-beam computed tomography with flat-panel detectors. *Med. Phys.*, 2004;31(12):3506–3520.
34. Endo, M., S. Mori, and T. Tsunoo, Magnitude and effects of X-ray scatter in a 256-slice CT scanner. *Med. Phys.*, 2006;33:3359–3368.
35. Barnes, G.T., H.M. Cleare, and M.S. Brezovich, Reduction of scatter in diagnostic radiology by means of a scanning multiple slit assembly. *Radiology*, 1976;120:691–694.
36. Moore, R., D. Korbuly, and K. Amplatz, A method to absorb scattered radiation without attenuation of the primary beam. *Radiology*, 1976;120:713–717.
37. Barnes, G.T. and I.A. Brezovich, The design and performance of a scanning multiple slit assembly. *Med. Phys.*, 1979;6(3):197–204.
38. King, M.A., G.T. Barnes, and M.V. Yester, A mammographic scanning multiple slit assembly: Design considerations and preliminary results. In *Reduced Dose Mammography*, W.W. Logan and E.P. Muntz (eds.). New York: Masson, 1979, pp. 243–252.
39. Yester, M.V., G.T. Barnes, and M.A. King, Experimental measurements of the scatter reduction obtained in mammography with a scanning multiple slit assembly. *Med. Phys.*, 1981;8(2):158–162.
40. Aslund, M. et al., Scatter rejection in scanned multi-slit digital mammography. *Proc. SPIE*, 2004;5368:478–487.
41. Aslund, M. et al., Scatter rejection in multislit digital mammography. *Med. Phys.*, 2006;33(4):933–940.
42. Shikhaliev, P.M., T. Xu, and S. Molloi, Photon counting CT: Concept and initial results. *Med. Phys.*, 2005;32(2):427–436.
43. Arfelli, F. et al., Silicon X-ray detector for synchrotron radiation digital radiology. *Nucl. Instrum. Methods Phys. Res. A*, 1994;353(1–3):366–370.
44. Arfelli, F. et al., At the frontiers of digital mammography: SYRMEP. *Nucl. Instrum. Methods Phys. Res.*, 1998;A409:529–533.
45. Arfelli, F., Synchrotron light and imaging systems for medical radiology. *Nucl. Instrum. Methods Phys. Res.*, 2000;A454:11–25.
46. Beuville, E. et al., An application specific integrated circuit and data acquisition system for digital X-ray imaging. *Nucl. Instrum. Methods Phys. Res.*, 1998;A406:337–342.
47. Aslund, M. et al., Physical characterization of a scanning photon counting digital mammography system based on Si-strip detectors. *Med. Phys.*, 2007;34(6):1918–1925.
48. Hilt, B., P. Fessler, and G. Prevot, The quantum X-ray radiology apparatus. *Nucl. Instrum. Methods*, 2000;A442:355–359.
49. Tumer, T.O. et al., Preliminary results obtained from a novel CdZnTe pad detector and read-out ASIC developed for an automatic baggage inspection system. In *2000 IEEE Nuclear Science Symposium Conference Record*, 2000, Vol. 1, Lyon, France, pp. 4/36–4/41.
50. Shikhaliev, P.M., Energy resolved computed tomography: First experimental results. *Phys. Med. Biol.*, 2008;53:5595–5613.
51. Shikhaliev, P.M., Projection X-ray imaging with photon energy weighting: Experimental evaluation with a prototype detector. *Phys. Med. Biol.*, 2009;54:4971–4992.
52. Shikhaliev, P.M. and S.G. Fritz, Photon counting spectral CT versus conventional CT: Comparative evaluation for breast imaging application. *Phys. Med. Biol.*, 2011;56:1905–1930.
53. Shikhaliev, P.M., Photon counting spectral CT: Improved material decomposition with K-edge filtered X-rays. *Phys. Med. Biol.*, 2012;57:1595–1615.

54. Schlomka, J.P. et al., Experimental feasibility of multi-energy photon-counting K-edge imaging in pre-clinical computed tomography. *Phys. Med. Biol.*, 2008;53(15):4031–4047.
55. Iwanczyk, J.S. et al., Photon counting energy dispersive detector arrays for X-ray imaging. In *2007 IEEE Nuclear Science Symposium Conference Record*, Honolulu, 2007, pp. 2741–2748.
56. Barber, W.C. et al., Characterization of a novel photon counting detector for clinical CT: Count rate, energy resolution, and noise performance. *Proc. SPIE*, 2009;7258:2401–2409.
57. Ballabriga, R. et al., Characterization of the Medipix3 pixel readout chip. *J. Instrum.*, 2011;6:C01052.
58. Koenig, T. et al., Imaging properties of small-pixel spectroscopic X-ray detectors based on cadmium telluride sensors. *Phys. Med. Biol.*, 2012;57:6743–6759.
59. Koenig, T. et al., Charge summing in spectroscopic X-ray detectors with high-Z sensors. *IEEE Trans. Nucl. Sci.*, 2013;60(6):4713–4718.
60. Poludniowski, G. et al., SpekCalc: A program to calculate photon spectra from tungsten anode X-ray tubes. *Phys. Med. Biol.*, 2009;54:433–438.
61. Johns, H.E. and J.R. Cunningham, *The Physics of Radiology*, 4th edn. Springfield, IL: Thomas, 1983, p. 785.
62. Hecht, K., Zum Mechanismus des lichtelektrischen Primarstromes in isolierenden Kristallen. *Z. Phys.*, 1932;77:235–245.
63. Akutagawa, W. and K. Zanio, Gamma response of semi-insulating material in the presence of trapping and detrapping. *J. Appl. Phys.*, 1969;40(9):3838–3854.
64. Barrett, H.H., J.D. Eskin, and H.B. Barber, Charge transport in arrays of semiconductor gamma-ray detectors. *Phys. Rev. Lett.*, 1995;75(1):156–159.
65. Eskin, J.D., H.H. Barrett, and H.B. Barber, Signals induced in semiconductor gamma-ray imaging detectors. *J. Appl. Phys.*, 1999;85(2):647–659.
66. Fritz, S.G. and P.M. Shikhaliev, CZT detectors used in different irradiation geometries: Simulations and experimental results. *Med. Phys.*, 2009;36(4):1098–1108.
67. Fritz, S.G., P.M. Shikhaliev, and K.L. Matthews, Improved X-ray spectroscopy with room temperature CZT detectors. *Phys. Med. Biol.*, 2011;56:5735–5751.
68. Soldner, S.A., D.S. Bale, and C. Szeles, Dynamic lateral polarization in CdZnTe under high flux X-ray irradiation. *IEEE Trans. Nucl. Sci.*, 2007;54(5):1723–1727.
69. Rundle, D., Private communication, 2006.
70. Szczykutowicz, T.P. and C.A. Mistretta, Design of a digital beam attenuation system for computed tomography: Part I. System design and simulation framework. *Med. Phys.*, 2013;40(2):021905-1–021905-12.
71. Hsieh, S.S. and N.J. Pelc, The feasibility of a piecewise-linear dynamic bowtie filter. *Med. Phys.*, 2013;40(3):031910-1–031910-12.
72. Silkwood, J.D., K.L. Matthews, and P.M. Shikhaliev, Photon counting spectral breast CT: Effect of adaptive filtration on CT numbers, CT noise, and CNR. *Med. Phys.*, 2013;40(5):051905-1–051905-15.
73. Lindqvist, M. et al., Computer simulations and performance measurements on a Si strip detector for edge-on imaging. *IEEE Trans. Nucl. Sci.*, 2000;47(4):1487–1492.
74. Nilsson, H.E., C. Fröjdh, and E. Dubaric, Monte Carlo simulation of charge sharing effects in silicon and GaAs photon-counting X-ray imaging detectors. *IEEE Trans. Nucl. Sci.*, 2004;51(4):1636–1640.
75. Nilsson, H.E. et al., Charge sharing suppression using pixel-to-pixel communication in photon counting X-ray imaging systems. *Nucl. Instrum. Methods Phys. Res.*, 2007;A576:243–247.
76. Norlin, B., C. Frojdh, and H.E. Nilsson, Spectral performance of a pixellated X-ray imaging detector with suppressed charge sharing. *Nucl. Instrum. Methods Phys. Res.*, 2007;A576:248–250.

77. Durst, J., G. Anton, and T. Michel, Discriminator threshold dependency of the zero-frequency DQE of photon-counting pixel detectors. *Nucl. Instrum. Methods Phys. Res.*, 2007;A576:235–238.
78. Korn, A. et al., Investigation of charge carrier transport and charge sharing in X-ray semiconductor pixel detectors such as Medipix2. *Nucl. Instrum. Methods Phys. Res.*, 2007;A576:239–242.
79. Chen, C.M.H. et al., Numerical modeling of charge sharing in CdZnTe pixel detectors. *IEEE Trans. Nucl. Sci.*, 2002;49(1):270–276.
80. Pellegrini, G. et al., Performance limits of a 55-m pixel CdTe detector. *IEEE Trans. Nucl. Sci.*, 2006;53(1):361–366.
81. d'Aillon, E.G. et al., Charge sharing on monolithic CdZnTe gamma-ray detectors: A simulation study. *Nucl. Instrum. Methods Phys. Res.*, 2006;A563:124–127.
82. Iniewski, K. et al., Modeling charge-sharing effects in pixellated CZT detectors. In *2004 IEEE Nuclear Science Symposium Conference Record*, 2007, Vol. NM2-3(6), pp. 4608–4611.
83. Kuvvetli, I. and C.B. Jørgensen, Measurements of charge sharing effects in pixilated CZT/CdTe detectors. In *2007 IEEE Nuclear Science Symposium Conference Record*, 2007, Vol. N47-6(3), pp. 2252–2257.
84. Shikhaliev, P.M., S.G. Fritz, and J.W. Chapman, Photon counting multienergy X-ray imaging: Effect of the characteristic x rays on detector performance. *Med. Phys.*, 2009;36(11):5107–5119.
85. Chang, C.H. et al., Computed tomography of the breast. *Radiology*, 1977;124:827–829.
86. Chang, C.H. et al., Specific value of computed tomographic breast scanner (CT/M) in diagnosis of breast diseases. *Radiology*, 1979;132:647–652.
87. Chang, C.H., J.L. Sibala, and S.L. Fritz, Computed tomographic evaluation of the breast. *Am. J. Roentgenol.*, 1978;131:459–464.
88. Chang, C.H.J. et al., Computed tomography in detection and diagnosis of breast cancer. *Cancer*, 1980;46:939–946.
89. Boone, J.M. et al., Dedicated breast CT: Radiation dose and image quality evaluation. *Radiology*, 2001;221:657–667.
90. Chen, B. and R. Ning, Cone-beam volume CT breast imaging: Feasibility study. *Med. Phys.*, 2002;29(5):755–770.
91. Karellas, A. and S. Vedantham, Breast cancer imaging: A perspective for the next decade. *Med. Phys.*, 2008;35(11):4878–4897.
92. Glick, S.J., Breast CT. *Annu. Rev. Biomed. Eng.*, 2007;9:501–526.
93. Shikhaliev, P.M., Photon counting computed tomography for breast imaging. NIH proposal R01 EB004563-01, January 26, 2004 (unpublished).

4 Pixelated Semiconductor and Parallel ASIC Design for Spectral Clinical Radiology

William C. Barber, Einar Nygard, and Jan S. Iwanczyk

CONTENTS

4.1 Introduction 81
4.2 Direct Conversion Semiconductor Sensor Development 87
4.3 ASIC Development 93
4.4 Module Development for Spectral Computed Tomography 95
4.5 Detector Development for Spectral Bone Mineral Densitometry 104
4.6 Module Development for Spectral Digital Mammography 107
4.7 Discussion 110
4.8 Conclusions 111
Acknowledgments 112
References 113

4.1 INTRODUCTION

This chapter focuses on the design of sensors and readouts for the development of X-ray imaging arrays for applications in clinical radiology. Clinical radiology is narrowly defined here as pertaining to the in vivo imaging of humans using X-ray photons. Silicon (Si)-based semiconductor and cadmium telluride (CdTe) and cadmium zinc telluride (CdZnTe) compound semiconductor sensor development using fast mixed signal application-specific integrated circuits (ASICs) as readouts that incorporate a portion of the digital readout are discussed with respect to their use in applications of spectral clinical radiology.

Soon after the discovery of X-rays, image capture and storage technologies were developed initially using photographic plates that were soon replaced with photographic film. Although currently being replaced by digital flat panel X-ray imaging arrays, X-ray photographic film is still widely used and consists of a transparent plastic substrate made of acetate or polyester that is coated a with gelatin emulsion layer, usually on both sides, which contains silver halide crystals. Exposure to photons splits ions producing atomic silver that appears black. The amount of blackening is proportional to the total energy deposited in the emulsion layers, which is equal to the X-ray intensity times the exposure time. Passing white light through the film then produces a negative image of the X-ray intensity. Since white light is passed through the film to view the image, the dynamic range, defined as the range between the minimum and maximum deposited energy discernible, is related to the film's optical density, which is a measure of film blackness for visible light. The optical density is related to the number of silver–bromine grains per unit area and the absorption cross section of the grains. There is therefore a trade-off between the film sensitivity and spatial resolution where increasing the grain size increases sensitivity and reduces spatial resolution. Also increasing the emulsion layer thickness increases sensitivity and reduces spatial resolution. Noise in the image, which limits the contrast, arises from fluctuations in the number of absorbed X-rays (quantum mottle), fluctuations in the amount of deposited energy (Swank noise), and fluctuations in the number of silver halide grains per unit area (random darkening). Quantum mottle and random darkening are the dominant sources of image noise in X-ray images captured with X-ray photographic film.

In attempts to overcome the limited dynamic range of X-ray photographic film, digital image capture and storage technologies were developed initially using phosphor plates that were eventually replaced with pixelated flat panel X-ray imaging arrays. Image capture is achieved in phosphor plates by exposure to X-rays that excite electrons that are trapped in the crystal lattice and thus stored. The image is then read out by scanning a visible light source across the plate and recording the luminescent signal emitted with a photomultiplier tube (PMT). Europium-doped barium fluorobromide is a commonly used phosphor. The luminescence generated is proportional to the number of trapped electrons that is proportional to the total deposited energy, as it was with X-ray photographic film; however, phosphor plates have a larger dynamic range and a much more linear response. The signal from the scanned PMT can be digitized producing a two-dimensional (2D) array of numbers where the numbers are proportional to the total deposited energy at that location, which in its most general sense is a digital 2D projection image. Flat panel X-ray imaging arrays provide image capture and storage in a single step by connecting a pixelated digital readout array to an X-ray sensor where the readout produces a 2D digital image that is easily stored and shared via computers. Separation of the sensor, responsible for X-ray absorption and signal generation, from the readout, responsible for signal amplification, processing, and storage, decouples the spatial resolution from the dynamic range; however, there remains a trade-off between the contrast resolution and the spatial resolution that is limited by image noise. In this case, noise in the image, which limits the contrast, arises from fluctuations in the number of absorbed X-rays (quantum mottle), fluctuations in the amount of deposited energy (Swank noise), and fluctuations in the number of charges due to the electronic noise in the readout and thermal current in the sensor under bias. Quantum mottle and electronic noise are the dominant sources of image noise in X-ray images captured with digital flat panel X-ray imaging arrays.

The sensors used in flat panel X-ray imaging arrays use either indirect or direct conversion methods to generate a signal from X-ray interactions. Indirect conversion involves the conversion of the X-ray energy deposited in a scintillator to optical photons that are read out by photodiodes (PDs) which produce a charge proportional to the number of optical photons that is proportional to the total energy deposited. Whereas, direct conversion involves the conversion of the X-ray energy deposited in a semiconductor to an electric charge that is proportional to the total energy deposited. Common indirect converters used for the sensors in flat panel X-ray imaging arrays include gadolinium oxysulfide (Gd_2O_2S) and thallium doped cesium iodide (CsI:Tl). These scintillators are either optically coupled to or grown directly on a thin-film transistor (TFT) array comprised of amorphous silicon (a-Si) on glass substrates. Each pixel contains a PD that generates the electrical signal. The most common direct converter used as a sensor in flat panel X-ray imaging arrays is amorphous selenium (a-Se). The top side of the a-Se layer, which is incident to the X-rays, is coated with a continuous thin metal film layer for applying a bias voltage between the top side of the a-Se layer and the pixels of the TFT array that are connected to the bottom side of the a-Se layer. X-ray photons absorbed in the a-Se layer generate electron–hole pairs. The charge generated travels along the electric field lines generated by the bias voltage and the induced signal is read out by a TFT array.

The sensors used in flat panel X-ray imaging arrays have a higher detective quantum efficiency (DQE) for the short wavelength X-rays used in radiology as compared to X-ray photographic film and phosphor plates, and this allows for a lower radiation dose for a given contrast and spatial resolution. Also the readouts used in flat panel X-ray imaging arrays allow for images to be captured and stored much faster than X-ray photographic film and phosphor plates, which require film development or storage phosphor readout, respectively.

Applications in radiology for digital X-ray imaging arrays are not limited to those served by flat panels such as general practice radiography for bone fracture and chest imaging, digital mammography (DM) and digital breast tomosynthesis for breast imaging, and cone beam computed tomography (CT) for brain imaging and treatment planning in image-guided radiation therapy, but are also used in scanning methods such as bone mineral densitometry (BMD) and whole body CT where in these cases the X-ray imaging arrays are tiled in multirow arrays and used in a fan beam geometry. TFT arrays that use a multiplexed readout where gate and drain lines are connected to the source and drain of the individual TFT's along rows and columns, respectively, to reduce the number of channel from N^2 to $2N$ for an $N \times N$ array are generally not used for CT due to a lower bound on the minimum integration (frame) time imposed by the speed of their circuits. Whereas the development of large-area TFT arrays coupled to or coated with large-area scintillator or a-Se layers allows for flat panel devices with a large field of view (FOV) for image capture and readout in radiography and fluoroscopy (cine mode), scanning arrays for BMD and fan beam CT require tiling of many modules to create a large FOV with readout schemes that allow for sequential frames to be read out with short frame times generally below 1 ms for clinical CT. These frame rates, in excess of 1000 frames per second (fps) for CT, require parallel channel readouts for all individual pixels.

All of these X-ray image capture and storage technologies, when applied to radiology, produce a spatially registered X-ray intensity map that provides an image of human anatomy because different tissues have different X-ray attenuation coefficients and the X-ray source uniformly irradiates the portion of the body being imaged for a fixed exposure time. Therefore an X-ray image is a spatial distribution of the linear attenuation coefficients of an object, μ, where μ at each location is determined by the chemical composition of the object, the mass density of the object, and the energy of X-ray photon.[1] In clinical radiology, the X-ray tube projects an X-ray beam with a broad energy spectrum and the number of photons with a particular energy obey Poisson statistics. The attenuation coefficients of materials are larger in general at lower photon energies, and X-ray photons with lower energy are more heavily attenuated than the ones with higher energy. Also, except near the k-edge energy of a particular element, the difference in attenuation coefficients between different tissue types is larger at lower energy.

All of these technologies, when used in X-ray imaging, are energy integrating (EI) in that they produce a signal that is proportional to the total deposited energy during a fixed exposure time. As mentioned, the X-rays used in clinical radiology projects a broad energy spectrum and the EI readout and/or display of these technologies places a *weight* proportional to the energy of each X-ray in the intensity signal that is not optimal due to the shape of the attenuation coefficient curve as a function of energy. Recognition of this has led to many dual-energy techniques usually using EI detectors and multiple exposures at different or modulated tube potentials (dual kVp) and/or with different filters between the source and patient.

The classic paper by Tapiovaara and Wagner[2] showed that broad-spectrum X-ray imaging with an energy modulating source and EI detectors could improve signal-to-noise ratios (SNRs) compared to a nonenergy-modulating source and EI detectors or a system using photon-counting (PC) detectors that count all photons above a single threshold but provide no energy information, provided the beam modulation is optimized for the specific imaging task such as the detection of a lesion in a surrounding tissue. The method of using an energy-modulating source has given rise to various dual-energy or dual-kVp methods where optimal coefficients (known as energy weights) are applied to data in different energy bands to increase the SNR for a specific imaging task. They also showed that single threshold PC detectors generally outperform conventional nonbeam-modulated radiography with EI detectors. Later, Giersch et al.[3] showed that optimal energy weighting could be implemented using PC detectors with high-energy-resolution pulse height analysis (PHA). In addition, they showed simulations that demonstrated substantial improvement in the SNR by using optimal energy weights. Niederlohner et al.[4] then showed that an improvement in SNR could still be obtained using PHA with a small number of bins. Alvarez[5] then proposed a method to produce near Tapiovaara–Wagner optimal SNR images with a variety of relatively low-energy-resolution detectors. The method uses the representation of the attenuation coefficients with a 2D linear space, followed by a solution of the nonlinear equations for the line integrals of the basis set coefficients as described by Alvarez and Macovski.[6] The method then transforms the line integrals of the coefficients, considered to be points in an abstract vector space, to a *whitened* space with uncorrelated data and unit variance of each coordinate.

An example is found in the use of dual-energy k-edge digital subtraction angiography (KEDSA) for the imaging of arterial and venous function using intravenous iodine contrast agent. In KEDSA, an exposure with the wavelength of maximum output from the tube below the k-edge of iodine is subtracted from an exposure with the wavelength of maximum output from the tube above the k-edge of iodine. Due to the large increase in the attenuation coefficient for iodinated blood that occurs just above the k-edge of iodine, the difference image yields high contrast for blood vessels with most of the signal from other tissues separated.

Another example is found in the technique of dual-energy X-ray absorptiometry (DEXA), which is commonly applied to the application of areal bone mineral density measurements. In DEXA, when using EI detectors, a low- and high-energy image is generated with exposures at different wavelengths of maximum output from the X-ray tube (dual kVp). The attenuation data at the two energies can decompose the ray path into two different materials differing in average atomic number. Thus, soft tissues can be decomposed into lean and adipose components and bone mineral can be separated from soft tissues.[7,8]

More recently, dual-energy methods are being applied to DM[9] and CT.[10–14] Good separation of the energies in the X-ray beams used to make the two measurements is desirable in order to increase the precision of the quantities estimated. To obtain good separation of the energy spectra, sometimes different anode materials are used and thicker and higher atomic number filters are used in order to drastically attenuate the low-energy photons in the beam. One limitation of the dual-kVp technique is that it requires two acquisitions and may increase the total dose delivered to the patient. Another limitation is the possible overlap in the energy spectra of the detected photons.

Methods to acquire dual-energy X-ray absorption data in single exposure have been developed using depth-segmented detectors with multiple detectors in the incident direction parallel to the beam (dual layer).[15–18] The front detector closest to the source is made thinner and is designed to detect largely low-energy X-rays. The back detector farther from the source is made thicker and designed to detect the higher-energy X-rays in the beam. These dual-layer detectors, in essence, take advantage of the beam hardening that takes place in the detector to obtain information about the energy dependence of the attenuation. The principal advantage of these systems is that they allow the use of a single measurement. However, there is fundamentally more overlap in the energy spectra of the detected photons than for dual kVp systems. This is partly because the optimal thickness of the front detector depends on the beam spectrum and will thus vary depending on the composition and thickness of the object imaged. This can be improved by putting filters between the two detectors, using bimodal energy X-ray beams or using detectors with multiple segments in the depth direction. However, filtering the X-ray beam after it passes through the object blocks photons that have passed through the patient and is not dose efficient.

As with single-kVp clinical radiology, the dual-kVp and dual-layer methods are limited by the energy EI detectors used. As mentioned, EI readout integrates both the signal and noise from the detector and electronics over time. When either the count rate or the X-ray energies are low, the signal from X-rays must exceed a noise level produced by the detector and readout electronics. Thus there is a minimum threshold in terms of X-ray flux that can be reliably detected, which increases as the X-ray energy decreases thereby placing a distinct non-zero lower limit on the dynamic range.[19] There are additional major deficiencies inherent with EI systems such as not taking advantage of statistical information carried by each photon (e.g., three photons of 30 keV carry the same signal as one photon of 90 keV), and since there are a small number of high-energy photons, they have a disproportionate contribution to image noise. Also, variation in deposited energy of each photon produces Swank noise.[20] And finally, because a polyenergetic X-ray spectrum is used, each detected photon also contributes different information to the resulting image depending on density and elemental composition of the examined tissue.

Despite these limitations, EI X-ray detectors are used in virtually all clinical X-ray systems including digital radiography (DR), DM, and CT. This is because high output count rates (OCRs) are required (in the order of tens of millions of counts per second per square millimeter [cps/mm^2]) and better detectors such as the energy-dispersive photon-counting (ED-PC) X-ray detectors being developed and described in this chapter have been previously unobtainable. The count rates incident on the detector depend on the various settings such as the tube current (mA), the tube potential (kVp), the size and the material of the object, and the distance between the X-ray focus and the detector.[21–23] The input count rates (ICRs) for the unattenuated beam in CT, which is the highest flux application in clinical radiology, can exceed 100×10^6 cps/mm^2 (Mcps/mm^2). ED-PC detectors have a number of potential advantages in noise reduction and contrast enhancement. Through the use of a threshold set above the noise floor, they eliminate electronic noise. Also the variation in the detected energy of each photon affects the energy resolution, but not the number of photons counted, thus eliminating Swank noise. This is important because it has been shown that one type of object decomposition analysis method required a noise level as low as 1/3 to 1/10 of the noise encountered in standard CT imaging.[24] Furthermore, ED-PC detectors have the potential to reduce the overlap in the spectra of the high- and low-energy detected photons compared to dual kVp or dual-layer systems since good energy resolution can be achieved and maintained at high flux as we demonstrate in the ED-PCs presented here.

Higher OCR is now obtainable due to the development of fast sensors connected to high-throughput ASICs, which read out the fast signals from the sensors. This allows for the development of ED-PC detectors with sufficient OCR for applications in clinical radiology, and they have been considered as alternatives to EI detectors conventionally used.[25–27] ED-PC detectors applied to DR and DM have shown dose reduction while maintaining sufficient contrast for these applications.[28,29] ED-PC detectors along with optimal energy weighting can increase dose efficiency up to 40% for DM as compared to a conventional integrating system.[30] ED-PC detectors fabricated by us and using our fast ASIC technology have been used to create a full FOV clinical CT system.[31] The system has collected the only simultaneous dual-energy patient images that were acquired with a single X-ray tube at one kVp setting. The conventional detectors of a clinical multislice scanner provided by GE Healthcare were replaced with ED-PC detectors. This ED-PC detector uses two independent energy bins per pixel. The image reconstruction generates *virtual-non-enhanced* images where the iodine-based contrast media is identified and removed. The patient CT imaging with the ED-PC detectors was a prospective study on patients with known carotid artery disease. The protocol was that of a CT angiography (CTA) scan of the carotid bifurcation region using 140 kVp but at a very low X-ray tube output of 40 mA (approximately 1/10 the mA setting as compared to a conventional 400 mA CTA exam). The clinical images demonstrated the vascular and stenotic elements with good image quality, which is particularly notable in regard to the low X-ray tube current technique used for these scans. A second patient study was performed in the abdomen at a tube current of 300 mA in patients with suspected renal structures. The images were acquired after contrast injection and the ability to display results without iodine precludes the need for a precontrast study. The results of these patient studies show equivalent image quality and the potential to significantly reduce dose for conventional single-energy scanning applications. The ED-PC detectors used to perform these trials have maximum OCRs above 5 Mcps/mm^2/pixel using CdTe with a 1 mm pixel pitch as a direct conversion sensor and provides dual-energy acquisition with a noise floor below 20 keV and an energy resolution of 10% (6 keV) at 60 keV using ASICs with 128 parallel channels with two thresholds per channel.[32]

There is little doubt that ED-PC detectors have the potential to significantly expand the diagnostic benefit of current clinical X-ray imaging applications provided they can achieve the required OCR while maintaining good energy resolution. However an additional advantage of the development of ED-PC based systems is in the potential to develop new functional and therapeutic applications using targeted high atomic number nanoparticles thus combining diagnosis and therapy (i.e., theranostics).

In considering the development of ED-PC detectors for clinical radiology, sufficient performance in terms of the ranges and resolutions required for the specific application must be achieved. Here we define two ranges, namely, the OCR range between the lowest and highest flux detectable and the dynamic range between the lowest and highest energy photons detectable with good DQE and SNR. And we define two resolutions, namely, the intrinsic spatial resolution as a number of line pairs per mm (lp/mm) and the energy resolution measured as the full width at half maximum (FWHM) for a monoenergetic input. The sensors and ASICs used, as well as the methods for interconnecting the sensor pixels to the ASIC inputs, need to be designed with the ranges and resolutions required by the application kept in mind. In this chapter, we discuss sensor, ASIC, and interconnect design for application in clinical CT, DEXA, and DM.

4.2 DIRECT CONVERSION SEMICONDUCTOR SENSOR DEVELOPMENT

In principle, indirect or direct converters could be used as sensors for high-flux ED-PC detectors. ED-PC X-ray imaging arrays using indirect converters may be possible by coupling fast scintillator arrays to arrays of silicon photomultipliers (SiPMs) to maintain spatial resolution and to provide information about the number of X-ray photons and their energy level. This is due to the combination of the very fast decay times and high light yield emission of certain scintillators and the highly light-sensitive and extremely fast response of SiPM devices. This method requires pixilation by optically coupling scintillator segments to PD arrays such as SiPMs. In recent years, several cerium (Ce)-doped scintillators have been explored such as lutetium oxyorthosilicate (LSO), lutetium yttrium oxyorthosilicate (LYSO), yttrium aluminum perovskite (YAP), lutetium aluminum perovskite (LuAP), and lanthanum bromide (LaBr$_3$). These scintillating materials have characteristics that include high light output, very fast scintillation decay, and high atomic number (high density). Table 4.1 summarizes physical properties of the scintillating materials listed earlier.[33] The requirement for an extremely low afterglow is not an important criterion for the selection of scintillation material in PC CT. In PC mode, the influence of the afterglow on the image can be eliminated by setting the electronic threshold above this signal or above an excess electronic noise due to this effect. The scintillator would consist of numerous segments, each of which are physically aligned with a corresponding SiPM element (pixel) to detect light produced by the interaction of X-rays with that element's scintillator. Optical reflectors between the scintillator segments can be used to reflect light back to the segment. The peak sensitivity of SiPMs is about 400 nm that matches reasonably well with all the scintillators listed in Table 4.1. LYSO offers certain advantages over YAP because of its higher light output and better match of the peak wavelength with the peak of the sensitivity of SiPM. However, YAP has faster decay time. The lower Z of YAP compared to LYSO will have sufficient efficiency at CT energies. Also, LSO and LuAP could be scintillators of interest. LaBr, despite having the highest light yield, is highly hydroscopic that makes it not very practical for dense segmentation applications. Additionally, scintillator-based detector coupled to a SiPM may not be able to achieve sufficient energy resolution due to the noise in the multiplication gain, and therefore indirect conversion based on high-flux X-ray imaging arrays may be suitable for PC only applications.

ED-PC X-ray imaging arrays using direct converters may be possible by coupling fast semiconductor arrays to ASICs to similarly maintain spatial resolution and to provide information about the number of X-ray photons and their energy level. This is due to the combination of the very fast signal generation and the extremely fast response of ASIC devices. In contrast to scintillators where the energy of an absorbed X-ray is converted to an electrical charge by an indirect method, first generating light photons in the scintillator, and then in turn the light photons are converted into electrical signal in the PD, a direct conversion semiconductor sensor makes use of

TABLE 4.1
Characteristics of Some Candidate Scintillator Materials for SiPM-Based Indirect Conversion ED-PC X-Ray Imaging Arrays

Scintillator	LSO	LYSO	YAP	LuAP	LaBr
Density (g/cc)	7.4	7.1	5.4	8.3	5.3
Light yield (photons/keV)	27	32	21	10	61
Effective, Z	66	64	31.4	65	46.9
Principal decay time (ns)	42	48	25	18	35
Peak wavelength (nm)	420	420	370	365	358
Index of refraction	1.82	1.8	1.94	1.95	1.88
Hygroscopic	No	No	No	No	Yes

the direct conversion of the X-ray photon energy deposited by each X-ray to charge. This method requires pixilation and electrical coupling of the semiconductor pixels to read out arrays such as can be achieve with ASICs. Counting and sorting X-ray photons with different energies is accomplished by the associated amplification and processing electronics for each pixel. In recent years, several semiconductors have been explored such as cadmium telluride (CdTe), cadmium zinc telluride (CdZnTe), mercuric iodide (HgI$_2$), thallium bromide (TlBr), lead iodide (PbI$_2$), and silicon (Si) for use as direct converters. Table 4.2 summarizes physical properties of the semiconductor materials listed previously.[34,35] With the exception of Si, for X-ray imaging applications, the significant properties of these materials are that they have high stopping power for X-rays, making the detectors very efficient, and they have low leakage current at room temperature, making intrinsically low-noise devices. An absorber with bulk resistivity larger than 10^9 Ω cm will minimize leakage current and inherent noise. Other than CdTe and CdZnTe, potential high-Z materials that are not yet available in large volume and material quality for applications in radiology include thallium bromide (TlBr),[36] lead iodide (PbI$_2$),[37,38] and mercuric iodide (HgI$_2$).[39,40] TlBr has a relatively large mean energy for the creation of an electron-hole pair as compared to CdZnTe and CdTe that is a very important material parameter, which is related to the efficiency in energy transfer of X-rays into ionized charges. A material with a small conversion energy will produce a larger number of ionized charges and therefore a signal with improved statistical characteristics. Additionally, TlBr, PbI$_2$, and HgI$_2$ have a lower electron mobility-lifetime product as compared to CdZnTe and CdTe, which is important especially at high count rates. The electron mobility-lifetime product is a measure of the charge collection efficiency in the material, and larger values of this parameter assure that more of the ionized charges are collected and larger resulting signals can be obtained.

The energy resolution can be significantly better with CdTe and CdZnTe than that achievable with indirect detectors utilizing PDs. Moreover, the efficiency of converting the X-ray signal to an electrical signal (mean energy necessary to produce an electron–hole pair) in CdTe and CdZnTe can be an order of magnitude smaller due to the basic underlying physics of the energy transfer process in the direct detection

TABLE 4.2

Characteristics of Some Candidate Semiconductor Materials for Direct Conversion ED-PC X-Ray Imaging Arrays

Material	Energy for e-h Creation (eV)	Density (g/cm³)	Atomic Number	Resistivity (Ω cm)	Electron Mobility-Lifetime Product (cm²/V)	Hole Mobility-Lifetime Product (cm²/V)
CdZnTe	5	6	48, 30, 52	10^{11}	1×10^{-3}	6×10^{-6}
CdTe	4.3	6.2	48, 52	10^9	3.3×10^{-3}	2×10^{-4}
HgI$_2$	4.2	6.4	80, 53	10^{13}	10^{-4}	4×10^{-5}
TlBr	6.5	7.56	81, 35	10^{12}	1.6×10^{-5}	1.5×10^{-6}
PbI$_2$	4.9	6.2	82, 53	10^{12}	8×10^{-8}	1.8×10^{-6}
Si	3.6	2.33	14	10^4	>1	1

approach. That is, the mean energy for creating an electron–hole pair in a semiconductor detector (4.43 eV per electron–hole pair in CdTe) is typically an order of magnitude smaller than the corresponding energy necessary to create an optical photon and consequently an electron–hole pair in the PD through the scintillation approach. This results directly in an order of magnitude larger signal for the same incident X-ray with a direct detector approach as compared to an indirect detector approach. Another important consideration is that charges generated by X-rays do not spread laterally (aside from negligible spread due to Coulomb repulsion and diffusion), but move along the applied electric field lines within the semiconductor. Spreading of light in the indirect conversion method is a more serious problem and can be limited by segmenting the scintillator. However, for small segments less than or equal to 1 mm^2, the physical segmentation and surface treatment for light reflection required contribute to appreciable loss in the efficiency of the scintillator (smaller volume). In addition, the segmentation introduces significant dead space.

When considering the dynamic range for the application, high-Z semiconductors can maintain high detection efficiency in thin sensors compared to lower Z semiconductors for higher energy ranges. However, the detection of X-ray photons with energies above the k-edges of the sensors can have deleterious effects on the spatial resolution and spectral response. When an X-ray photon interacts with the semiconductor through the photoelectric effect, an electron in the K shell of the sensor atoms is ejected, leaving the atom ionized. The vacancy of the ejected electron is then filled by an electron from a higher shell, and the transition energy can be realized as emission of either a characteristic *secondary* fluorescent X-ray photon or an electron (Auger effect). The photoelectric interactions with the detector material and X-rays with energies above the k-edges of the detector material are the prevailing type of events in high dynamic range applications such as CT and the production of characteristic radiation is a dominant effect due to the high fluorescent yields of cadmium (Cd) and tellurium (Te). Because the photons are emitted in a random direction, they may either be absorbed by the pixel with the primary interaction again, be detected by an adjacent pixel affecting the spatial resolution and spectral response), or leave the sensor completely (affecting the spectral response). In the second and third cases, the recorded energy is lower by the energy of the characteristic radiation, as the characteristic X-ray photon *escapes* from the pixel. In the first case, the two charge clouds generated by the primary and secondary X-ray photons may result in quasi-coincident events, which may be detected as two separate counts if the detector electronics is fast or as a single count if not. The K fluorescent yield is the number of photons of all lines in the K series emitted in a unit time divided by the number of K-shell vacancies formed during the same time. The fluorescence yields of the K shells in Cd and Te are 84% and 87.5%, respectively.[41] This means that the photoelectric interactions with a CdTe or CdZnTe detector produce predominantly characteristic radiation with an absorption range that can be quite long compared to the shorter range and highly absorbed Auger electrons. The mean range of the characteristic radiation can be expressed as the inverse of the linear attenuation coefficient. These values correspond to 124.4 μm for Cd–K X-rays at 23.1 keV and 61.6 μm for Te–K X-rays at 27.4 keV. The K-shell energy of zinc is 8.6 keV; thus, this will be absorbed in a short distance. In addition, all the charge clouds generated by such secondary photons are subject to charge sharing at each location, resulting in a multiplicative effect. Our analysis showed that, at the absence of anticharge sharing circuits, the previously discussed effects set the lower limit of the pixel size to about 500 μm in order to preserve reasonably good spectral characteristics and therefore achieve good energy resolution. Note that the lower limit can be smaller with an anticharge sharing circuit. Although Si has a much lower energy of X-ray characteristic radiation that can be reabsorbed in close proximity to the primary interaction site avoiding excess charge sharing from characteristic escape, it suffers from low detection efficiency and Compton scatter effects. For Si to be used in the development of ED-PC X-ray imaging arrays applied to radiology, stacked arrays or an edge illumination geometry must be used to provide sufficient detection efficiency as the thickness of fully depleted junctions required for charge collection is usually limited to below 1 mm. Having multiple layers along the depth direction significantly decreases the count rates each layer needs to handle. This approach has a number of challenges related to the need for corrections for Compton scatter and in the implementation of the interconnections between the various layers and readout electronics. In order to keep the detector layers close together, connections to the readout electronics need to be elongated that might cause additional stray capacitances and vulnerability to cross talk, if not carefully designed. The increased stray capacitance is particularly harmful at high count rates, contributing to high electronic noise and/or high-power requirements for readout electronics. Also, detectors with multiple layers may be more expensive to produce. Edge illumination avoids these problems; however, this geometry will limit the FOV to 1D arrays and scanning methods.

When considering the OCR for applications in spectral radiology, sensors with rapid signal generation are required. The high OCR combined with a need for high detection efficiency using two 2D pixelated arrays requires the development of detector structures that can provide formation of the response signal much faster than the transit time of carriers over the whole detector thickness. A number of strategies have been investigated for this purpose including detectors utilizing the *small pixel effect*[42] and parallel drift structures.[43] The small pixel effect is achieved in detectors with pixel dimensions smaller than the detector thickness. The electrode configuration creating the small pixel effect is very effective in reducing the duration of the fast portion of an induced signal because the induced signal is almost entirely due to the motion of the electrons as they approach the vicinity of the pixel anode. The pixel size, however, cannot be too small. When pixels are very small (much less than 0.5 mm) with a sensor thickness of 2–3 mm required for detection efficiency in CT using CdTe or CdZnTe, there are significant penalties to be paid due to charge sharing between pixels. When an X-ray photon is absorbed in a semiconductor, a charge cloud is initially created in the sensor. These charges (electrons and holes) drift to their respective electrodes due to the electric field generated by the applied high voltage bias. The charge cloud grows in size due to diffusion effects and the Coulomb force. If the electron charge cloud is generated near a pixel boundary, it may be divided and detected by multiple pixels at energies lower than the original energy, causing distortions in the spectral response. The significance of this effect depends on the sensor material that governs the mobility of the charge carriers, the pixel sizes, the applied bias voltage, and the depth of interaction (DOI) in the sensor.[44] Another strategy to shorten the signal duration is the implementation of parallel drift structures. Drift structures allow for the collection of electrons from larger volumes on a small anode.[45] A good energy spectral response is preserved with this method and fast signal formation is achieved due to the small anode dimensions. However, drift structures require a larger number of electrodes and smaller anodes than a corresponding electrode configuration using the small pixel effect for the same detector size. More electrodes and smaller anode dimensions may increase the difficulties involved in electrically connecting the sensor's pixels to the ASIC's inputs. Si direct conversion sensors applied to spectral radiology can achieve rapid charge collection and hence good signal formation at high output counting ranges since the charge is collected over the relatively thin detector thickness albeit with the limitations of stacking or edge illumination.

Other effects in the sensor that can limit the ranges and resolutions include Compton scattering, charge trapping, polarization, and stability. When an X-ray photon interacts with the semiconductor by Compton scattering, the scattered photon has changed direction and less energy, and the lost energy is deposited at the interaction site. This is called the Compton effect or Compton scatter. The scattered photon may be reabsorbed by the incident pixel, detected by an adjacent pixel, or leave the sensor completely. Contrary to the discrete energy loss with K-escape X-rays, the energy loss with Compton scatter depends on the scattered angle and is therefore continuous, resulting in a long tail at low energies in the recorded spectrum. The internal charge trapping process is one in which an electron or hole is captured by a trapping center and then, after a delay, is thermally reemitted into the conduction or valence band. The trapping centers are usually related to impurities and lattice defects in semiconductors. The trapping effect deteriorates spectral responses by reducing amplitudes of pulses toward lower energies than the original and by creating low-energy tailing in the spectral characteristics. Polarization and long-term stability of CdTe and CdZnTe detectors can be a very serious problem for detectors exposed to high-intensity X-ray beams. The polarization under certain operational conditions may lead to a decrease in the OCRs and charge collection efficiency, which are dependent on time or incoming flux intensity.[46,47] The polarization phenomenon is due to the existence of deep trapping levels in the detector material, and several models have been proposed to explain this complex effect.[48,49] However, with proper selection of the starting material, surface preparation, contact deposition, and good surface passivation, CdTe and CdZnTe detectors can operate stably and reliably for a very long time. This has already been proven in commercial applications such as DEXA for BMD measurements. Two systems using semiconductor ED-PC arrays, the Lunar iDXA systems (GE Healthcare) and the Stratos DR systems (DMS-APELEM) with detectors from DxRay, Inc., have been introduced into the medical market several years ago. Good performance of the CdTe and CdZnTe detectors under very high X-ray fluxes has been reported. However, their long-term reliability in CT applications still needs to be proven. Unlike the compound semiconductors CdTe and CdZnTe, Si has fewer problems with quality, uniformity, and cost of production. The majority of the interactions of diagnostic X-ray photons with Si is Compton scattering, not the photoelectric effect, due to low atomic number of Si, which along with the low detection efficiency may limit the use of Si as a direct conversion ED-PC detector to lower dynamic (energy) range applications in radiography.

4.3 ASIC DEVELOPMENT

The development of ED-PC arrays for applications in spectral radiology requires dense parallel channels of amplification and processing electronics that provide a dedicated electronic channel for each sensor pixel. There are a number of approaches being investigated to develop unique ASIC readout electronics to achieve this including using common digital electronic schemes, anticharge sharing schemes, and schemes that simultaneously count photons in parallel with an EI readout. Here we outline the basic architecture and detection mechanism of ED-PC ASIC readouts using pulse height discrimination where the sensors' individual pixelated anodes are connected to the inputs of ASICs containing parallel channels. Each channel consists of a preamplifier, a pulse shaper, N pulse height comparators for implementing adjustable energy windows, and N counters per sensor pixel. Each pulse height comparator is followed by a dedicated counter that is iterated up when a pulse arrives with a larger amplitude than the comparator threshold is set to. The number of counts between two energies (energy window) can be obtained by subtracting the number of counts from a comparator with a higher threshold setting from the counts of a comparator with a lower threshold setting for a single sensor pixel and within a single exposure time. The number of energy windows per pixel is N only if the energy windows are contiguous with the highest energy window having no upper bound, and the number of energy windows per pixel is N/2 if the energy windows are separated with upper and lower bounds. For calibration, data can be acquired in frames where the threshold values are stepped across the dynamic range between each frame. In each frame the number of counts above the comparator discriminator level is read out from the digital counters and recorded. The number of counts plotted as a function of threshold level represents the integrated spectrum. An energy spectrum can be obtained by differentiating these data. During a patient scan, the threshold levels are fixed creating up to N energy binned images.

As mentioned previously, due to charge sharing and fluorescent X-ray K-escape, a photon may be counted by adjacent pixels at wrong energies especially when the pixel size is small. A network of charge summing circuits that communicate between adjacent pixels for detecting coincidences and reconstructing charges can be used to ameliorate the effects of charge sharing and K-escape. The reconstructed charge will then be exclusively allocated to a pixel with the largest charge once it exceeds a set energy threshold.[50] There are challenges due to the complicated coincidence electronics for allocating charges that can reduce the count rate capabilities and generate additional heat. The count rate capabilities will be a strong function of the peaking time used in the preamplifier circuit, which should match fairly well to the rise time of the signal produced by the sensor. In order to detect the total signal generated in a photoelectric or Compton scattering event, while considering K-escape of the characteristic X-rays of Cd and Te, it is necessary to extend the peaking time of the preamplifier to collect all of the signals generated by primary and secondary X-ray photons. This sets the lower limit on the minimum peaking time of the amplification electronics to account for the full charge to avoid spectral distortion. In order for an electron cloud to travel a distance that corresponds to the mean absorption ranges of Cd–K X-rays (124.4 µm), it takes about 2.9 ns assuming a typical electrical field in the detector. In addition, in order to take into account the spread of the primary and secondary charge clouds during their transport, it is necessary to extend the peaking time and the processing electronics should be no less than 5 ns.[51]

Given the finite lower limit on the peaking time imposed by the sensor, pulse pileup will always be present in direct conversion semiconductor PC detector systems and is a function of the ICR and detector deadtime. Detector systems with longer deadtimes exhibit this effect at lower count rates. Multiple pulses generated by quasi-coincident photons may be piled up and observed as one pulse, resulting in a loss of counts and a wrong recorded energy. With OCRs as high as those required for radiology and when using a bipolar pulse shape, two types of pulse pileup effects are observed, namely, peak pileup and tail pileup. Coincidences during the initial part of a pulse are recorded as a single count at a higher energy than the original pulse's energies. This is called peak pulse pileup. The long tail of the pulse affects the recorded energy of subsequent events. For bipolar-shaped pulses, a peak overlapping the tail of a preceding pulse results in a lower recorded energy and for unipolar-shaped pulses in a higher recorded energy. This is called tail pulse pileup. Both peak pileup and tail pulse pileup distort the recorded spectrum, and the amount of distortion depends strongly on the count rate.

The electrical interconnections between semiconductor sensor arrays and silicon ASIC electronics are challenging due to the large density of connections between dissimilar materials and in some cases a different pitch of the sensor pixels and individual ASIC channels. Standard wire bonding or regular solder reflow technologies cannot be used with CdTe or CdZnTe. Instead, low-temperature solder reflow, silver epoxy, or other bump bonding technologies must be utilized. The assembly often involves interposer boards to assist the interconnections. A dense parallel channel method to increase throughput of a detection system is an approach utilizing an increased number of parallel detection and signal processing channels within a given area. However, densely packed multichannel fast electronics lead to large power consumption, and the resulting heat needs to be dissipated to the ambiance without negatively affecting the detection system. Sensors may have a pixel pitch smaller than the intrinsic spatial resolution requirement for the particular application in order to decrease count rates per detector pixel. In that case several sensor pixels could use common electronics such as discriminators and counters to reduce power consumption.[52] In general, in order to reduce power consumption, the shortest possible connections to the detector pixels for the lowest stray capacitance should be implemented.

With these constraints in mind and considering the ranges and resolutions required for the applications targeted, we have developed ASICs for CT, BMD, and DM applications. Although differing in geometry, bonding method, number of channels, and/ or peaking time, all the ASICs contain four threshold comparators for pulse height discrimination. Figure 4.1 shows the basic architecture of an individual pixel detection channel in the ASICs. In general each channel consists of a charge-sensitive preamplifier, a shaper/gain stage, four discriminators (level-sensitive comparators), and four 18-bit counters (static ripple type). The threshold levels of all discriminators can individually be fine-tuned through separate 6-bit digital-to-analog converters (DACs) attached to each discriminator. These four threshold main levels are

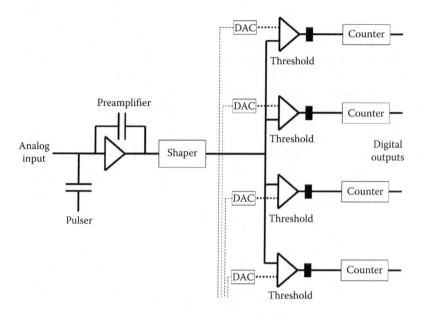

FIGURE 4.1 Schematic of a single channel within the ASICs where each channel has four parallel thresholds per pixel.

externally adjustable. There is also a test input capacitor at the input to the preamplifiers to inject electronic pulses for testing and calibration. The discriminator fine-tuned DACs are programmable through series shift registers (i.e., all DAC registers are linked in a series daisy chain). The readout of the chip (dumping of the counters' content) is done by reading each channel counter one after the other in a fixed sequence (from top to bottom). All 18 bits of the counter are read out fully in parallel. Except for the channel-specific, digitally programmable threshold fine-tuned DACs (which adjust the threshold main levels for each pixel), the biasing of all other analog functions inside a channel is generated by a single bias-generator unit that is common to all channels (including the biasing of the range of the fine-tuned DACs). This bias-generator unit is again biased via external connections, and adjustments of these make changes to the parameters of the amplifier/discriminator chain (shape, gain, etc.), with all channels changed equally. The external biases, voltage, and current levels must be provided by external circuitry.

4.4 MODULE DEVELOPMENT FOR SPECTRAL COMPUTED TOMOGRAPHY

Although used for several decades, clinical CT performance continues to improve as new technologies are incorporated and the effort to improve CT detectors continues to be an active area of research. Room temperature X-ray imaging arrays based on high-Z direct conversion compound semiconductors are being developed for CT.[53–60] The CT detector arrays presented here operate in an ED-PC mode by connecting a multichannel ASICs to the pixels of a single-crystal semiconductor where the pixels are formed with patterned anode contacts. The direct interaction of photons within semiconductor detectors can generate a large signal with good energy and spatial resolution in a compact design as compared to indirect conversion scintillator-based systems. If arrays of this type can achieve sufficient OCRs while maintaining sufficient energy resolution, then optimal energy weighting and material decomposition methods can be used to increase material-specific contrast and/or reduce dose for specific imaging tasks.[61–63] CT applications require very high detector OCR capabilities. These high count rates can be achieved by using very short shaping times in the amplification circuitry.

For CT, the X-ray imaging arrays are tiled in multirow arrays and used in a fan beam geometry. This requires that detector modules must be developed with a vertically integrated readout that fits within the active area of the sensors.[64] Figure 4.2a shows a schematic of our ED-PC CT module design. ASICs are packaged into a ball grid array (BGA) that provides electrical connection between the detector's pixels and the inputs of the ASICs as well as acts as an enclosure to facilitate encapsulating the ASICs. The ASICs are flip-chip bump bonded to an array of gold pads within the BGA with interior pads for connections to the sensor pixels and the exterior pads for input/output (I/O) connections. CdZnTe or CdTe sensors are then coupled to the ASICs (through the BGAs) by a low-temperature reflow of solder balls, a method we optimized for high yield. A single module is formed by single-crystal

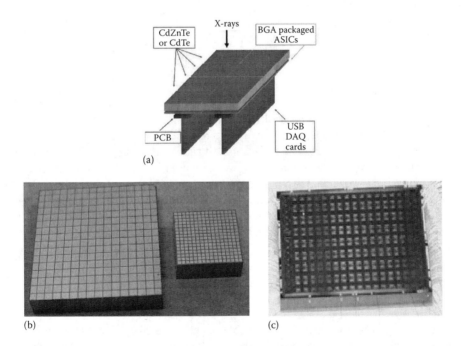

FIGURE 4.2 Schematic diagram of the CT modules where pixelated CdZnTe or CdTe is coupled to 2D ASICs and vertical integration allows blocks in 2D to scale to any FOV (a). Images of 3 mm thick CdTe crystals with a 16 × 16 array of pixels with a 1.0 mm pitch and 0.5 mm pitch (b). An image of a 2D ASIC with a 16 × 16 array of amplifier inputs with a 0.4 mm pitch with 10 or 20 ns peaking times for use with 0.5 and 1.0 mm pixels, respectively (c).

sensors with a pixelated grid anode interconnected to an ASIC constituting parallel channels with fast PC with four energy bin functionality. The individual modules can then be mounted to a printed circuit board (PCB) that forms a mechanical and electrical substrate thus producing any FOV in principle since the modules are four-side buttable and can be tiled in 2D with a vertically integrated high-speed readout. The main risks associated with obtaining good yield with this technology is the interconnection of the sensors and ASICs in a completely vertical package. Finally, the edge pixels are slightly smaller allowing tiling in 2D with virtually no dead space and with preserved pixel pitch.

Here we demonstrate that equivalent results can be obtained using CdTe or CdZnTe for spectral CT using ED-PC arrays. One difference between CdTe and CdZnTe, which have similar attenuation (dense high-Z materials) and band gaps (room temperature operation), is in the bulk resistivity, which is on average more than one order of magnitude higher for CdZnTe as compared to CdTe for good material. At the relatively long shaping times used in low flux applications, such as single-photon emission computed tomography (SPECT) and positron emission tomography (PET), the dark current can be the dominant source of noise contribution from the sensor, and CdZnTe usually produces better energy resolution as compared to CdTe when the sensors are fabricated as conduction counters and long shaping times are used. For the high-flux incident on the detector in CT, very short shaping times are required to achieve a high OCR and the dominant electronic source of noise contribution is due to the sensor capacitance. Therefore, it is expected that CdTe and CdZnTe could produce essentially the same results in this application because they have similar capacitances and good electron charge collection. In order to demonstrate equivalent results in both materials, we have fabricated CdZnTe and CdTe sensors with arrays of 1 and 0.5 mm pixels connected to the inputs of parallel channel ASICs with fast peaking times. The results can then be compared directly. Figure 4.2b shows a CdTe crystals configured with 256 pixel anodes at a 1.0 and 0.5 mm pitch where surface preparation and film deposition were performed. The perimeter pixels are slightly smaller than the interior pixels so that when several detectors are tiled together a 0.5 mm pixel pitch can be maintained. The same method of using smaller perimeter pixels to preserve pitch upon tiling is used in all the CdZnTe and CdTe sensors at a 1.0 and 0.5 mm pitch. The 1 and 0.5 mm pixelated sensors have been connected to ASICs with 20 and 10 ns peaking times, respectively. Figure 4.2c shows a 2D ASIC configured with 256 amplifier input pads that fit within the active area of the 1.0 and 0.5 mm pitch sensors. In all cases, the single-crystal sensors are 3 mm thick with a continuous thin-film metal cathode on one side and a pixelated thin-film metal anode with an array of 16 × 16 pixels on the other side. We use a method of passivating the lateral surfaces of CdZnTe and CdTe detectors and eliminated the need for guard rings. Reducing surface current on these crystals without the use of guard rings allows our pixel maps to be extended to the edge of the crystal, thus making more effective use of the material and allowing tiling with little or no dead space. In order to extend the pixelated anode to the edge of the active area of the crystals, the lateral surfaces are passivated with an insulating material that also acts as a chemical getter to remove impurities that can increase surface current.

Both the 20 and 10 ns ASICs contain a two 2D array of input bond pads surrounded by I/O connections for flip-chip bump bonding to BGAs, thus eliminating wire bonding and shortening the connections to the sensor pixels for the lowest stray capacitance. The 2D ASICs contain a 16 × 16 array of readout channels where each channel is contained within a 0.3 mm × 0.3 mm cell. Four thresholds per pixel allows for either four continuous energy windows with no upper bound on the highest energy window, three continuous energy windows with an upper bound on the highest energy window, or two energy windows with independent upper and lower bounds. Although used with fixed thresholds for imaging, the ASICs contain a feature to sweep the thresholds across the dynamic range producing integrated spectra for calibration.

In all cases, the CdZnTe and CdTe sensors and readout ASICs are mounted to a substrate PCB, which provides mechanical support for both, as well as electrical contacts between the pixels and the ASIC inputs. The substrate PCB is then connected to a field programmable gate array data acquisition card that parses the data from the counters and provides I/O support for the system. The OCRs are measured by exposure to increasing flux controlled by increasing the tube current on an X-ray generator at fixed source to detector distance and recording the counts above a pulse height threshold setting for each pixel corresponding to about 30 keV. There are no counts observed above a pulse height corresponding to about 20 keV when no source is present. The pulse height spectra results are obtained by sweeping a threshold for each pixel from high to low energy and differentiating the integrated spectra produced by plotting the counts above the thresholds as a function threshold setting under constant flux from radionuclide sources placed on the incident (cathode) side of the sensor.

In order to develop a modular design for CT clinical, a CdZnTe sensor with 1 mm pixels was connected to an ASIC with a 20 ns peaking time in order to match the peaking time of the amplifier with the signal rise time from the sensor. The sensors are 3 mm thick for sufficient DQE up to 140 kVp and make use of the *small pixel effect* for rapid signal generation with a three to one ratio between the sensor thickness and the pixel size. Measurements of the OCR as a function of ICR have been obtained. Figure 4.3a shows the OCR as a function of increasing ICR using CdZnTe with 1 mm pixels for a tube setting of 120 kVp at various tube currents and at a distance of about 20 cm from the tube's focal spot and with a 1 mm Cu filter between the source and

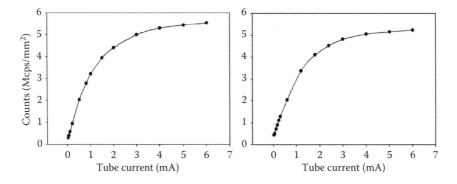

FIGURE 4.3 The output count rate as a function of increasing X-ray tube current for a typical 1 mm CdZnTe pixel (a) and typical 1 mm CdTe pixel connected to a fast 20 ns peaking time parallel channel 2D ASIC optimized for use with the 1 mm pixels (b).

detector. Note that the X-ray tube current is proportional to the ICR. A CdTe sensor with 1 mm pixels was connected to the same ASIC and measurements of the OCR as a function of ICR have been obtained under the same conditions. Figure 4.3b shows the OCR as a function of increasing ICR using CdTe with 1 mm pixels. In both cases, data are acquired in a 1 ms frame time. For both the CdZnTe and CdTe, 1 mm pixels exhibit a linear OCR response to between 1.5 million counts per second (Mcps) and 2 Mcps/pixel using preamplifiers with 20 ns peaking time within the ASIC. Also for both materials the OCR saturates above 5 Mcps/pixel due to pulse pileup. In both cases, results are shown for a typical pixel.

The CdZnTe and CdTe detectors with 1 mm pixels have also been used to obtain spectra from radionuclide sources. Figure 4.4a shows an ^{241}Am spectrum taken with the CdZnTe sensor with 1 mm pixels. An approximately 10 µCi ^{241}Am source was placed directly above the sensor on the cathode (incident) side and data were taken over several minutes. A single threshold for each pixel is swept from approximately 140 keV down to approximately 10 in 0.5 keV steps. The resulting plot of the number of counts above the threshold setting as a function of threshold setting is then passed through a smoothing filter and differentiated to produce the pulse height spectrum. An ^{241}Am spectra from the CdTe sensor with 1 mm pixels has been obtained under the same conditions. Figure 4.4b shows the ^{241}Am spectrum taken with the CdTe sensor with 1 mm pixels. In order to determine the noise floor and resolution in terms of energy, the detector needs to be calibrated. This is done by sweeping all the thresholds in the presence of ^{109}Cd, ^{133}Ba, ^{241}Am, and ^{57}Co sources and differentiating the resulting curves to obtain spectra in terms of the pulse height response for the detector. Figure 4.5 shows spectra from all these sources where a linear calibration is used from the corresponding peaks to convert from pulse height to energy demonstrating a noise floor of 20 keV. The photopeaks, from 22 to 135 keV, show a FWHM of approximately 7 keV and is independent of energy indicating good charge collection across the entire dynamic range for CT. The response is linear across the entire dynamic range to 140 keV and the energy resolution is the same across dynamic range as well indicating that the energy resolution is dominated by electronic noise.

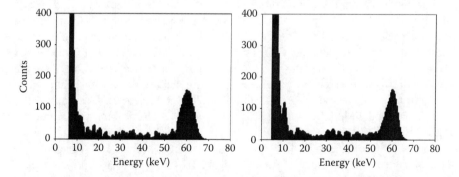

FIGURE 4.4 Pulse height spectra from a ^{241}Am source taken with a typical pixel from an array of 1 mm CdZnTe pixels (a) and from an array of 1 mm CdTe pixels (b) both connected to fast 20 ns peaking time parallel channel 2D ASICs optimized for use with the 1 mm pixels.

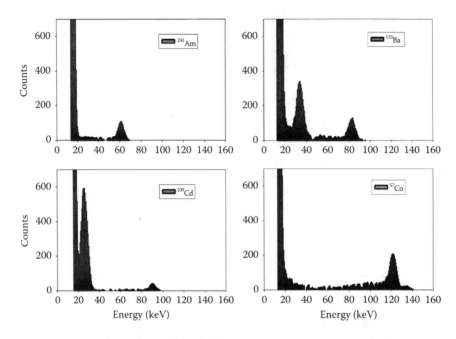

FIGURE 4.5 Energy spectra from ^{109}Cd, ^{133}Ba, ^{241}Am, and ^{57}Co sources taken with a typical pixel from an array of 1 mm CdTe pixels connected to a fast 20 ns peaking time parallel channel 2D ASIC optimized for use with the 1 mm pixels.

In order to develop a better modular design with a higher OCR better suited for CT, a CdZnTe sensor with 0.5 mm pixels was connected to an ASIC with a 10 ns peaking time. The sensors are 3 mm thick for sufficient DQE at CT dynamic ranges up to 140 kVp. The sensors also make use of the *small pixel effect* for rapid signal generation with a six to one ratio between the sensor thickness and the pixel size. Measurements of the OCR as a function of ICR have been obtained. Figure 4.6a shows the OCR as a function of increasing X-ray tube current, which is proportional to the ICR, using CdZnTe with 0.5 mm pixels for a tube setting of 120 kVp at various tube currents and at a distance of about 10 cm from the tube's focal spot with a 0.5 mm Cu filter between the source and detector. A CdTe sensor with 0.5 mm pixels was connected to the same ASIC and measurements of the OCR as a function of ICR have been obtained under the same conditions. Figure 4.6b shows the OCR as a function of increasing ICR using CdTe with 0.5 mm pixels. In both cases, data are acquired in a 1 ms frame time. For both the CdZnTe and CdTe, 0.5 mm pixels exhibit a linear OCR response up to between 3.5 and 4 Mcps/pixel using preamplifiers with 10 ns peaking time within the ASIC. A 15% OCR deficit due to pulse pileup occurs at about 5.5 Mcps/pixel for both materials. The CdZnTe and CdTe sensors with 0.5 mm pixels have also been used to obtain spectra from radionuclide sources. Figure 4.7 shows an ^{241}Am spectrum taken with the CdZnTe sensor with 0.5 mm pixels connected to a 2D ASIC with a 10 ns peaking time. An approximately 10 µCi ^{241}Am source was placed directly above the sensor on the cathode (incident) side and data were taken over several minutes by sweeping thresholds across the dynamic range and differentiating the resulting curve

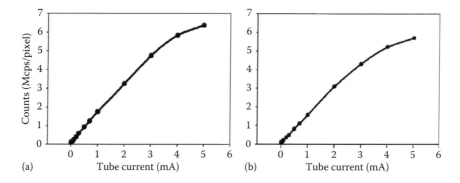

FIGURE 4.6 The output count rate as a function of increasing X-ray tube current for a typical 0.5 mm CdZnTe pixel (a) and a typical 0.5 mm CdTe pixel (b) both connected to a fast 10 ns peaking time parallel channel 2D ASIC optimized for use with the 0.5 mm pixels.

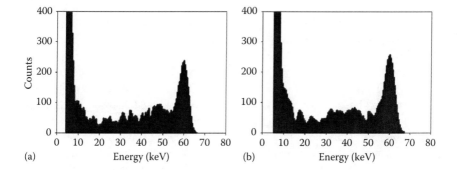

FIGURE 4.7 Pulse height spectra from a ^{241}Am source taken with a typical pixel from an array of 0.5 mm CdZnTe pixels using the 2D ASIC with a 10 ns peaking time optimized for use with the 0.5 mm pixels (a). Similar results are obtained with 0.5 mm CdTe pixels using the 2D ASIC with a 10 ns peaking time (b).

as described previously. Charge deficit and charge sharing between nearest neighbor pixels resulting from the escape of characteristic X-rays of Cd and Te and other effects described earlier is expected to increase as a function of decreasing pixel size. This will produce increased tailing (counts below the photopeak) particularly in response to X-rays with energies above the k-edge of Cd or Te. Notice there is increased tailing as compared to results obtained with 1 mm pixels (Figure 4.4) as expected from an increase in charge sharing effects. The FWHM energy resolution of the photopeak is approximately 7 keV similar to what was obtained with 1 mm pixels and a 20 ns peaking time. The CdTe sensor with 0.5 mm pixels was connected to the same 2D ASIC with a 10 ns peaking time, and a measurement of the FWHM energy resolution has been obtained under the same conditions with essentially the same results as seen in the graph of Figure 4.7b.

For applications in clinical CT, which requires an intrinsic spatial resolution of 1 mm, four nearest neighbor 0.5 mm pixels in a 1 mm^2 block are summed with four global thresholds levels fixed to generate four energy windows during patient scanning. Since each individual threshold (4 per pixel) contains a DAC voltage adjustment, the offset and gain differences are compensated for at the energy that each of the four global levels are set to. For this reason there is no spectral distortion by summing the data from the four nearest neighbor pixels. Figure 4.8a shows the OCR and a function of ICR for a typical 1 mm^2 block of 0.5 mm pixels from CdZnTe. All the counts above the threshold set to 30 keV is plotted. The OCR is linear to between 14 and 16 Mcps/mm^2 and has a 15% deficit at about 23 Mcps/mm^2 for typical pixels. Figure 4.8a shows spectra generated with the 0.5 mm pixel CdTe detector. The spectra are normalized by plotting the number of counts divided by the tube current from an X-ray tube set to 120 kVp operated from 0.1 to 3.0 mA of tube current at approximately 20 cm with no filter for an incident flux of about 0.4 Mcps/mm^2 (at 0.1 mA) to 27 Mcps/mm^2 (at 3 mA). At a threshold value (pulse height) corresponding to 30 keV, no counts are seen without X-rays or sources present. Notice that at very high flux (27 Mcps/mm^2 ICR corresponding to 23 Mcps/mm^2 OCR) pulse pileup distorts the X-ray spectra. However, the spectra up to 20 Mcps/mm^2 ICR corresponding to 18 Mcps/mm^2 OCR (2 mA) are essentially the same showing very little pileup. Pulse pileup effects eventually limit the upper bound on the flux linearity of the detector system. The high-flux rate of this device extends the range for PC detectors as compared to previous devices of this type.

Applications in CT require stability and uniformity without hysteresis from the detectors. The temporal stability and uniformity under the high flux expected in CT applications has been measured using the CdZnTe and CdTe sensors with 0.5 mm pixels. One thousand temporally contiguous 1 ms frames are acquired with an average OCR of 3 Mcps/pixel that give approximately 3000 counts/frame/pixel. Six typical pixels each are taken from the 0.5 mm pitch CdZnTe and the 0.5 mm pitch CdTe sensors whose OCRs are shown in Figure 4.6. Figure 4.9 shows the difference in the number of counts in an individual frame divided by the mean versus the mean. The top two rows of Figure 4.9 are from CdZnTe pixels and the bottom two rows are

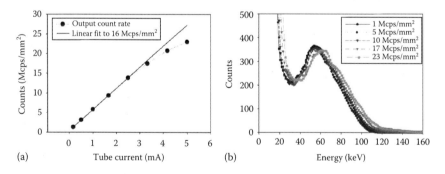

FIGURE 4.8 Output count rate (OCR) as a function of increasing X-ray tube current (ICR) from four nearest neighbor 0.5 mm × 0.5 mm pitch pixels patterned on 3 mm thick CdZnTe and connected to a 2D ASIC with a 10 ns peaking time optimized for use with the 0.5 mm pixels. Similar results are obtained with 0.5 mm CdTe pixels (a). X-ray spectra taken with a 120 kVp X-ray tube setting and no filter at approximately 20 cm and increasing X-ray tube current with a CdTe sensor with 0.5 mm pixels connected to a 2D ASIC and operated at a 10 ns peaking time (b).

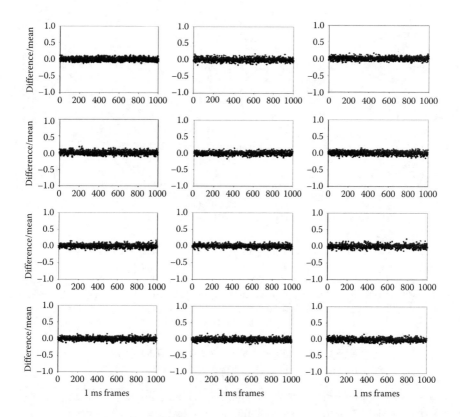

FIGURE 4.9 Graphs of the difference divided by the mean versus time as determined by acquiring data in 1000 successive 1 ms frame times from a 2D ASIC with a 10 ns peaking time under constant flux from a clinical CT X-ray generator using both CdZnTe and CdTe sensors with an 0.5 mm pixel pitch.

from CdTe pixels. No drift is detected and additionally the variance versus the mean falls within the counting statistics of the mean. We have also measured the variance versus the mean as a function of increasing ICR. Figure 4.10a shows the variance, calculated as the mean squared difference over 1000 successive 1 ms air exposures under constant flux from a clinical CT X-ray generator and readout with a 2D ASIC operating with a 20 ns peaking time versus the mean for a typical 1 mm CdZnTe pixel at OCRs from 0.03 to 6 Mcps/mm^2. The graph follows a trendline with a slope of one up to between 1.5 and 2.0 Mcps/mm^2, which is in the linear portion of the OCR curve shown in Figure 4.3. Similar results are obtained from typical 1.0 mm CdTe pixels using the same 2D ASICs operating with a 20 ns peaking time. Figure 4.10b shows the variance versus the mean over 1000 successive 1 ms air exposures for a typical 0.5 mm CdZnTe pixel using a 2D ASIC operating with a 10 ns peaking time at OCRs from 0.04 to 23 Mcps/mm^2. The graph follows a trendline with a slope of one up to between 14 and 16 Mcps/mm^2, which is in the linear portion of the OCR curve shown in Figure 4.8. Similar results are obtained from typical 0.5 mm CdTe pixels using the same 2D ASICs operating with a 10 ns peaking time.

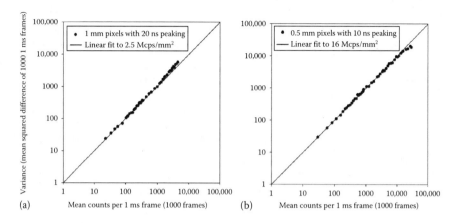

FIGURE 4.10 Graph of the variance versus the mean as determined by acquiring data in 1000 successive 1 ms frame times as a function of increasing flux from a 1.0 mm pixel CdZnTe sensor and a 2D ASIC with a 20 ns peaking time (a). Graph of the variance versus the mean as determined by acquiring data in 1000 successive 1 ms frame times as a function of increasing flux from a 0.5 mm pixel CdZnTe sensor and a 2D ASIC with a 10 ns peaking time (b).

4.5 DETECTOR DEVELOPMENT FOR SPECTRAL BONE MINERAL DENSITOMETRY

The standard for the diagnosis of osteoporosis is a measurement of areal bone mineral density (BMD) at the proximal femur acquired with a DEXA scanner. Although more sophisticated methods have been developed and introduced, no alternative bone measurement method has yet shown to be superior in the ability to identify groups of those at future risk of fracture from a population of controls. Fundamentally a DEXA scanner is designed to separate materials that differ in atomic number by using X-ray attenuation measurements at two substantially different photon energies. In medical applications, the method is mostly used to measure BMD and to measure body composition. Similar to CT, the intrinsic spatial resolution requirement for the detector is about 1 mm; however, the OCR range is significantly less. The ideal DEXA method would use two discrete energies in a dichromatic spectrum, but X-ray sources are intrinsically polychromatic with broad energy spectra. Most practical methods involve two relatively broad energy bands achieved by using appropriate filters typically with some spectral overlap that reduces the efficiency of the method. The two bands can be achieved by spitting the spectrum into high and low-energy regions that are measured separately with a counting detector using pulse height discrimination. Alternatively two different spectra are generated with different kVp settings on the X-ray tube and beam filters and measured separately with EI detectors. Usually counting detectors can be made to be more efficient than kVp switching with EI detector designs. The efficiency of a medical X-ray system is best described in terms of SNR per unit radiation dose that can be optimized for a given system design. ED-PC detectors can provide superior performance by providing better separation between the low- and high-energy images provided the energy resolution is good.

Pixelated Semiconductor and Parallel ASIC Design 105

In order to develop a superior detector for BMD using DEXA, CdTe sensors are connected to ASICs with an 80 ns peaking time. The sensors are 1 mm thick for sufficient DQE at BMD dynamic ranges up to 80 kVp. The sensors make use of a Schottky barrier structure for reducing the dark current. Here, the junction structure limits the dark current's contribution to the electronic noise allowing the longer peaking time to accommodate the lower OCR range requirement as compared to CT and gaining better energy resolution to improve accuracy in BMD measurements using DEXA. Although the intrinsic spatial resolution required for CT and DEXA is similar, the dynamic range and OCR range required are different. DEXA is typically performed with up to an 80 kVp X-ray tube setting, as compared to CT that can operate up to 140 kVp, and for this application we have created detector arrays with 1 mm thick CdTe sensors for good detection efficiency in this range. Figure 4.11a shows a schematic of our ED-PC BMD detector design. The sensors employ a continuous cathode on the incident side and a pixelated anode on the

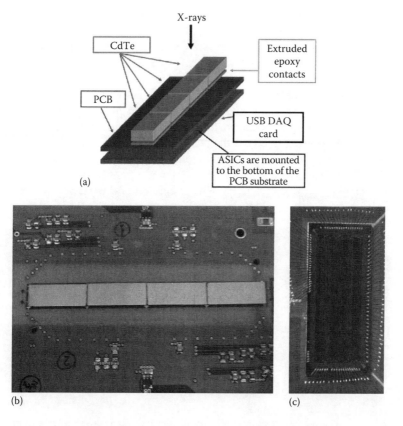

FIGURE 4.11 Schematic of the BMD detectors that are translated during scanning and do not require vertical integration (a). Tiling of crystals allows blocks in 1D to scale to any 1D FOV. Multiple rows reduce scan time. An image of a 1D tiling of CdTe crystals each with a 16 × 4 array of pixels (b). An image of a 1D ASIC with a 32 × 1 array of amplifier inputs with an 0.16 mm pitch (c).

other side. The pixel size is 1.1 mm in the coronal or sagittal plane and 1.4 mm in the transverse plane, which is along the axial or scan direction. Each sensor has a 16 × 4 grid of 1.1 mm × 1.4 mm pixels that are electrically and mechanically interconnected to a 32-channel 1D ASIC. A CdTe crystal configured with 64-pixel anodes where surface preparation and film deposition were performed has been fabricated. When several sensors are tiled together, a 1.1 mm pixel pitch can be maintained in the coronal or sagittal plane. The lower dynamic range and significantly lower OCR requirements allow for the development of CdTe detector arrays with very low electronic noise and excellent energy resolution allowing for increased sensitivity in low-energy X-ray imaging applications. The BMD detector uses CdTe sensors and ASICs with a 1D array of wire bond pads, which are both mounted to a PCB that forms a mechanical and electrical substrate, thus producing elements that can be tiled in 1D or is two-side buttable. Figure 4.11b shows the tiling of four CdTe sensors in 1D. Figure 4.11c shows a 1D ASIC configured with 32 amplifier input pads for the BMD detector. The four 16 × 4 pixel CdTe sensor elements each using two 32-channel ASICs are tiled to create a large area 64 × 4 pixel multislice fan beam array for whole body DEXA scanning. Figure 4.12a shows the OCR as a function of increasing ICR for a tube setting of 80 kVp at various tube currents and at a distance of about 20 cm from the tube's focal spot and with a 1 mm Cu filter between the source and detector for a typical ~1.5 mm² area pixel of the BMD detector. The output is linear to 0.4 Mcps/mm² and saturates above 1 Mcps/mm². Figure 4.12b shows an individual pixel's spectrum from the DEXA detector. A small 10 μCi ^{133}Ba source was placed directly above the sensor, and 2000 ms frames (i.e., images of the number of counts above the thresholds) were acquired at threshold levels stepped in 0.2 keV increments from below the noise floor to above the maximum energy response at 100 keV. The FWHM resolution of the peaks is measured to be approximately 2 keV. The response is linear across the entire dynamic range

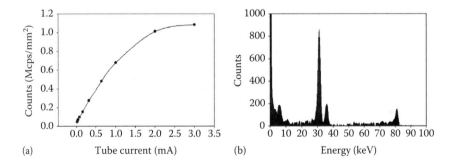

FIGURE 4.12 The output count rate as a function of increasing X-ray tube current for an array of 1.1 mm × 1.4 mm CdTe pixels connected to an 80 ns peaking time parallel channel ASIC (a). An energy spectrum from ^{133}Ba taken with the BMD detector and the 1D ASIC optimized for use with the CdTe barrier contact pixels and operating at a peaking time of approximately 80 ns. The full width at half maximum energy resolution is about 2 keV across the entire dynamic range for DEXA (b).

Pixelated Semiconductor and Parallel ASIC Design

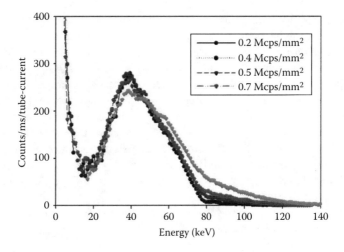

FIGURE 4.13 X-ray spectra taken with a 80 kVp X-ray tube setting and no filter at approximately 20 cm and increasing X-ray tube current with a CdTe sensor with 1.1 mm × 1.4 mm pixels connected to a 1D ASIC and operated at a 80 ns peaking time.

to 80 keV and the energy resolution is the same across dynamic range as well indicating that the energy resolution is dominated by electronic noise as with the modules designed for CT. Figure 4.13 shows energy spectra generated with the BMD detector and an X-ray tube operating at 80 kVp. The energy spectra in Figure 4.13 are normalized by plotting the number of counts divided by the tube current from an X-ray tube set to 80 kVp operated from 0.2 to 1.0 mA of tube current at approximately 20 cm with no filter for an incident flux of about 0.2 Mcps/mm^2 (at 0.2 mA) to 0.7 Mcps/mm^2 (at 1 mA). At a threshold value (pulse height) corresponding to 10 keV, no counts are seen without X-rays or sources present. Notice that at the higher flux (0.7 Mcps/mm^2 OCR), pulse pileup distorts the X-ray spectra. However, the spectra up to 0.5 Mcps/mm^2 OCR, which is sufficient for the application, are essentially the same showing very little pileup.

4.6 MODULE DEVELOPMENT FOR SPECTRAL DIGITAL MAMMOGRAPHY

Breast cancer screening has been performed for many years using X-ray projection mammography using 2D screen-film systems where the examined breast tissue is compressed.[65–67] Mammography is the current standard of care for breast cancer screening because it can detect soft tissue lesions and microcalcifications in short imaging times with cost-effective systems. However, there are significant limitations to the detectability of cancer, especially for women with dense breasts.[68] Recently, DM systems with increased dynamic range using pixelated 2D detectors have begun to replace screen-film systems due to significant SNR advantages among women who are pre- or perimenopausal and/or have dense breast tissue.[69] In order to make further improvements, contrast-enhanced DM is currently being developed using iodine contrast injection. There are, however, additional significant limitations including detector noise, Compton scatter, and the superposition of the breast anatomy on the 2D projection image, all of which result in reduced SNR.[70] Also, the overlap of normal breast parenchyma can obscure tumor identification and it can lead to false-positive detections by the radiologist.[71–76] In an effort to remove detector noise and the effect of Compton scatter in the breast in the recorded image, a system with a stack of 1D PC detector panels where each pixel has a single threshold comparator allowing PC above the noise level has been developed (Phillips MicroDose™). The stack of panels are translated across the breast to perform DM. Fore and aft collimators, with respect to the examined breast, along with the 1D geometry of the pixelated detector arrays rejects virtually all the Compton scatter originating in the breast and single PC eliminates dark noise in the detector. However, to record energy information for the individual X-rays, an upgraded system has been announced that contains two comparators per pixel for dual-energy imaging.

For a film screen the spatial resolution can be as good as 25 μm (20 lp/mm); however, the actual pixel size and spatial resolution requirements for DM are believed to be in the range of 50–100 μm. Studies have shown improved detectability of low-contrast objects on digital systems at a resolution less than that of film systems (at 100 μm pixel size or 5 lp/mm).[77,78] Figure 4.14a shows a schematic of our ED-PC DM module design. The Si direct conversion sensor is edge illuminated with a 5° tilt to create a 6 mm thick absorber for the incident X-rays overcoming the low-Z properties of Si for DM; however, the thickness of the Si sensor is 0.5 mm, which is dictated by the technology used to implement Si pin structure. Another advantage to this design is that the tilted edge illumination avoids dead regions in the sensor related to a guard ring surrounding the strip contacts (anodes) that pixelate the

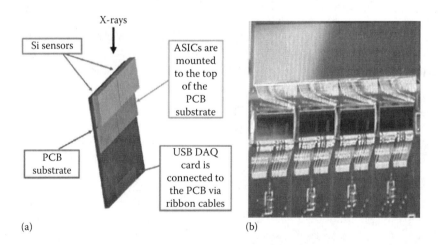

FIGURE 4.14 Schematic of a side view along the edge of the edge-illuminated DM detector (a). An image of a 0.5 mm thick Si crystal with a 256 × 1 array of strips that can be wire bonded directly to the inputs of ASICs with a 1D array of input pads (b).

sensor. Although modules based on edge-illuminated Si strip detectors can be tiled in 1D, this cannot be done without small gaps, given the high spatial resolution requirements for DM, due to spacing between the Si sensors. Figure 4.14b shows a Si sensor with an array strips wire bonded directly to 1D arrays of input pads of the parallel channel ASICs for DM. Figure 4.15a shows the OCR of the ED-PC DM module as a function of increasing ICR for a tube setting of 50 kVp at various tube currents and at a distance of about 10 cm from the tube's focal spot and with a 1 mm Al filter between the source and detector for a typical 100 μm² area pixel of the DM module where a 100 μm parallel slit collimator is used to restrict the effective pixel size perpendicular to the 1D tiling direction. The output is linear to 0.4 Mcps/μm² and saturates above 1 Mcps/μm². This corresponds physically to a linear OCR up to 40 Mcps/mm² and that saturates above 100 Mcps/mm². Figure 4.15b shows an individual pixel's spectrum from the DM module. A small 10 μCi ^{109}Cd source was placed directly above the sensor and 40,000 ms frames were acquired at threshold levels stepped in 0.1 keV increments from below the noise floor to above the maximum energy response at 60 keV. The FWHM resolution of the peak at 22 keV is measured to be approximately 2.2 keV.

Applications in DM require stability and uniformity from the detectors. The temporal stability and uniformity under the flux expected in DM has been measured using the Si sensors with 0.1 mm pixels. We have tiled four 256 pixel modules in a linear array of 1024 pixels and acquired five temporally separated 30 ms exposures to a flat X-ray field taken at 50 kVp over 45 min at room temperature. The images are acquired with an average of approximately 1800 counts/exposure/pixel. Figure 4.16 shows all five images plotted together. No drift is detected, and additionally, the variance of the number of counts versus the mean falls within the counting statistics of the mean.

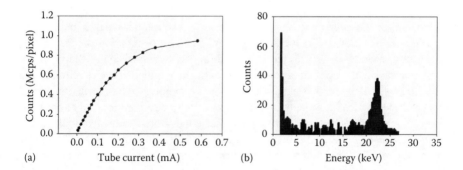

FIGURE 4.15 Graph showing the output count rate as a function of increasing X-ray tube current for an array of 100 μm strip pixels connected to a fast 100 ns peaking time parallel channel 1D ASIC optimized for use with the 100 μm Si strip contacts (a). An energy spectrum from ^{109}Cd taken with the DM module and the 1D ASIC optimized for use with the Si strip pixels and operating at a peaking time of approximately 100 ns. The full width at half maximum energy resolution is about 2.2 keV at 22 keV (b).

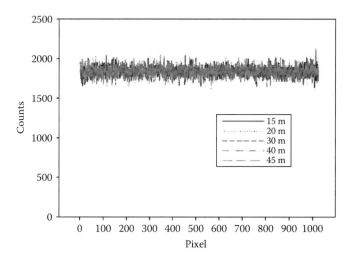

FIGURE 4.16 Graphs of the flat field response in time as determined by acquiring data in 530 ms exposures in a linear array of 1024 0.1 mm edge-illuminated Si pixels under constant flux from an X-ray generator set to 50 kVp.

4.7 DISCUSSION

When developing sensors and associated ASICs for applications in spectral radiology, the ranges and resolutions required must be considered. The OCR range must accommodate the incident flux on the detector minimizing pulse pileup. The dynamic range, defined in ED-PC X-ray imaging arrays as the range of X-ray energies for which high DQE is required rather than as the range of intensities over which a linear response is achievable as with EI X-ray imaging arrays, needs to be good enough to demonstrate a clinical advantage over existing EI technologies that may be application dependent. Of course the best possible energy resolution is desired and requires rapid signal formation in direct conversion semiconductors to be achieved so that the energy resolution is preserved across the useful OCR range. Finally the OCR and dynamic range with good energy resolution must be obtained at the required spatial resolution, which places physical limitations on the speed of signal formation and the density of mechanical and electrical interconnections coupling the sensor and ASIC.

For CT, relatively thick (>1 mm) high-Z semiconductors sensors can make use of the *small pixel effect* to achieve signal formation faster than the charge transit time across the sensor. With guard ring elimination achieved by lateral surface passivation, 2D arrays can be achieved with minimal dead space, and with the ASIC vertically integrated within the active area of the sensor, a four-side buttable design can be achieved. For these reasons we use CdTe or CdZnTe sensors coupled to a BGA encapsulated ASIC with a 2D array of bond pads. In this design we achieve rapid output signal from the sensor matching the fast peaking time of the ASIC achieving tiling in 2D with a high OCR. The fast peaking time limits the dark current's contribution to the noise permitting the use of conduction structures for the sensor with essentially equivalent results from CdTe and CdZnTe. The vertical integration requirement for the readout forces an intimate coupling of the sensor and ASIC, which has the advantage of reducing the input capacitance to the ASIC that is the dominant source of electronic noise at very fast peaking times. There is therefore synergy between the OCR, the dynamic range (thick sensors), and vertical integration (2D tiling) in this design for spectral clinical CT application that together provide good energy resolution limited by the electronic noise in the ASIC. In addition to range and resolution requirements, tiling of many segmented modules is required to achieve a FOV suitable for the application. This means the front-end sensor and ASIC combination must accommodate a back plane readout, without a bandwidth limitation truncating the OCR, that can be integrated with the front-end electronics. For CT, a vertically integrated back plane readout is required to achieve a multislice fan beam geometry.

For BMD, the OCR and dynamic range requirement is less as compared to CT but with the same intrinsic spatial resolution requirement of about 1 mm. In this case, thinner high-Z semiconductors can be used with an ASIC with longer peaking time. In this design rapid signal formation is achieved across the thin sensor and we use CdTe with Schottky structures to limit the dark current's contribution to the noise at the longer peaking time. BMD images are acquired by translation of either an array or single pixel in either a fan beam or pencil beam geometry, respectively. In our BMD detector design, ASICs with a 1D array of bond pads are interconnected by wire bonding (as opposed to the flip-chip bump bonding used with our 2D ASICs). The BMD modules have a two-side buttable geometry sufficient for 1D tiling and suitable for fan beam scanning for BMD measurements. In this design there is synergy between the lower OCR and dynamic range (thin sensors) and the nonvertical integration (1D tiling), which together provide good energy resolution limited by the electronic noise in the ASIC.

For DM, the dynamic range is lower than for both CT and BMD but the intrinsic spatial resolution requirement is an order of magnitude higher at 100 μm. Although the dynamic range for DM is relatively small for an application in radiology, it requires 6 mm to 1 cm of the low-Z semiconductor Si to achieve 75%–90% DQE, respectively, for a 50 kVp source. Our design for DM uses edge-illuminated SI strip sensors to achieve sufficient DQE for DM but requires a high OCR due to the need to translate the 1D FOV X-ray imaging arrays and maintain a reasonable image acquisition time. In this case with this design there is synergy between the edge illumination and the OCR requirement since the charges generated are collected across the thin detector for all DOI absorption. The edge illumination where the pixel size in the scanning direction is controlled by the width of a parallel slit collimator rejects virtually all of the scatter that is a major benefit in DM. In this case it is desirable to vertically integrate the ASICs on the side of the edge-illuminated Si strip sensors that is away from the X-ray tube in order to place the sensor very close to the lower breast compression plate to avoid magnification, but the back plane readout does not also need to be vertically integrated within the active area of the sensor as with the CT modules.

4.8 CONCLUSIONS

The example ED-PC X-ray imaging arrays designed for applications in spectral radiology presented here demonstrate the role that geometry plays in the design of technologies with sufficient ranges and resolution for a particular application. Sensor geometry such as the thickness in the incident direction for a given sensor material must achieve the required dynamics, while sensor thickness and pixel pitch play roles in determining the possible OCR range as well as the spatial and energy resolutions. Also ASIC geometry such as the minimum line width process used in fabrication of the Si wafer and the pitch and layout of the bonding pads need to match the sensors pixel pitch so that modules can be tiled to achieve the required FOV for the particular application. And finally, concerning the role of geometry in the design of these ED-PC X-ray imaging arrays, the need to electrically interconnect the sensor's pixels to the ASIC's inputs while also achieving the required integration of the back plane readout to provide I/O control and data readout from the ASIC can be constrained by the density of connections that can be achieved within PCBs and in the connections between PCBs.

ED-PC X-ray imaging has the potential to address both of the two major problems inherent in EI X-ray imaging, namely, dose and contrast. However, in radiology there are very demanding requirements on the detector for particularly high OCRs as well as good DQE and reasonable energy resolution across the useful portion of the OCR range. To meet these range and resolution requirements, there is a need for the development of novel detector structures, customized fast low-noise and low-power ASIC electronics, and appropriate interconnections between detector pixels and electronics for the construction of tilable detector modules.

The high X-ray photon OCRs required impose a number of other requirements on an ED-PC detector arrays such as the design of the processing electronics (back plane readout), the selection of detector material, and the method of electrode fabrication. To avoid effects in CdTe or CdZnTe that would lead to distortion and eventual collapse of the electric field, care must be taken that the charge generated by the radiation is removed from the device at a sufficiently fast rate.[79] For this reason, hole collection (holes being less mobile than electrons) is important. In this respect CdTe has some advantages over CdZnTe as typically the lifetimes of holes are significantly longer in CdTe than in CdZnTe.

This chapter presents our novel detector designs that have been specifically developed for and evaluated for spectral X-ray imaging applications in clinical radiology. Most of our work has concentrated on the development CdTe, CdZnTe, and Si detector arrays, fast low-noise ASIC readout electronics, and solving the interconnecting problems for coupling the detectors and electronics in order to create modules that can be tiled together. The developed detectors based on semiconductor sensors provide a fast signal response time over the whole dynamic range of the application. At the same time the individual detector pixels exhibit good spectral performance at very short peaking times at room temperature for all the detectors presented here.

ACKNOWLEDGMENTS

We would like to thank the National Institute of Biomedical Imaging and Bioengineering (NIBIB) at the National Institutes of Health (NIH) for partial support at DxRay Inc., for the development of the CdTe and CdZnTe detectors (grants numbered R44EB007873 and R44EB012379), and the National Cancer

Institute (NCI) at the NIH for partial support at DxRay Inc., for the development of the Si strip detectors (grant number R43CA177093).

REFERENCES

1. Hu H, He HD, Foley WD, and Fox SH, Four multidetector-row helical CT: Image quality and volume coverage speed, *Radiology* 215(1): 55–62 (2000).
2. Tapiovaara MJ and Wagner RF, SNR and DQE analysis of broad-spectrum X-ray-imaging, *Physics in Medicine and Biology* 30(6): 519–529 (1985).
3. Giersch JD, Niederlohner, and Anton G, The influence of energy weighting on X-ray imaging quality, *Nuclear Instruments & Methods in Physics Research Section A—Accelerators Spectrometers Detectors and Associated Equipment* 531(1–2): 68–74 (2004).
4. Niederlohner D, Karg J, Giersch J, Firsching M, and Anton G, Practical aspects of energy weighting in X-ray imaging, *Nuclear Science Symposium Conference Record, IEEE*, pp. 3191–3654 (2004).
5. Alvarez RE, Near optimal energy selective X-ray imaging system performance with simple detectors, *Medical Physics* 37(2): 822–841 (2010).
6. Alvarez RE and Macovski A, Energy-selective reconstructions in X-ray computerized tomography, *Physics in Medicine and Biology* 21(5): 733–744 (1976).
7. Brody WR, Butt G, Hall AL, and Macovski A, A method for selective tissue and bone visualization using dual energy radiography, *Medical Physics* 8(3): 353–357 (1981).
8. Lehmann LA, Alvarez RE, Macovski A, Brody WR, Pelc NJ, Riederer SJ, and Hall AL, Generalized image combinations in dual KVP digital radiography, *Medical Physics* 8(5): 659–667 (1981).
9. Bisogni M, Gambaccini M, Fabbri S, Gambaccini M, Marziani M, Novelli A, Quattrocchi M et al., A digital system based on a bi-chromatic X ray source and a single photon counting device: A single exposure dual energy mammography approach, *Proceedings of SPIE* 4682: 620–632 (2002).
10. Kalender WA, Perman WH, Vetter JR, and Klotz E, Evaluation of a prototype dual-energy computed tomographic apparatus, *Medical Physics* 13(3): 340–343 (1986).
11. Marshall W, Hall E, and Doost-Hoseini A, An implementation of dual energy CT scanning, *Journal of Computed Assisted Tomography* 8(4): 745–749 (1984).
12. Marshall WH, Alvarez RE, and Macovski A, Initial results with prereconstruction dual-energy computed tomography (PREDECT), *Radiology* 140(2): 421–430 (1981).
13. Drost DJ and Fenster A, Experimental dual xenon detectors for quantitative CT and spectral artifact correction, *Medical Physics* 7(2): 101–107 (1980).
14. Kelcz F, Joseph P, and Hila SI, Noise considerations in dual energy CT scanning, *Medical Physics* 6(5): 418–425 (1979).
15. Alvarez RE, Seibert JA, and Thompson SK, Comparison of dual energy detector system performance, *Medical Physics* 31(3): 556–565 (2004).
16. Brooks RA and Di Chiro G, Split-detector computed tomography: A preliminary report, *Radiology* 126(1): 255–257 (1978).
17. Stevens GM and Pelc NJ, Depth-segmented detector for X-ray absorptiometry, *Medical Physics* 27(5): 1174–1184 (2000).
18. Shikhaliev PM, Tilted angle CZT detector for photon counting/energy weighting X-ray and CT imaging, *Physics in Medicine and Biology* 51(17): 4267–4287 (2006).
19. Whiting BR, Massoumzadeh P, Earl OA, O'Sullivan JA, Snyder DL, and Williamson JF, Properties of preprocessed sinogram data in X-ray computed tomography, *Medical Physics* 33(9): 3290–3303 (2006).
20. Swank RK, Absorption and noise in X-ray phosphors, *Journal of Applied Physics* 44(9): 4199–4203 (1973).

21. Taguchi K, Srivastava S, Kudo H, and Barber WC, Enabling photon counting clinical X-ray CT, *Nuclear Science Symposium Conference Record, IEEE*, Orlando, FL, pp. 3581–3585 (2009).
22. Toth T and Hoffman DM, System and method of determining a user-defined region-of-interest of an imaging subject for X-ray flux management control, U.S. Patent 6,990,171 (2006).
23. Kudo H, Courdurier M, Noo F, and Defrise M, Tiny a priori knowledge solves the interior problem in computed tomography, *Physics in Medicine and Biology* 53: 2207–2231 (2008).
24. Heismann BJ, Signal-to-noise Monte-Carlo analysis of base material decomposed CT projections, *Nuclear Science Symposium Conference Record, IEEE*, San Diego, CA, pp. 3174–3175 (2006).
25. Johns PC, Dubeau J, Gobbi DG, and Dixit MS, Photon-counting detectors for digital radiography and X-ray computed tomography, *Regional Meeting on Optoelectronics, Photonics, and Imaging, SPIE*, Vol. TD01, pp. 367–369 (2002).
26. Shikhaliev PM, Tong X, and Molloi S, Photon computed tomography: Concept and initial results, *Medical Physics* 32(2): 427–436 (2005).
27. Maidment A, Ullberg C, Lindman K, Adelow L, Egerstrom J, Eklund M, Francke T et al., Evaluation of a photon-counting breast tomosynthesis imaging system, *Medical Imaging 2005, Proceedings of SPIE* 5745: 572–582 (2005).
28. Lundqvist M, Cederstrom B, Chmill V, and Danielsson M, Evaluation of a photon-counting X-ray imaging system, *Transactions on Nuclear Science* 48(4): 1530–1536 (2001).
29. Chmeissani M, Frojdh C, Gal O, Llopart X, First experimental tests with a CdTe photon counting pixel detector hybridized with a Medipix2 readout chip, *Transactions on Nuclear Science* 51(5): 2379–2385 (2004).
30. Danielsson M, Bornefalk H, Cederstroem B, Chmill V, Hasegawa BH, Lundqvist M, Nygren DR, and Tabar T, Dose-efficient system or digital mammography, *Proceedings of SPIE* 3977(1): 239–250 (2000).
31. Iwanczyk JS, Nygard E, Meirav O, Arenson J, Barber WC, Hartsough NE, Malakhov N, and Wessel JC, Photon counting energy dispersive detector arrays for X-ray imaging, *Transactions on Nuclear Science* 56: 535–542 (2009).
32. Barber WC, Nygard E, Wessel JC, Malakhov N, Hartsough NE, Gandhi T, Wawrzyniak G, and Iwanczyk JS, Photon-counting energy-resolving CdTe detectors for high-flux X-ray imaging, *Nuclear Science Symposium Conference Record, IEEE*, Knoxville, TN, pp. 3953–3955 (2010).
33. Wojtowicz AJ, Drozdowski W, Wisniewski D, Lefaucheur JL, Galazka, Gou Z, Lukasiewicz T, and Kisielewski J, Scintillation properties of selected oxide monocrystals activated with Ce and Pr, *Optical Materials* 28(1–2): 85–93 (2006).
34. Squillante M and Shah K, Other materials: Status and prospects. In *Semiconductors for Room Temperature Nuclear Detector Applications* (San Diego, CA: Academic Press, Schiesinger TE, James RB (eds.), 1995), pp. 470–471.
35. Zentai G, Partain LD, Pavlyuchkova R, Proano C, Virshup G, Melekhov L, Zuck A et al., Mercuric iodide and lead iodide X-ray detectors for radiographic and fluoroscopic medical imaging, *Proceedings of SPIE* 5030: 77–91 (2003).
36. Bennett PR, Shah KS, Cirignano LJ, Klugerman LP, Moy LP, and Squillante MR, Characterization of polycrystalline TlBr films for radiographic detectors, *Transactions on Nuclear Science* 46: 266–270 (1999).
37. Shah KS, Bennett P, Dimitriyev Y, Cirignanno L, Klugerman M, Squillante MR, Street RA, Rahn JT, and Ready SE, PbI2 for high resolution digital X-ray imaging, *Proceedings of SPIE* 3770: 164–173 (1999).

38. Street RA, Rahn JT, Ready SE, Shah K, Bennett PR, Dimitriyev Y, Mei P et al., X-ray imaging using lead iodide as a semiconductor detector, *Proceedings of SPIE* 3659: 36–47 (1999).
39. Schieber M, Hermon H, Zuck A, Vilensky A, Melekhov L, Shatunovsky R, Meerson E, and Saado H, Polycrystalline mercuric iodide detectors, *Proceedings of SPIE* 3770: 146–155 (1999).
40. Schieber M, Zuck A, Melekhov L, Shatunovsky R, Hermon H, and Turchetta R, High flux X-ray response of composite mercuric iodide detectors, *Proceedings of SPIE* 3768: 296–309 (1999).
41. Bambynek W, Crasemann B, Fink RW, Freund HU, Mark H, Swift CD, Price RE, and Rao PV, X-ray fluorescence yields, Auger, and Coster-Kronig transition probabilities, *Reviews of Modern Physics* 44: 716–813 (1972).
42. Barrett HH and Myers K, *Foundations of Image Science*, 1st edn. (New York: Wiley-Interscience, 2003).
43. Iwanczyk JS, Patt BE, Segal J, Plummer J, Vilkelis G, Hedman B, Hodgson KO, Cox AD, Rehn L, and Metz J, Simulation and modelling of a new silicon X-ray drift detector design for synchrotron radiation applications, *Nuclear Instruments and Methods in Physics Research Section A: Accelerators, Spectrometers, Detectors and Associated Equipment* 380: 288–294 (1996).
44. Myronakis ME and Darambara DG, Monte Carlo investigation of charge-transport effects on energy resolution and detection efficiency of pixelated CZT detectors for SPECT/PET applications, *Medical Physics* 38: 455–467 (2011).
45. Patt BE, Iwanczyk JS, Vilkelis G, and Wang YJ, New gamma-ray detector structures for electron only charge carrier collection utilizing high-Z compound semiconductors, *Nuclear Instruments and Methods in Physics Research Section A: Accelerators, Spectrometers, Detectors and Associated Equipment* 380: 276–281 (1996).
46. Siffert P, Berger J, Scharager C, Cornet A, Stuck R, Bell RO, Serreze HB, and Wald FV, Polarization in cadmium telluride nuclear radiation detectors, *Transactions on Nuclear Science* 23: 159–170 (1976).
47. Szeles C, Soldner SA, Vydrin S, Graves J, and Bale DS, Ultra high flux 2-D CdZnTe monolithic detector arrays for X-ray imaging applications, *Transactions on Nuclear Science* 54: 1350–1358 (2007).
48. Malm HL and Martini M, Polarization phenomena in CdTe: Preliminary results, *Journal of Applied Physics* 51: 2336–2340 (1973).
49. Bell RO, Entine G, and Serreze HB, Time-dependent polarization of CdTe gamma-ray detectors, *Nuclear Instruments and Methods* 117: 267–271 (1974).
50. Gimenez EN, Ballabriga R, Campbell M, Horswell I, Llopart X, Marchal J, Sawhney KJS, Tartoni N, and Turecek D, Characterization of Medipix3 with synchrotron radiation, *Transactions on Nuclear Science* 58: 323–332 (2011).
51. Iwanczyk JS and Taguchi K, Vision 20/20: Single photon counting X-ray detectors in medical imaging, *Medical Physics* 40(10): 100901 (2013).
52. Ballabriga R, Campbell M, Heijne EHM, Llopart X, and Tlustos L, The Medipix3 prototype, a pixel readout chip working in single photon counting mode with improved spectrometric performance, *Transactions on Nuclear Science* 54: 1824–1829 (2007).
53. Barber WC, Iwata K, Hasegawa BH, Bennett PR, Cirignano LJ, and Shah KS, Current mode operation of a CdZnTe for CT imaging, *Penetrating Radiation Systems and Applications IV, SPIE* 4786: 144–150 (2002).
54. Mikulec B, Campbell M, Heijne E, Llopart X, and Tlustos L, X-ray imaging using single photon processing with semiconductor pixel detectors, *Nuclear Instruments and Methods A* 511: 282–286 (2003).

55. Tomita Y, Shirayanagi Y, Matsui S, Misawa M, Takahashi H, Aoki T, and Hatanaka Y, X-ray color scanner with multiple energy differentiate capability, *Nuclear Science Symposium Conference Record,* Vol. 6, pp. 3733–3737 (2004).
56. Le HQ and Molloi S, Segmentation and quantification of materials with energy discriminating computed tomography: A phantom study, *Medical Physics* 38: 228–237 (2011).
57. Taguchi K, Zhang M, Frey EC, Wang X, Iwanczyk JS, Nygard E, Hartsough NE, Tsui BMW, and Barber WC, Modeling the performance of a photon counting X-ray detector for CT: Energy response and pulse pileup effects, *Medical Physics* 38: 1089–1102 (2011).
58. Barber WC, Nygard E, Wessel JC, Malakhov N, Hartsough NE, Gandhi T, Wawrzyniak G, and Iwanczyk JS, Photon-counting energy-resolving CdTe detectors for high-flux X-ray imaging, *Nuclear Science Symposium Conference Record, IEEE,* Vol. R18(1), pp. 3953–3955 (2011).
59. Cammin J, Srivastava S, Barber WC, Iwanczyk JS, Hartsough NE, Nygard E, Wessel JC, Malakhov N, and Taguchi K, A tabletop clinical X-ray CT scanner with energy-resolving photon counting detectors, *Medical Imaging 2011: Physics of Medical Imaging, SPIE,* Lake Buena Vista, FL, pp. 79611S1–79611S7 (2011).
60. Kappler S, Henning A, Kreisler B, Schoeck F, Stierstorfer K, and Flohr T, Photon counting CT at elevated X-ray tube currents: Contrast stability, image noise and multi-energy performance, *Proceeding of SPIE* 9033: 90331C (2014).
61. Taguchi K, Frey EC, Wang X, Iwanczyk JS, and Barber WC, An analytical model of the effects of pulse pileup on the energy spectrum recorded by energy resolved photon counting X-ray detectors, *Medical Physics* 37(8): 3957–3969 (2010).
62. Leng S, Yu L, Wang J, Fletcher JC, Mistretta CA, and McCollough CH, Noise reduction in spectral CT: Reducing dose and breaking the trade-off between image noise and energy bin selection, *Medical Physics* 38: 4946–4957 (2011).
63. Shikhaliev PM, Beam hardening artifacts in computed tomography with photon counting, charge integrating and energy weighting detectors: A simulation study, *Physics in Medicine and Biology* 50(24): 5813–5827 (2005).
64. Barber WC, Malakhov N, Damron MQ, Hartsough NE, Moraes D, Weilhammer P, Nygard E, and Iwanczyk JS, Guard ring elimination in CdTe and CdZnTe detectors, *Nuclear Science Symposium Conference Record, IEEE,* San Diego, CA, pp. 2414–2416 (2006).
65. Schmitt EL and Threatt BA, Effective breast-cancer detection with film-screen mammography, *Journal of the Canadian Association Radiologists* 36: 304–307 (1985).
66. Dorsi CJ, Early detection of breast cancer: Mammography, *Breast Cancer Research and Treatment* 18: S107–S109 (1991).
67. Nishikawa RM, Mawdsley GE, Fenster A, and Yaffe MJ, Scanned-projection digital mammography, *Medical Physics* 14: 717–727 (1987).
68. Sala M, Comas M, Macia F, Martinez J, Casamitjana M, and Castells X, Implementation of digital mammography in a population-based breast cancer screening program: Effect of screening round on recall rate and cancer detection, *Radiology* 252: 31–39 (2009).
69. Pisano ED, Gatsonis C, Hendrick E, Yaffe M, Baum JK, Acharyya S, Conant EF et al., Diagnostic performance of digital versus film mammography for breast-cancer screening, *New England Journal of Medicine* 353: 1773–1783 (2005).
70. Bird RE, Wallace TW, and Yankaskas BC, Analysis of cancers missed at screening mammography, *Radiology* 184: 613–617 (1992).
71. Swann CA, Kopans DB, Mccarthy KA, White G, and Hall DA, Mammographic density and physical assessment of the breast, *American Journal of Roentgenology* 148: 525–526 (1987).
72. Jackson VP, Hendrick RE, Feig SA, and Kopans DB, Imaging of the radiographically dense breast, *Radiology* 188: 297–301 (1993).

73. Pisano ED, Gatsonis C, Hendrick E, Yaffe M, Baum JK, Acharyya S, Conant EF etal., Diagnostic performance of digital versus film mammography for breast-cancer screening, *New England Journal of Medicine* 353: 1773–1783 (2005).
74. Diekmann F, Freyer M, Diekmann S, Fallenberg EM, Fischer T, Bick U, and Pollinger A, Evaluation of contrast-enhanced digital mammography, *European Journal of Radiology* 78: 112–121 (2011).
75. Pinker K, Perry N, Milner S, Mokbel K, and Duffy S, Accuracy of breast cancer detection with full-field digital mammography and integral computer-aided detection correlated with breast density as assessed by a new automated volumetric breast density measurement system, *Breast Cancer Research and Treatment* 12(3): P4 (2010).
76. Pinker K, Perry N, Milner S, Mokbel K, and Duffy S, Sensitivity of integral computer-aided detection with full-field digital mammography for detection of breast cancer according to different histopathological tumor types and appearances, *Breast Cancer Research and Treatment* 12(3): P13 (2010).
77. Karssemeijer N, Frieling J, and Hendricks HCC, Spatial resolution in digital mammography, *Investigative Radiology* 28: 413–419 (1993).
78. Shaw de Paredes E, Fatouros PP, Thunberg S, Cousins J, Wilson J, and Sedgewick T, Evaluation of a digital spot mammographic unit using a contrast detail phantom. In *Digital Mammography* (Amsterdam, the Netherlands: Kluwer Academic Publishers, Karssemeijer N, Thijssen M, Hendricks J, Nijmegen (eds.), 1998), pp. 47–50.
79. Szeles C, Soldner SA, Vydrin S, Graves J, and Bale DS, Ultra high flux 2-D CdZnTe monolithic detector arrays for X-ray imaging application, *Transactions Nuclear Science* 54(4): 1350–1358 (2007).

5 Silicon Photomultipliers in Detectors for Nuclear Medicine

Martyna Grodzicka, Marek Moszyński, and Tomasz Szczęśniak

CONTENTS

5.1 Introduction 119
5.2 Operation Principles 121
    5.2.1 Basic Parameters of a SiPM 122
        5.2.1.1 Gain 122
        5.2.1.2 Photon Detection Efficiency 124
        5.2.1.3 After-Pulses 126
        5.2.1.4 Optical Crosstalk 129
        5.2.1.5 Dark Noise 130
        5.2.1.6 Linearity of an SiPM 131
        5.2.1.7 Excess Noise Factor 133
        5.2.1.8 Transit Time Spread 133
    5.2.2 Digital Silicon Photomultipliers 136
    5.2.3 SiPMs in Gamma Spectrometry with Scintillators 137
    5.2.4 SiPMs in Fast Timing with Scintillators 138
5.3 SiPMs in Medical Instrumentations 141
References 143

5.1 INTRODUCTION

Scintillation detectors are one of the most commonly used types of radiation detectors in nuclear medicine. These detectors consist of a photodetector and a dense crystalline scintillation material that absorbs gamma quanta and emits light as a result. The scintillation light is emitted isotropically in short pulses, lasting typically a couple of hundred nanoseconds (Nassalski et al. 2007, Lewellen 2008). In positron emission tomography (PET) scanners, the typical number of light photons emitted from a single 511 keV gamma scintillation is up to 20,000 photons, depending on the scintillator used (Conti et al. 2009). The number of photons detected or "seen" by a photodetector is the main parameter deciding about its performance. These photons are the "information carriers" about the detected gamma radiation. The higher is their number, the more detailed information about the radiation is possible.

Therefore, the first requirement for a PET photodetector is to possess a very high sensitivity in order to achieve a good signal-to-noise ratio (SNR).

Another important requirement for a photodetector concerns its timing performance. The recent development of bright and fast scintillators such as lutetium orthosilicate (LSO), lutetium yttrium orthosilicate (LYSO), and lanthanum(III) bromide ($LaBr_3$) has enabled the usage of time-of-flight PET (TOF-PET), which explores the difference between the arrival times of the gamma pair to estimate the position along the line of response where the annihilation took place (Moses and Derenzo 1999). Therefore, to actually improve the SNR and image contrast with TOF-PET, the employed detectors must feature subnanosecond timing performance.

Historically, the most commonly used photodetectors in PET scanners were photomultiplier tubes (PMTs) (Del Guerra et al. 2009, Szczesniak et al. 2009). This was mainly due to their very high gain, low noise, and fast response. However, PMTs are formed by a vacuum tube and, as such, they are somewhat bulky and fragile. In addition, they also require power supplies of up to thousands of volts and are sensitive to magnetic fields. Due to these disadvantages, solid-state detectors, like avalanche photodiodes (APD), have long been proposed as an alternative to PMTs (Moszynski et al. 2002).

At present, the silicon photomultiplier (SiPM) has been showing promising results in the field of PET (Otte et al. 2005). A SiPM is a relatively new kind of a photodetector that is made up of a matrix of parallel connected micro-APDs (APD cells), working in Geiger mode (Antich et al. 1997, Buzhan et al. 2003, Golovin and Saveliev 2004). Each APD cell generates a constant signal after detection of a single photon, and a sum of the signals from all the APD cells gives a SiPM output pulse that is proportional to the number of detected photons. However, this proportionality is disrupted because of such phenomena as crosstalk and after-pulse, which give additional, false pulses that are not directly related to the detected photon. In consequence, the number of fired APD cells is larger than the number of detected photons. Moreover, each APD cell possesses an internal recovery time that lasts from a few to tens of ns after detection of a single photon (this value depends on the SiPM and its manufacturer). Therefore, if at the same time two or more photons are incident on one APD cell, the signal will be lost and the linearity of photon detection will deteriorate. This is because the number of incident photons that hit the photosensitive area and could be potentially detected is larger than the number of fired APD cells. Summarizing, the sum of signals from all of the fired APD cells of the entire array is an output signal of a SiPM, but the total number of fired APD cells does not directly correspond to the number of incident photons. A SiPM's response depends on its total number of APD cells, its effective recovery time and the duration of the detected pulse, or the decay time of the light pulses in the scintillators (Dolgoshein et al. 2006, Musienko et al. 2006, Renker 2006, 2009, Du and Retiere 2008, Finocchiaro et al. 2009).

SiPMs belong to a group of highly sensitive detectors with single-photon detection capability, which is in contrast to other popular silicon photodiodes, such as single APD or P-type, Intrinsic, N-type (PIN) diode. SiPMs possess high internal gain (typically 10^5–10^6), which is comparable to the gain of PMTs, are insensitive to magnetic fields, are resistant to damage, require a low bias voltage (below 100 V), have a compact size, and possess high quantum efficiency (QE). The drawbacks of SiPMs are small active area, sensitive to temperature and bias voltage, high capacitance (especially for large active areas), and generation of false pulses caused by phenomena such as crosstalk and after-pulses.

SiPMs are widely tested in high-energy physics, in neutrino physics, and also in commercial applications such as nuclear medicine or border monitoring devices (Andreev et al. 2005, Otte et al. 2005, 2006, Yamamoto et al. 2005, Raylman et al. 2006, Renker 2006, 2009, Danilov 2007, Korpar et al. 2008, Achenbach et al. 2009, Schaart et al. 2009, Yokoyama et al. 2010, Musienko 2011). Until recently, the small total active area of the individual detector (from 1×1 mm^2 to 2×2 mm^2) limited their potentiality of being used in gamma spectrometry with scintillators. Presently, a slightly larger total active area of single detectors (3×3 mm^2 and 6×6 mm^2) and their matrices (from 6×6 mm^2 to 57×57 mm^2), also on a common substrate, allow for the efficient use of SiPMs in scintillation detectors for gamma spectrometry.

The rapid development of SiPMs has facilitated construction of compact, efficient, and magnetic response (MR)-compatible PET scanners (Yamamoto et al. 2005, Raylman et al. 2006). Moreover, the use of TOF information in PET has been demonstrated to enable significant improvement in image noise properties and, therefore, lesion detection, especially in heavier patients (Allemand et al. 1980, Mullani et al. 1981, Laval et al. 1982). It has recently been shown that a very good timing resolution can be achieved with SiPM-based scintillation detectors (Kim et al. 2009, Schaart et al. 2010), opening a new way for TOF-PET scanners.

5.2 OPERATION PRINCIPLES

Nowadays, SiPMs are manufactured by several companies: Hamamatsu Photonics (Japan), SensL (Ireland), Zecotek Photonics (Singapore), Fondazione Bruno Kessler (FBK) (Trento, Italy), STMicroelectronics (Catania, Italy), Amplification Technologies (New Jersey, United States), Ketek GmbH (Munich, Germany), and others. The name *SiPM* is the most commonly used, but in the literature, this type of photodetector can also be referred to as a multipixel photon counter (MPPC), micropixel APD, multipixel Geiger-mode APD, solid-state photomultiplier, single-photon avalanche diode array, or pixelated Geiger-mode avalanche photon detector.

A SiPM consists of many small cells of APDs (APD cells) that are fabricated on a common Si substrate. Nowadays, depending on the manufacturer's technology, commercially available SiPMs can be a single device with a total active area from 0.18×0.18 mm^2 (manufactured by Amplification Technologies) up to 6×6 mm^2 (manufactured by SensL). Detectors with a larger active area are based on SiPM arrays of different formats composed of these single elements. The array elements can be built on one substrate or multiple separate small devices can be stacked together to form the array. Available single SiPM devices consist of many APD cells from ~100 to 15,000 per mm^2. The size of the each individual cell varies from 15×15 μm^2 up to a maximum of 100×100 μm^2 and is constant for a given device.

Each APD cell operates in a Geiger mode, which means that each APD cell is reverse biased above the electrical breakdown voltage (V_{bd}). In these conditions, the electric field in the depletion region of the APD cell is high enough for free carriers, that is, the electrons and holes (which were produced by light absorption), to produce additional carriers by impact ionization, thus resulting in a self-sustaining avalanche

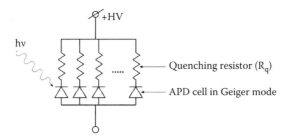

FIGURE 5.1 A simplified electric structure of a SiPM composed of several APD cells in a series with a quenching resistor.

(Vacheret et al. 2011). This Geiger discharge is stopped due to a drop in the voltage below the breakdown value, either passively by an external resistor connected in a series with the diode (R_q, quenching resistor, typically from about 100 kΩ to several MΩ; see Figure 5.1) or actively by the special quenching electronics (Bondarenko et al. 2000). After a certain effective recovery time, which lasts several tens of nanoseconds, the voltage is restored again to the operating value, and the cell is ready to detect the arrival of another photon (Spinelli and Lacaita 1997). Before restoring the full voltage at a given cell, this cell generates signal with reduced gain (or amplitude).

The independently operating APD cells are connected in parallel to the same readout line; therefore, the combined output signal corresponds to the sum of all fired APD cells, which is a measure of the light flux (Dolgoshein et al. 2006).

Due to the device operation principle, the avalanche can be triggered not only by photogenerated carriers but also by carriers that are thermally generated or emitted as a result of phenomena such as after-pulses and crosstalk (discussed in Sections 5.2.1.3 and 5.2.1.4, respectively).

5.2.1 Basic Parameters of a SiPM

One of the key parameters of a SiPM is the size of the APD cell. It determines both the gain and the dynamic range. For geometrical considerations, for a given border technology, the smaller the microcell size, the smaller the fill factor (the percentage of the sensitive area in respect to the total area of the device). Thus, from the efficiency point of view, large cells are desirable. On the other hand, the main positive aspects of a small microcell (e.g., ≤ 30 μm) are a higher dynamic range (due to a high number of microcells on the same substrate area), a shorter recovery time, and a smaller correlated noise (optical crosstalk and after-pulsing, both proportional to the gain of the detector) (Piemonte et al. 2013).

5.2.1.1 Gain

The gain of one APD cell (G) is defined as the ratio between the charge produced in a single avalanche and the elementary charge, and it can be expressed as (Dolgoshein et al. 2006, van Dam et al. 2010, Vacheret et al. 2011)

$$G = \frac{Q}{e} = \frac{C_{APDcell} \cdot V_{ob}}{e} \tag{5.1}$$

where

> Q is the charge of a single avalanche in an APD cell, typically created by a single carrier (unit charge) and can be triggered either by a photon or by thermal noise
> e is the elementary charge, equal to 1.602×10^{-19} C
> $C_{APDcell}$ is the APD cell capacitance
> V_{ob} is the voltage over breakdown also referred to as overvoltage, defined as

$$V_{ob} = (V_b - V_{bd}) \tag{5.2}$$

where

> V_b is the operating voltage or bias voltage
> V_{bd} is the breakdown voltage of a SiPM

The gain of a single APD cell in a SiPM increases linearly with the overvoltage, as opposed to the exponential voltage dependence of the gain in standard APDs. The gain is independent on the temperature if V_{ob} is the same, although V_{bd} is dependent on the temperature. It means that when the bias voltage is constant and the temperature has changed, the gain will also change due to shift of the breakdown voltage. Figure 5.2 shows a typical gain versus bias voltage characteristic for the Hamamatsu MPPC with three different APD cell sizes: 25 × 25 μm, 50 × 50 μm, and 100 × 100 μm. These characteristics are straight lines that, for the gain equal to 0, intersect the x-axis at values being the breakdown voltages. Similar plots are observed for SiPMs from other producers. However, in the case of FBK SiPMs, the nonlinear behavior of the relative gain curve is observed due to the partial depletion of the epilayer at the breakdown and up to approximately 27 V of bias voltage

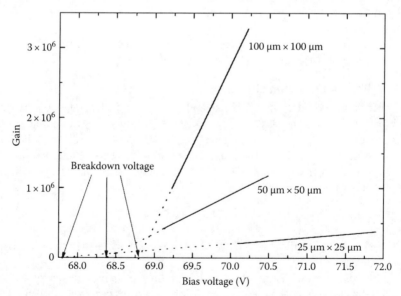

FIGURE 5.2 A typical gain characteristic of a 1 × 1 mm² Hamamatsu MPPC for three types of sensors, following the manufacturer's data. (From Hamamatsu, MPPC Multi-pixel photon counter technical information, 2009.)

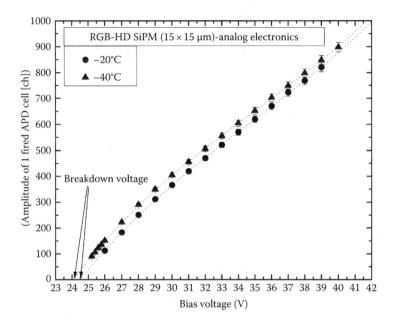

FIGURE 5.3 A typical relative gain characteristics of 2.2 × 2.2 mm² FBK SiPM (15 × 15 μm). (From Grodzicka, M. et al., *JINST*, P08004, 2014.)

(Figure 5.3). After this voltage, the epilayer rapidly gets fully depleted, increasing the depletion width of the junction and thus lowering the cell capacitance. This results in the two different slopes of the gain versus overvoltage curve clearly visible in Figure 5.3.

In Table 5.1, the main parameters of 3 × 3 mm² Hamamatsu MPPC, SensL (Ireland), and FBK (Trento, Italy) SiPMs are compared.

5.2.1.2 Photon Detection Efficiency

Like other Si photodiodes, the SiPM has high QE, which is dependent on the wavelength of the incident photons (λ, nm), and the expected QE is more than 80% at 500 nm (Dolgoshein et al. 2006, Vacheret et al. 2011). However, the overall photon detection efficiency (PDE) for the present-state SiPMs is smaller due to two additional contributions apart from the QE. The PDE can be defined as (Dolgoshein et al. 2006, Renker and Lorenz 2009)

$$PDE = QE \cdot FF \cdot THR \quad (5.3)$$

where
   FF is a geometrical fill factor, that is, the ratio of the active area to the total area of the device
   QE is the quantum efficiency, that is, the probability that an incident photon will generate an electron–hole pair in a region in which carriers can produce an avalanche
   THR is the combined probability of electrons and holes to initiate a Geiger discharge, which is a strong function of the electric field

TABLE 5.1
Comparison of the Main Parameters of 3 × 3 mm² SiPMs of Hamamatsu, SensL, and FBK

Manufacturer	FBK		SensL	Hamamatsu
Type	ASD-RGB3S-P ASD-NUV3S-P	B-series 30020, 30035, 30050	M-series 30035, 30050	S12572–010C S12572–015C S12572–025C S12572–050C S12572–100C
Active area		3 × 3 mm		
Number of APD cells	5520	30020: 10998 30035: 4774 30050: 2668	30035: 4774 30050: 2668	–010C: 90000 –015C: 40000 –025C: 14400 –050C: 3600 –100C: 900
APD cell size (µm²)	40 × 40	30020: 20 × 20 30035: 35 × 35 30050: 50 × 50	30035: 35 × 35 30050: 50 × 50	–010C: 10 × 10 –015C: 15 × 15 –025C: 25 × 25 –050C: 50 × 50 –100C: 100 × 100
Fill factor (%)	60%	30020: 48% 30035: 64% 30050: 72%	30035: 64% 30050: 72%	–010C: 33% –015C: 53% –025C: 65% –050C: 62% –100C: 78%
Spectral resp. range (λ)	350–900 nm	300–800 nm	400–1000 nm	320–900 nm
Peak sensitivity wavelength (nm)	RGB3S-P: 550 NUV3S-P: 420	420 nm	500 nm	–010C: 470 nm –015C: 460 nm –025C: 450 nm –050C: 450 nm –100C: 450 nm
Photon detection efficiency	32.5% at 550 nm	30020: 24% 30035: 31% 30050: 35% at 420 nm and 2.5 V (overvoltage)	30035: 20% 30050: 23% at 500 nm and 2 V (overvoltage)	–010C: 10% at 470 nm –015C: 25% at 460 nm –025C: 35% at 450 nm –050C: 35% at 450 nm –100C: 35% at 450 nm
Breakdown voltage (V_{br})	RGB3S-P: 27 ± 2 NUV3S-P: 26 ± 2	24.5 ± 0.5	27.5 ± 0.5	~65 ± 10

(*Continued*)

TABLE 5.1 (*Continued*)
Comparison of the Main Parameters of 3 × 3 mm² SiPMs of Hamamatsu, SensL, and FBK

Manufacturer	FBK		SensL	Hamamatsu
Gain	RGB3S-P: 2.7×10^6	30020: $\sim 1 \times 10^6$	30035: $\sim 2.3 \times 10^6$	−25C: $\sim 5.15 \times 10^5$
	NUV3S-P: 2.1×10^6	30035: $\sim 3 \times 10^6$	30050: $\sim 4 \times 10^6$	−50C: $\sim 1.25 \times 10^6$
		30050: $\sim 6 \times 10^6$		−100C: $\sim 2.8 \times 10^6$
Dark count (kcps/mm²)	RGB3S-P: <400	30020: <733	<1112	<223
	NUV3S-P: <200	30035: <744		
		30050: <833		
Capacitance	—	30020: 770 pF	30035: 870 pF	320 pF
		30035: 850 pF	30050 : 990 pF	
		30050: 920 pF		
Temperature coefficient	$\sim 27\,\text{mV/°C}$	$\sim 21.5\,\text{mV/°C}$	$\sim 21.5\,\text{mV/°C}$	$\sim 60\,\text{mV/°C} \pm 10$

The typical curves of a PDE (as a function of a wavelength) for Hamamatsu MPPCs with three different APD cell sizes, 25 μm × 25 μm, 50 μm × 50 μm, and 100 μm × 100 μm, are presented in Figure 5.4. The curves include such effects as crosstalk and after-pulses. It is worth noting that this figure shows only an approximate curve because the manufacturer does not provide the voltage at which the curve was measured. It can only be assumed that the curves were measured for the recommended voltage suggested by the producer.

Because FF is constant for a given MPPC and the QE of the device is constant for a given scintillator or a given wavelength of the laser diode pulser, the PDE of a single device depends only on the THR and in consequence is a function of the MPPC bias voltage (V). The PDE (without effects of crosstalk and after-pulses) for MPPC with an APD cell size of 50 μm × 50 μm, as a function of overvoltage at three temperatures, is shown in Figure 5.5, according to Vacheret et al. (2011). PDE increases with the voltage over breakdown (V_{ob}); however, for a fixed overvoltage, there is no observable dependence on the temperature (Vacheret et al. 2011).

5.2.1.3 After-Pulses

The after-pulses of the SiPM are false pulses that are caused by the trapping of charge carriers created during an avalanche, by defect or impurity in the silicon lattice. The trapped carriers are released after a certain time, triggering a new avalanche inside the same APD cell as the original avalanche. The after-pulse avalanche is delayed by the trapping time. Because the traps in the silicon may be of various types with a different trap lifetime, the after-pulsing can exhibit more than one time constants and can be different for various technologies (or fabrication processes). For example, two time components, 15 ns (short) and 83 ns (long),

Silicon Photomultipliers in Detectors for Nuclear Medicine

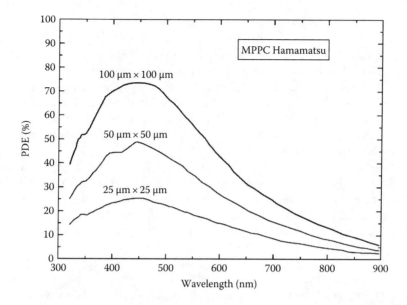

FIGURE 5.4 Photon detection efficiency including effects of crosstalk and after-pulses, according to the manufacturer. (From Hamamatsu, MPPC Multi-pixel photon counter technical information, 2009.)

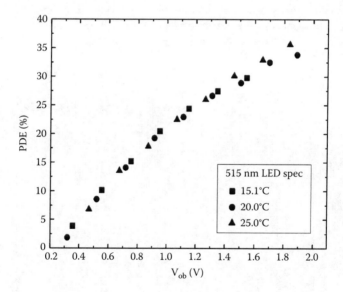

FIGURE 5.5 Photon detection efficiency of an MPPC (with an APD cell size of 50 μm × 50 μm) for green light (~515 nm) as a function of overvoltage at three temperatures. (From Vacheret, A. et al., *Nucl. Instrum. Methods A*, 656, 69, 2011.)

for an MPPC with an APD cell size of 50 × 50 µm were observed by Du and Retiere (2008). For comparison, with a similar MPPC (Vacheret et al. 2011), there are three time components observed: 17 ns (first), 70 ns (second), and 373 ns (third). They also observed that both the first and second components decreased as the temperature increased. Operation in low temperature elongates the delay release because of the trapping center deexcitation time (Renker and Lorenz 2009). Moreover, the after-pulse probability for the first component decreases with temperature, in contrast to the probability of the after-pulse for the second component, which increases with temperature.

After-pulsing may also be partially suppressed due to existence of the effective dead time of APD cell. If a carrier is released while the APD cell voltage has not reached the nominal value, then the charge produced in the avalanche will be lower than for nominal avalanches (Figure 5.6). Only if the delay is larger than the APD cell's effective recovery time, a standard avalanche signal is triggered (Piemonte 2006, Eckert et al. 2010, Yokoyama et al. 2010, Vacheret et al. 2011).

During an avalanche only a very small fraction of the trapping level is filled. Therefore, the trap population is always well below saturation, and the carrier trapping probability remains constant during all of the avalanche pulses. The probability that an after-pulse will occur increases with the amount of charge that flows through the diode during a Geiger discharge. Thus, the after-pulsing probability increases exponentially with the increasing bias voltage and, consequently, with the overvoltage. The after-pulse probability is also a function of the APD cell size and increases for a larger APD cells. The characteristics of the after-pulse probability as a function of the overvoltage for a Hamamatsu MPPC with three different APD cell sizes, 25 µm × 25 µm, 50 µm × 50 µm, and 100 µm × 100 µm, are presented in Figure 5.7 (Eckert et al. 2010).

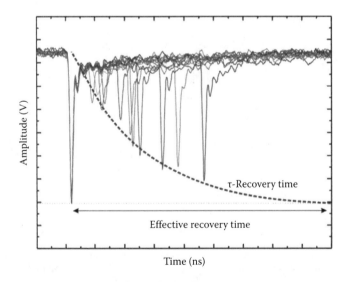

FIGURE 5.6 Examples of after-pulses with different delays. (From Piemonte, *FNAL*, 2006.)

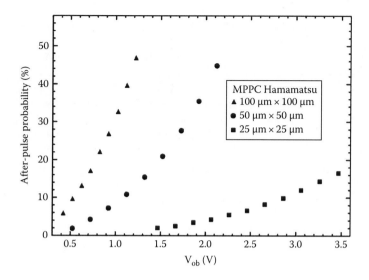

FIGURE 5.7 After-pulse probability as a function of the overvoltage for three different sensor types. (Reprint from Eckert, P. et al., *Nucl. Instrum. Methods A*, 620, 217, 2010.)

5.2.1.4 Optical Crosstalk

The optical crosstalks in SiPM are false pulses that are caused by optical photons emitted during an avalanche discharge in a single APD cell. These photons can trigger another Geiger discharge in the same or neighboring APD cell (Musienko et al. 2006). The aforementioned photons are emitted mainly because of spontaneous, direct carrier relaxation in the conduction band, and this effect is known as hot-carrier luminescence (Lacaita et al. 1990, Renker and Lorenz 2009, Eckert et al. 2010). In accordance with Lacaita et al. (1993), 2.9×10^{-5} photons with an energy higher than 1.14 eV and with a wavelength less than 1 μm are emitted per carrier crossing the junction. Optical crosstalk can manifest itself in the three different ways (Piemonte et al. 2013):

1. *Direct*, when the emitted photon generates a carrier in the active region of a neighboring APD cell, thus producing a second avalanche in coincidence with the first one.
2. *Delayed*, when the emitted photon is absorbed in the nondepleted region beneath the same or neighboring APD cell, thus generating a carrier able to reach the active region by diffusion. The diffusion process is relatively slow, so the second avalanche can be delayed in time with respect to the first one.
3. *External*, when the emitted photon tries to escape from the device but is reflected by structures placed on top of the device, such as a scintillator wrapped with reflecting/diffusing material.

The number of optical crosstalks mainly depends on the APD cell size, the distance between the high-field regions, and gain. The characteristics of crosstalk probability as a function of the gain for a Hamamatsu MPPC with three different APD cell sizes, 25 μm × 25 μm, 50 μm × 50 μm, and 100 μm × 100 μm, are presented in Figure 5.8, according to Eckert et al. (2010). It is worth noting that devices with a smaller APD cell size have a larger crosstalk probability as compared to devices with larger APD cells. This is because photons have to travel a longer average distance in the case of larger cells before reaching a neighboring APD cell, where they can cause a second avalanche.

Presently, the most popular method of reducing the number of false pulses resulting from direct crosstalks is implementation of trenches between the cells to provide electrical and partial optical isolation (Piemonte et al. 2013).

5.2.1.5 Dark Noise

The main source of dark noise in SiPMs is charge carriers generated thermally within the depletion region, which subsequently enter the Geiger multiplication area and trigger avalanches. It is worth remembering that any avalanche can, in turn, initiate secondary avalanches through after-pulsing and crosstalks (Vacheret et al. 2011).

The dark-noise rate of the SiPM is mainly dependent on the capacitance of an APD cell and is proportional to its total active area. The noise increases almost linearly with increasing overvoltage and exponentially with increasing temperature. Nowadays, depending on the manufacturer, commercially available SiPMs possess a dark-noise rate from a few, typically 1–2 MHz/mm^2, to a dozen of kHz/mm^2 at room temperature and gain of ~10^6. Such a dark-noise rate limits SiPM performance at room temperatures, especially for a large sensitive area (~100 mm^2), however, only in

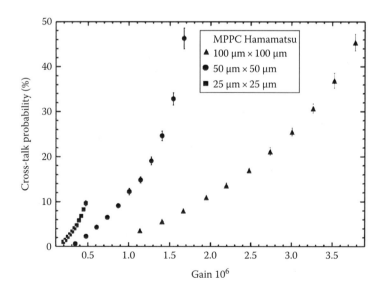

FIGURE 5.8 Crosstalk probability as a function of the SiPM gain for three different sensor types. (From Eckert, P. et al., *Nucl. Instrum. Methods A*, 620, 217, 2010.)

detection of very small light intensities (one, a few, or a dozen of photoelectrons). It does not strongly affect the measurements in the case of larger light signals (Buzhan et al. 2001, Yokoyama et al. 2010).

5.2.1.6 Linearity of a SiPM

In order to define the basic parameters of a detector used for gamma spectrometry (such as energy resolution or number of photoelectrons per MeV), its response should be linear in the studied energy range or the characteristics of its nonlinear response should be known in order to correct the nonlinear data. The construction and principle of operation of SiPMs make this type of device nonlinear by definition. The detector consists of a finite number of sensitive elements (APD cells), and this number limits the number of photons that can be detected. Moreover, even when the number of incident photons is much lower than the total number of APD cells, two photons may interact with the same cell. In such a case, the second photon is lost and *invisible* for the SiPM. As a result, the response of the detector (number of fired APD cells) stops to be proportional to the number of incident photons. The linear range of SiPM response mainly depends on the total number of APD cells, the number of illuminating photons, and the effective dead time of APD cells in relation to the width (or decay time) of the light pulse (Musienko et al. 2006, Grodzicka et al. 2013b, 2014b).

For light pulses shorter than the effective recovery time and for an *ideal* SiPM (without phenomena such as crosstalk and after-pulsing), the response of the SiPM is described by a well-known equation (Renker and Lorenz 2009):

$$N_{fired} = N_{total} \cdot \left[1 - \exp\left(\frac{-(N_{photon} \cdot PDE)}{N_{total}} \right) \right] \quad P_w \leq t_d \quad (5.4)$$

where
N_{fired} is the number of excited APD cells
N_{total} is the total number of APD cells
N_{photon} is the number of incident photons
PDE is the photon detection efficiency
P_w is the pulse width
t_d is the effective recovery time

The product of the number of incident photons (N_{photon}) and the PDE can be considered as the number of photons having the potential to be detected (N_{pd}):

$$N_{pd} = N_{photon} \cdot PDE \quad (5.5)$$

Figure 5.9 presents the theoretical response of three types of 1×1 mm² MPPCs, according to Equation 5.4, where N_{total} equals to 100, 400, and 1600 APD cells and the APD cell size is equal to 100×100 µm², 50×50 µm², and 25×25 µm², respectively.

Phenomena such as optical crosstalk and after-pulses produce optical photons and electrons resulting from additional discharges of APD cells. Thus, for the same number of incident photons, the number of fired APD cells is higher than is apparent from Equation 5.3. This difference is related to the probability of crosstalk and

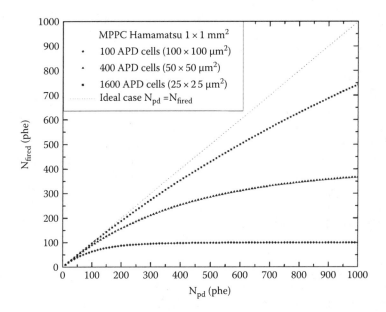

FIGURE 5.9 Theoretical lines—the number of fired APD cells versus the number of photons having the potential to be detected on the MPPC for three different sensor types.

after-pulses. These probabilities become higher with an increase of the bias voltage and depend on the type and production technology of the SiPM used. For a constant bias voltage, this probability is constant and independent on the number of incident photons. Taking into account the aforementioned considerations, the response of a SiPM, for light pulses shorter than the effective recovery time, can be described by the following equation:

$$N_{fired} = N_{total} \cdot \left[1 - \exp\left(\frac{-\left((1+P) \cdot N_{photon} \cdot PDE\right)}{N_{total}}\right)\right] \quad P_w \leq t_d \quad (5.6)$$

where
P_w is the width of rectangular pulse
t_d is the effective recovery time
P is the probability of crosstalk and after-pulses

The total number of events having potential to be detected (N_{ted}) includes optical crosstalk photons and electrons causing after-pulses and is given by

$$N_{ted} = \left((1+P) \cdot N_{photon} \cdot PDE\right) \quad (5.7)$$

In Grodzicka et al. (2014b), it has been proven experimentally that for pulses longer than the effective recovery time, the number of N_{fired} is increased, and in consequence, an extension of the SiPM linear range is observed. For these situations, Equation 5.6 has to be modified by taking into account the probability of multiple firings of the single APD cell during the light pulses with duration (P_w) longer than the effective recovery time of the SiPM (t_d). In order to include this effect in Equation 5.6, a coefficient P_w/t_d has to be introduced in the following way:

$$N_{fired} = N_{total} \cdot \left(\frac{P_w}{t_d}\right) \cdot \left[1 - \exp\left(\frac{-\left((1+P) \cdot N_{photon} \cdot PDE\right)}{N_{total} \cdot (P_w/t_d)}\right)\right] \quad P_W > t_d \quad (5.8)$$

The effective recovery time was calculated from (5.8) using Mathematica software and is given by

$$t_d = \frac{N_{total} \times P_w \times \begin{pmatrix} N_{fired} \times W(-((1+P) \times N_{photon} \times PDE) \times \exp(-((1+P) \times N_{photon} \times PDE)/N_{fired}/N_{fired}) \\ + ((1+P) \times N_{photon} \times PDE) \end{pmatrix}}{N_{fired} \times ((1+P) \times N_{photon} \times PDE)}$$

(5.9)

where W is the Lambert W Function.

5.2.1.7 Excess Noise Factor

The excess noise factor (ENF) for all photodetectors is described by a pulse-to-pulse fluctuation of the charge of the output signal. The ENF for photomultipliers and APDs is a result of fluctuation of the multiplication process. In the case of the SiPM, the relative gain fluctuation has only little impact on the ENF (Figure 5.10), and the ENF is mainly affected by crosstalk and after-pulses. Figure 5.11 presents the dependence of the ENF as a function of overvoltage. The ENF increases with increasing overvoltage due to an increase in the crosstalk and after-pulsing, which introduce additional avalanches in a stochastic manner. The ENF for the 3×3 mm^2 MPPC (with a 25×25 µm^2 APD cell size) increases exponentially with overvoltage and reaches about 2 at V_{ob} of about 4 V. It is worth noting that the ENF of SiPMs can be significantly larger than in the case of PMTs, where the ENF is equal to about 1.1–1.2.

5.2.1.8 Transit Time Spread

The time resolution capabilities of any photodetector are described mainly by three factors: the number of photoelectrons due to detected light flux, ENF, and time jitter (Moszyński and Bengtson 1979). Time jitter or transit time spread (TTS) is defined by the distribution of arrival times of pulses induced by single photons illuminating a whole surface of a photodetector. The full width at half maximum (FWHM) of this distribution describes an intrinsic timing resolution of a photodetector. Since the active layer of silicon in a SiPM is very thin (2–4 µm) and the process of the breakdown development is fast, very good timing properties even for single photons can be expected.

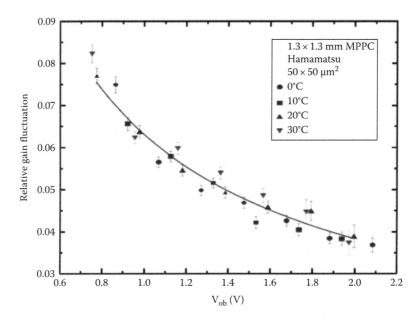

FIGURE 5.10 Relative gain fluctuation versus overvoltage at different temperatures. (From Vacheret, A. et al., *Nucl. Instrum. Methods A*, 656, 69, 2011.)

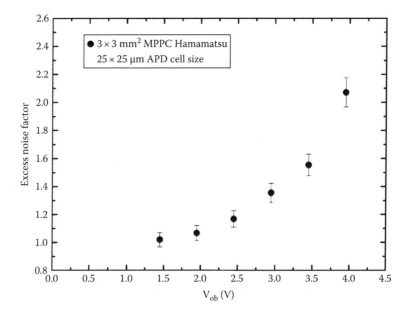

FIGURE 5.11 The excess noise factor versus the overvoltage measured for the Hamamatsu MPPC 33-025. (From Grodzicka, M. et al., *JINST*, 8, P02017, 2013b.)

Fluctuations in the avalanche development are mainly due to a lateral spreading by diffusion and by the photons emitted in the avalanche (Lacaita et al. 1990, 1993, Buzhan et al. 2001). A large spread in the experimental data concerning the time jitter of SiPMs is observed. For 3×3 mm^2 MPPC, values vary between 200 and 500 ps at FWHM. In the case of 1 mm devices, the situation is similar (Mazzillo et al. 2010, Ronzhin et al. 2010, Szczesniak et al. 2012, Gundacker et al. 2013). Moreover, in most cases, the TTS is lower for red photons than for blue ones (Ronzhin et al. 2010).

Table 5.2 summarizes the recent measurements of TTS in 1×1 mm^2 and 3×3 mm^2 SiPMs.

A large spread of the TTS reported in different papers is the effect of an influence of different factors on the measured quantity. The measurements are mostly carried out with illumination of SiPMs by fast light pulses of 50 ps wide or less from a laser pulser. During the measurements, it is important to attenuate the detected light by at least 10–20 times to assure detection of only single photons. In Szczesniak et al. (2012), the single-photon regime was checked in two ways: first, by a comparison of average light pulses recorded by the digital oscilloscope with laser on and off—in both cases, the pulses were indistinguishable—second, by setting the number of coincidences between the laser photons (emitted with 10 kHz rate) and the SiPM response at a level below 5%. A contribution of double photons in the detected light improves artificially the measured time jitter.

Time resolution measurements with a low signal, corresponding to the single photoelectron, particularly for 3 mm SiPMs, are affected by a dark noise of electronics and a high counting rate of SiPM dark pulses. Both effects may destroy the measured time spectra of single photoelectrons showing a poorer TTS. Thus, a correction for the noise contribution has to be applied (Szczesniak et al. 2012), see Figure 5.12. In this respect, the measurements done with 1 mm SiPMs are less affected by the noise.

TABLE 5.2
Transit Time Spread of Some SiPMs

SiPM	APD Cell Size	TTS	References
Hamamatsu 1×1 mm^2	25×25 μm^2	235 ps	Ronzhin et al. (2010)
Hamamatsu 1×1 mm^2	50×50 μm^2	280 ps	Ronzhin et al. (2010)
Hamamatsu 1×1 mm^2	100×100 μm^2	380 ps	Ronzhin et al. (2010)
Hamamatsu 1×1 mm^2	50×50 μm^2	300 ps[a]	Szczesniak et al. (2012)
IRST 1×1 mm^2	40×40 μm^2	140 ps	Mazzillo et al. (2010)
Hamamatsu 3×3 mm^2	50×50 μm^2	520 ps	Ronzhin et al. (2010)
Hamamatsu 3×3 mm^2	50×50 μm^2	410 ps[a]	Szczesniak et al. (2012)
Hamamatsu 3×3 mm^2	50×50 μm^2	340 ps	Mazzillo et al. (2010)
Hamamatsu 3×3 mm^2	50×50 μm^2	190 ps	Gundacker et al. (2013)
Hamamatsu 3×3 mm^2	50×50 μm^2	500 ps	Hamamatsu (2009)
STM 3.5×3.5 mm^2	32×32 μm^2	395 ps	Mazzillo et al. (2010)

[a] corrected for contribution of noise; see Szczesniak et al. (2012).

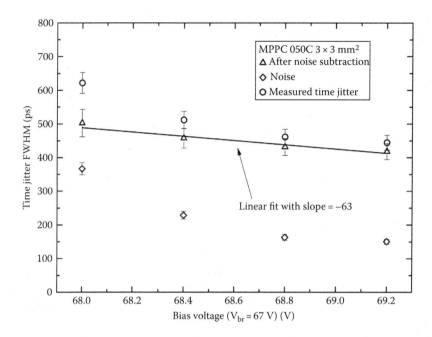

FIGURE 5.12 The dependence of the transit time spread on bias voltage in 3 × 3 mm² MPPC. (From Szczesniak, T. et al., Time jitter of silicon photomultipliers, *IEEE NSS-MIC Conference Records on CD*, 2012, p. K2467.)

The observed improvement of the time jitter for smaller SiPM pixel size (Table 5.2) is likely due to the smaller drift distance for the photoelectrons/holes in the smaller pixels, before they get to the high field gain region (Ronzhin et al. 2010). A final crucial question is still addressed to the further study: Is the poorer time jitter of 3 × 3 mm² SiPMs in comparison to the 1 × 1 mm² devices the effect of a lower and slower voltage signal because of a much larger capacitance?

5.2.2 Digital Silicon Photomultipliers

Around 10 years ago, SiPMs gained a lot of interest as a replacement for PMTs. SiPMs overcome some of the drawbacks of other solid-state detectors but still do not exploit the intrinsic performance of the Geiger-mode cells as their building blocks due to parasitic capacitances and inductances of the interconnect, the influence of electronic noise, and the sensitivity to temperature drifts. Thus, the digital SiPM (dSiPM) was recently developed by Philips Digital Photon Counting to overcome these problems by early digitization of the Geiger-cell output and integrated electronics on chip (Degenhardt et al. 2009, 2011, Frach et al. 2009, 2010, Haemisch et al. 2012, Mandai and Charbon 2012).

The dSiPM consists of an array of Geiger-mode cells with integrated electronics. Each detected photon is converted into a digital signal as early as possible in each of the Geiger-mode cell of the sensor. In addition, the complete trigger logic, a time-to-digital converter, and a controller are integrated into the photodetector. The output of the detector consists of data packets containing the number of detected photons

Silicon Photomultipliers in Detectors for Nuclear Medicine

and the corresponding timestamp. For a detailed description of the operating principle and the intrinsic detector performance, see Degenhardt et al. (2009) and Frach et al. (2010). Some reports on the performance and applications of dSiPMs are collected by Degenhardt et al. (2011), Schaart et al. (2011), Braga et al. (2012, 2014), Degenhardt et al. (2012), Haemisch et al. (2012), and Mandai and Charbon (2012). The newest development of dSiPMs is addressed to build arrays of single photodetectors (Braga et al. 2012, Haemisch et al. 2012, Mandai and Charbon 2012), particularly for detectors in PET scanners (Schaart et al. 2011, Degenhardt et al. 2012, Braga et al. 2014).

5.2.3 SiPMs in Gamma Spectrometry with Scintillators

The wide studies of SiPMs in gamma spectrometry with scintillators were reported by Grodzicka et al. (2012, 2013a,b,c, 2014a). The studies covered tests of 3×3 mm^2 SiPMs and 2×2 ch and 4×4 ch arrays of 3×3 mm^2 devices, working as the single photodetector. During the tests with 3×3 mm^2 MPPC, lutetium fine silicate (LFS-3) and CsI/Tl crystals were used, and in the case of 6×6 mm^2 MPPC, CsI/Tl, LSO/Ce/Ca, LaBr$_3$/Ce, and BGO scintillators were applied. The largest amount of crystals was used in the case of 12×12 mm^2 MPPC; it means CsI/Tl, BGO, LSO, LaBr$_3$, NaI/Tl, CsI/Na, LaCl$_3$, CaF$_2$/Eu, and CdWO$_4$.

Figure 5.13 presents the energy spectrum of 662 keV gamma rays from a ^{137}Cs source measured with $12 \times 12 \times 12$ mm^3 CsI(Tl) crystal fitted to the 12×12 mm^2 MPPC array (Grodzicka et al. 2013c).

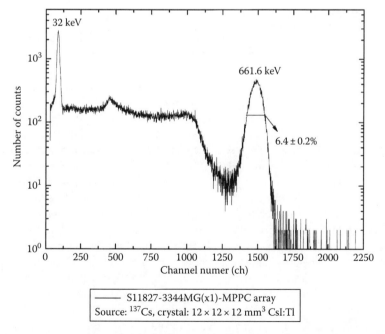

FIGURE 5.13 The energy spectrum of 661.6 keV gamma rays, as measured with a $12 \times 12 \times 12$ mm^3 CsI/Tl scintillator coupled to the MPPC array (S11827-3344MG(X1)). (From Grodzicka, M. et al., *JINST*, 8, P09020, 2013c.)

The study showed that SiPMs and particularly their large area arrays, like the measured 4 × 4 ch SiPM with the total active area of 12 × 12 mm^2, are working well as the scintillation photodetector; however, special attention has to be paid to the optimization of such detector. The optimization should cover verification of the breakdown voltage for each channel of an MPPC array, selection of the optimum operating voltage, verification of the linearity of the MPPC response for the used crystals, verification of the number of photoelectrons, and gamma-ray spectrometry with crystals. In Table 5.3, the energy resolution results obtained with MPPC array readout and nine scintillators are compared to those obtained for the same scintillators and XP2020Q or XP5212 PMTs, following Grodzicka et al. (2013c).

The energy resolution measured with small scintillators, fully fitted to the sensitive area of MPPC array, is comparable or better than that measured with the XP2020Q PMT. The spectra measured with fast scintillators have to be corrected for the nonlinear response of the MPPC array; see examples of linearity characteristics in Figure 5.14.

A worse but still reasonable energy resolution is obtained also for the large scintillators, with larger size of the optical surface comparing to the effective active area of the MPPC; see Table 5.3. A poorer energy resolution is caused mainly by some losses in the collected light.

A good energy resolution of detectors based on the SiPM light readout is of importance in a potential application to gamma cameras and in PET scanners limiting a contribution of false events due to scattered gamma quanta.

5.2.4 SiPMs in Fast Timing with Scintillators

The successful development of TOF-PET scanners with improved image quality, based on the photomultipliers, triggered intense studies of fast timing with SiPMs (Otte et al. 2005, Yamamoto et al. 2005, Raylman et al. 2006, Musienko et al. 2007, Kolb et al. 2008, Nishikido et al. 2008, Pestotnik et al. 2008, Schaart et al. 2008a,b, 2009, Chagani et al. 2009, Jarron et al. 2009, Kadrmas et al. 2009, Kim et al. 2009, Schaart et al. 2009, 2010, Vinke et al. 2009, Llosá et al. 2010, Szczesniak et al. 2010, Yamamoto et al. 2010, Llosa et al. 2011, Piemonte et al. 2011, Powolny et al. 2011, Roncali and Cheery 2011, Gola et al. 2012, Gundacker et al. 2012, 2013, 2014, Seifert et al. 2012, Szczesniak et al. 2012, Wang et al. 2012, Yeom et al. 2013). Timing properties of SiPMs are reported to be extremely good, although the literature data are very inconsistent. The superior time resolution of LYSO crystals, coupled to the MPPC 050C, of 138 ps for two detectors, was measured by the Delft group (Seifert et al. 2012) using digital timing based on the Acqiris DC282 digitizer. An even better time resolution of 108 ps was reported in Gundacker et al. (2013), as measured using NINO chips (Jarron et al. 2009) working as the leading discriminators. In contrast, the best time resolution of 240 ps for two detectors, measured with analog timing setups, was reported by the GE group (Kim et al. 2009). The timing resolution of 190 ps was reported by C. Piemonte et al. measured by means of the differential leading edge discriminator (DLED) method, in which the arrival time of the event is not extracted directly from the signal but rather from the difference between the signal and its delayed replica (Gola et al. 2012). The DLED method allows to compensate effectively the dark events allowing very low thresholds to be set.

TABLE 5.3
Comparison of MPPC Array and XP2020Q and XP5212 PMTs in Gamma Spectrometry with Scintillators

				4 × 4 ch MPPC		XP2020Q PMT	
Crystal	Size (mm³)	Decay Time (ns)	Wavelength of Peak Emission	Energy Resolution at 661.6 keV (%)	phe/MeV	Energy Resolution at 661.6 keV (%)	phe/MeV
CsI:Tl	12 × 12 × 12	1000	550	6.4 ± 0.2	12 200 ± 610	6.6 ± 0.2	4 500 ± 135
	Ø 1 × 1 in.			7.1 ± 0.2	7 700 ± 385	6.8 ± 0.2	4 160 ± 125
LSO	12 × 12 × 12	40	420	8.8 ± 0.3 (non-linear)	8 360 ± 240 (non-linear)	9.4 ± 0.3	5 740 ± 170
				9.4 ± 0.3	10 850 ± 320		
BGO	12 × 12 × 12	300	480	9.5 ± 0.3	2 200 ± 110	10.5 ± 0.3	1 085 ± 30
NaI:Tl	Ø 1 × 1 in.	250	420	8.7 ± 0.3	4 290 ± 210	6.6 ± 0.2	8 550 ± 260
CsI:Na	Ø 1 × 1 in.	630	420	8.9 ± 0.3	4 000 ± 200	7.4 ± 0.2	6 050 ± 180
LaCl₃	Ø 1 × 1 in.	28	350	9.6 ± 0.3	1 900 ± 95	5.6 ± 0.2	5 260 ± 160
CaF₂:Eu	Ø 1 × 1 in.	940	435	7.2 ± 0.2	5 190 ± 260	6.3 ± 0.2	5 670 ± 170
CdWO₄	Ø 20 × 20 mm	14000	475	15.6 ± 0.5	915 ± 45	12.1 ± 0.4	824 ± 25
				4 × 4 ch MPPC		XP 5212 PMT	
LaBr₃	Ø 1 × 1 in.	16	380	5.3 ± 0.2	6 430 ± 320	3.2 ± 0.1	16 070 ± 490

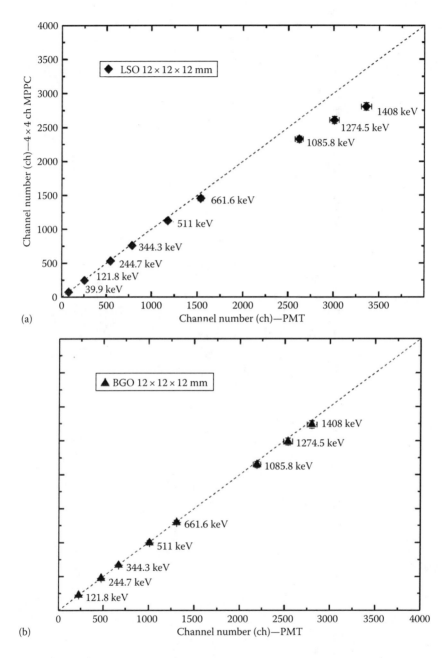

FIGURE 5.14 Linearity of the 4 × 4 ch MPPC array in the case of readout of different scintillation materials: (a) 12 × 12 × 12 mm³ BGO and (b) 12 × 12 × 12 mm³ LSO scintillators. (From Grodzicka, M. et al., *JINST*, 8, P09020, 2013c.)

TABLE 5.4
Time Resolution Measured with LSO/LYSO Crystals Coupled to SiPMs

SiPM	Crystal	Time Resolution	Reference
Hamamatsu MPPC-S10362-33-050C	LYSO $3 \times 3 \times 5\,mm^3$	138 ± 2 ps	Vinke et al. (2009)
Hamamatsu MPPC-S10931-050P	LSO:Ce, Ca(0.4%) $2 \times 2 \times 3\,mm^3$	108 ps	Gundacker et al. (2014)
Hamamatsu MPPC-S10362-33-050C	LSO $2 \times 2 \times 3\,mm^3$	125 ± 2 ps	Yeom et al. (2013)
Hamamatsu MPPC-S10362-33-050C	LYSO $3 \times 3 \times 5\,mm^3$	147 ± 3 ps	Yeom et al. (2013)
Hamamatsu MPPC-S10362-33-050C	LYSO $3 \times 3 \times 20\,mm^3$	186 ± 3 ps	Yeom et al. (2013)
Hamamatsu MPPC-S10362-33-050C	LYSO $3 \times 3 \times 10\,mm^3$	240 ps	Kim et al. (2009)
FBK, Trento, SiPM-4 $\times$ 4 mm^2-67 μm	LYSO $3 \times 3 \times 5\,mm^3$	190 ps	Piemonte et al. (2011)
FBK, Trento, SiPM-4 $\times$ 4 mm^2-67 μm	LYSO $3 \times 3 \times 15\,mm^3$	230 ps	Piemonte et al. (2011)

In Table 5.4, some of the best results of SiPM tests in timing with LSO/LYSO crystals for 511 keV annihilation quanta are collected. The time resolution of 108 ps was measured by CERN group using Hamamatsu MPPC of 3×3 mm^2 size and $2 \times 2 \times 3$ mm^3 LSO codoped with 0.4 mol% Ca (Powolny et al. 2011). Such good value was possible due to the fastest decay time of 30.3 ± 1 ns of the codoped crystal (Powolny et al. 2011). Assuming a comparable light output of standard LSO crystal used in Gundacker et al. (2012) and a typical decay time of the light pulse of 40 ns, the measured time resolution follows the square root of the decay time constant.

An even better time resolution of 101 ± 2 ps was reported by Delft group measured with a $3 \times 3 \times 5$ mm^3 LaBr$_3$ crystal (Schaart et al. 2010). This improvement was possible due to a larger light output and faster decay time of LaBr$_3$ light pulse. On the other hand, the time resolution of the LaBr$_3$ detector is limited by a finite rise time of 0.93 ns of the light pulse in typical crystals doped with 5 mol% of Ce (Levin et al. 2013).

The data collected in Table 5.4 point out an important contribution of the light transport in the longer pixel crystals. It is the effect of the time spread of the collected light and light attenuation in longer crystals.

5.3 SiPMs IN MEDICAL INSTRUMENTATIONS

An intense study of SiPM performance in PET detectors has led to the development of prototype PET scanners (Degenhardt et al. 2012, Levin et al. 2013, Schneider et al. 2013) and the first PET/MR commercial SIGNA™ PET/MR scanner proposed by GE Healthcare (GE Healthcare 2014).

The first prototype of a PET scanner based on dSiPMs was developed by Philips; see Figure 5.15. The detection system of the scanner, based on $4 \times 4 \times 22$ mm^3 LYSO

FIGURE 5.15 Prototype PET scanner using the dSiPMs, developed by Philips. (From Degenhardt, C. et al., Performance evaluation of a prototype positron emission tomography scanner using digital photon counters [DPC], *NSS-MIC Conference Record*, 2012, p. 2820.)

crystals, is characterized by a coincidence timing resolution of 266 ps at FWHM and an energy resolution of 10.7% at FWHM (Degenhardt et al. 2012).

Another prototype of the TOF-PET system based on SiPMs has been developed at Stanford University for simultaneous whole body PET/MR imaging (Levin et al. 2013). This PET system comprises of five rings of 112 detector blocks (each a 4 × 9 array of LYSO crystals, 3.95 × 5.3 × 25 mm^3) coupled to 1 × 3 arrays of SiPM devices. Using a Ge-68 pin source, the measured PET energy resolution was 10.5% FWHM at 511 keV for both radiofrequency (RF) on and RF off. The per crystal timing resolution was 390 ps with RF off and 399 ps with RF on. The transaxial spatial resolution of 3.9 mm FWHM was measured at 1 cm from the isocenter using an F-18 capillary tube point source.

The Technical University of Munich has developed a PET prototype based on digital SiPMs and GAGG scintillators (Schneider et al. 2013) consisting of two facing modules of dSiPMs assembled in combination with a rotational stage for the objects. Using GAGG, coincidence timing was 430 ps FWHM, and the energy resolution was 9.0% FWHM at 511 keV, while LSO coupled to the dSiPM resulted in coincidence timing of 200 ps FWHM and energy resolution of 9.7% FWHM at 511 keV.

The first commercial PET/MR scanner based on SiPMs in PET detectors was proposed on August 4, 2014, by GE Healthcare; see Figure 5.16. It is the first integrated, simultaneous, TOF-capable, whole body SIGNA PET/MR scanner.

Besides PET detectors with TOF capability and applicable to the dual modality PET/MR scanners, SiPMs were recently proposed to be used in the handheld gamma camera for intraoperative imaging (Popovic et al. 2014). The camera incorporates a cerium-doped lanthanum bromide (LaBr/Ce) plate scintillator, an array of 80 SiPM photodetectors, and a two-layer parallel-hole collimator. The disk-shaped camera housing is 75 mm in diameter, approximately 40.5 mm thick, and has a mass of only

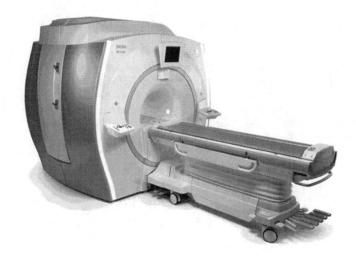

FIGURE 5.16 GE Healthcare's SIGNA PET/MR.

1.4 kg, permitting either handheld or arm-mounted use. The field of view is circular with a 60 mm diameter. The gamma camera has an intrinsic spatial resolution of 4.2 mm FWHM, an energy resolution of 21.1% FWHM at 140 keV, and a sensitivity of 481 and 73 cps/MBq when using the single- and double-layer collimators, respectively.

Recent studies of SiPM applications in gamma spectrometry with scintillators (Grodzicka et al. 2013a,b,c, 2014a,b, 2014) open a new field of SiPM use in medical instrumentation.

REFERENCES

Achenbach, P., A. S. Lorente, S. S. Majos, and J. Pochodzalla. Future use of silicon photomultipliers for KAOS at MAMI and PANDA at FAIR. *Nucl. Instrum. Methods Phys. Res. A*, 2009;610:358–361.

Allemand, R., C. Gresset, and J. Vacher. Potential advantages of a cesium fluoride scintillator for time-of-flight positron camera. *J. Nucl. Med.*, 1980;21:153–155.

Andreev, V. et al. A high-granularity scintillator calorimeter readout with silicon photomultipliers. *Nucl. Instrum. Methods Phys. Res. A*, 2005;540:368–380.

Antich, P. P., E. N. Tsyganov, N. A. Malakhov, and Z. Y. Sadygov. Avalanche photo diode with local negative feedback sensitive to UV, blue and green light. *Nucl. Instrum. Methods Phys. Res.*, 1997;A389:491–498.

Bondarenko, G. et al. Limited Geiger-mode microcell silicon photodiode: New results. *Nucl. Instrum. Methods*, 2000;A442:187–192.

Braga, L. H. C. et al. A fully digital 8 16 SiPM array for PET applications with per-pixel TDCs and real-time energy output. *IEEE J. Solid State Circuits*, 2014;49:301.

Braga, L. H. C., L. Gasparini, and D. Stoppa. A time of arrival estimator based on multiple timestamps for digital PET detectors. *Proceedings of IEEE Nuclear Science Symposium and Medical Imaging Conference*, Anaheim, CA, 2012.

Buzhan, P. et al. An advanced study of silicon photomultiplier. *ICFA Instrum. Bull.*, 2001;23:28–41.

Buzhan, P. et al. Silicon photomultiplier and its possible applications. *Nucl. Instrum. Methods Phys. Res. A*, 2003;504:48–52.

Chagani, H. et al. Tests of silicon photomultiplier PET modules. *IEEE Nuclear Science Symposium and Medical Imaging Conference Record*, Orlando, FL, 2009, pp. 1518–1520.

Conti, M., L. Eriksson, H. Rothfuss, and C.L. Melcher. Comparison of fast scintillators with TOF PET potential. *IEEE Trans. Nucl. Sci.*, 2009;56:926–933.

Danilov, M. CALICE Collaboration, Scintillator tile hadron calorimeter with novel SiPM readout. *Nucl. Instrum. Methods Phys. Res. A*, 2007;A584:451–456.

Degenhardt, C. et al. The digital silicon photomultiplier—A novel sensor for the detection of scintillation light. *Nuclear Science Symposium Conference Record (NSSIMIC)*, Orlando, FL, 2009, pp. 2383–2386.

Degenhardt, C. et al. Performance evaluation of a prototype positron emission tomography scanner using digital photon counters (DPC). *NSS-MIC Conference Record*, Anaheim, CA, 2012, p. 2820.

Degenhardt, C., B. Zwans, T. Frach, and R. de Gruyter. Arrays of digital silicon photomultipliers—Intrinsic performance and application to scintillator readout. *NSS-MIC Conference Record*, Orlando, FL, 2011.

Del Guerra, A., N. Belcari, M. G. Bisogni, G. Llosá, S. Marcatili, and S. Moehrs. Advances in position-sensitive photodetectors for PET applications. *Nucl. Instrum. Methods Phys. Res. A*, 2009;604:319–322.

Dolgoshein, B. et al. Status report on silicon photomultiplier development and its applications. *Nucl. Instrum. Methods Phys. Res. A*, 2006;563:368–376.

Du, Y. and F. Retiere. After-pulsing and cross-talk in multi-pixel photon counter. *Nucl. Instrum. Methods A*, 2008;596:396–401.

Eckert, P., H. C. Schultz-Coulon, W. Shen, R. Stamen, and A. Tadday. Characterisation studies of silicon photomultipliers. *Nucl. Instrum. Methods A*, 2010;620:217–226.

Finocchiaro, P. et al. Features of silicon photo-multipliers: Precision measurements of noise, cross-talk, after pulsing, detection efficiency. *IEEE Trans. Nucl. Sci.*, 2009;56:1033–1041.

Frach, T., G. Prescher, C. Degenhardt, R. de Gruyter, A. Schmitz, and R. Ballizany. The digital silicon photomultiplier—Principle of operation and intrinsic detector performance. *Nuclear Science Symposium Conference Record (NSSIMIC)*, Orlando, FL, 2009, pp. 1959–1965.

Frach, T., G. Prescher, C. Degenhardt, and B. Zwans. The digital silicon photomultiplier—System architecture and performance evaluation. *NSS-MIC Conference Record*, Knoxville, TN, 2010.

GE Healthcare's. New SIGNA™ PET/MR. 2014.

Gola, A., C. Piemonte, and A. Tarolli. The DLED algorithm for timing measurements on large area SiPMs coupled to scintillators. *IEEE Trans. Nucl. Sci.*, 2012;59:358–365.

Golovin, V. and V. Saveliev. Novel type of avalanche photodetector with Geiger mode operation. *Nucl. Instrum. Methods Phys. Res. A*, 2004;A518:560–564.

Grodzicka, M. et al. Performance of FBK high-density SiPMs in scintillation spectrometry. *JINST*, 2014a;9:P08004.

Grodzicka, M. et al. Characterization of CsI:Tl at a wide temperature range (−40°C to +22°C). *Nucl. Instrum. Methods Phys. Res. A*, 2013a;707:73–79.

Grodzicka, M., M. Moszyński, T. Szczęśniak, M. Kapusta, M. Szawłowski, and D. Wolski. Energy resolution of small scintillation detectors with SiPM light readout. *JINST*, 2013b;8:P02017.

Grodzicka, M., M. Moszyński, T. Szczęśniak, M. Szawłowski, and J. Baszak. Characterization of 4 × 4 ch MPPC Array in Scintillation Spectrometry. *JINST*, 2013c;8:P09020.

Grodzicka, M., M. Moszyński, T. Szczęśniak, M. Szawłowski, D. Wolski, and J. Baszak. MPPC array in the readout of CsI(Tl), LSO:Ce(Ca), LaBr$_3$, and BGO scintillators. *IEEE Trans. Nucl. Sci.*, 2012;59:3294–3303.

Grodzicka, M., M. Moszyński, T. Szczęśniak, M. Szawłowski, D. Wolski, and K. Grodzicki. Characterization of silicon photomultipliers: Effective dead time, new method of evaluating the single photoelectron response, gamma spectrometry with BGO scintillator. *Nucl. Instrum. Methods Phys. Res. A*, 2014b;783:58–64.

Gundacker, S. et al. A systematic study to optimize SiPM photodetectors for highest time resolution in PET. *IEEE Trans. Nucl. Sci.*, 2012;59:1798.

Gundacker, S. et al. Time of flight positron emission tomography towards 100 ps resolution with L(Y)SO: An experimental and theoretical analysis. *JINST*, 2013a;59(5):1798–1804.

Gundacker, S. et al. SiPM time resolution: From single photon to saturation. *Nucl. Instrum. Methods Phys. Res. A*, 2013b;718:569–572.

Gundacker, S., A. Knapitsch, E. Auffray, P. Jarron, T. Meyer, and P. Lecoq. Time resolution deterioration with increasing crystal length in a tof-pet system. *Nucl. Instrum. Methods Phys. Res. A*, 2014;737:92–100.

Haemisch, Y., T. Frach, C. Degenhardt, and A. Thon. Fully digital arrays of silicon photomultipliers (dSiPM)—A scalable alternative to vacuum photomultiplier tubes (PMT). *Phys. Procedia*, 2012;37:1546–1560.

Hamamatsu. MPPC Multi-pixel photon counter technical information. Hamamatsu Photonics K.K., Solid State Division, 2009.

Jarron, P. et al. Time based readout of a silicon photomultiplier (SiPM) for time of flight positron emission tomography (TOF-PET). *Nuclear Science Symposium Conference Record (NSS/MIC)*, Orlando, FL, 2009, pp. 1212–1219.

Kadrmas, D. J., M. E. Casey, M. Conti, B. W. Jakoby, C. Lois, and D. W. Townsend. Impact of time-of-flight on PET tumor detection. *J. Nucl. Med.*, 2009;50:1315–1323.

Kim, C. L., G. C. Wang, and S. Dolinsky. Multi-pixel photon counters for TOF PET detector and its challenges. *IEEE Trans. Nucl. Sci.*, 2009;56:2580–2585.

Kolb, A., M. S. Judenhofer, E. Lorenz, Renker D., and B. J. Pichler. PET block detector readout approaches using G-APDs. *IEEE Nuclear Science Symposium Conference Records*, Dresden, Germany, 2008.

Korpar, S. et al. Measurement of Cherenkov photons with silicon photomultipliers. *Nucl. Instrum. Methods Phys. Res. A*, 2008;A597:13–17.

Lacaita, A., M. Mastrapasqua, M. Ghioni, and S. Vanoli. Observation of avalanche propagation by multiplication assisted diffusion in p-n junctions. *Appl. Phys. Lett.*, 1990;57:489–491.

Lacaita, A. L., F. Zappa, S. Bigliardi, and M. Manfredi. On the Bremsstrahlung origin of hot-carrier-induced photons in silicon devices. *IEEE Trans. Electron Devices*, 1993;40:577–582.

Laval, M. et al. Contribution of the time-of-flight information to the positron tomographic imaging. *Proceedings of the Third World Congress Nuclear Medicine and Biology*, Paris, France. Pergamon Press, Paris, France, 1982, p. 2315.

Levin, C., G. Glover, T. Deller, D. McDaniel, W. Peterson, and S. H. Maramraju. Prototype time-of-flight PET ring integrated with a 3T MRI system for simultaneous whole-body PET/MR imaging. *J. Nucl. Med.*, 2013:148.

Lewellen, T. K. Recent developments in PET detector technology. *Phys. Med. Biol.*, 2008;53:R287–R317.

Llosá, G. et al. Characterization of a PET detector head based on continuous LYSO crystal and monolithic 64-pixel silicon photomultiplier matrices. *Phys. Med. Biol.*, 2010;53:R287–R317.

Llosa, G. et al. Development of a PET prototype with continuous LYSO crystals and monolithic SiPM matrices. *IEEE Nuclear Science Symposium and Medical Imaging Conference Record*, Valencia, Spain, 2011, pp. 3631–3634.

Mandai, S. and E. Charbon. Multi-channel digital SiPMs: Concept, analysis and implementation. *Proceedings of IEEE Nuclear Science Symposium and Medical Imaging Conference*, Anaheim, CA, 2012, pp. 1840–1844.

Mazzillo, M. et al. Timing performances of large area silicon photomultipliers fabricated at STMicroelectronics. *IEEE Trans. Nucl. Sci.*, 2010;57:2273–2279.

Moses, W.W. and S.E. Derenzo. Prospects for time-of-flight PET using LSO scintillator. *IEEE Trans. Nucl. Sci.*, 1999;46:474–478.

Moszyński, M. and B. Bengtson. Status of timing with plastic scintillation detectors. *Nucl. Instrum. Methods*, 1979;158:1–31.

Moszynski, M., M. Szawłowski, M. Kapusta, and M. Balcerzyk. Large area avalanche photodiodes in scintillation and X-rays detection. *Nucl. Instrum. Methods Phys. Res. A*, 2002;485:504–521.

Mullani, N. A., D. C. Ficke, R. Hartz, J. Markham, and G. Wong. System design of a fast PET scanner utilizing time-of-flight. *IEEE Trans. Nucl. Sci.*, 1981;NS-28:104–107.

Musienko, Y. State of the art in SiPM's. *CERN, SiPM Workshop*, Zurich, Switzerland, February 16, 2011.

Musienko, Y., E. Auffray, P. Lecoq, S. Reucroft, Swain J., and J. Trummer. Study of multi-pixel Geiger-mode avalanche photodiodes as a read-out for PET. *Nucl. Instrum. Methods Phys. Res. A*, 2007;571:362–365.

Musienko, Y., S. Reucroft, and J. Sawin. The gain, photon detection efficiency and excess noise factor of multi-pixel Geiger-mode avalanche photodiodes. *Nucl. Instrum. Methods Phys. Res. A*, 2006;567:57–61.

Nassalski, A. et al. Comparative study of scintillators for PET/CT detectors. *IEEE Trans. Nucl. Sci.*, February 2007;54:3–10.

Nishikido, F. et al. Four-layer DOI-PET detector with a silicon photomultiplier array. *IEEE Nuclear Science Symposium Conference Record*, Dresden, Germany, 2008, pp. 3923–3925.

Otte, A.N. et al. A test of silicon photomultipliers as readout for PET. *Nucl. Instrum. Methods Phys. Res. A*, 2005;545:705–715.

Otte, A. N. et al. Prospects of using silicon photomultipliers for the astroparticle physics experiments EUSO and MAGIC. *IEEE Trans. Nucl. Sci.*, 2006;53:636–640.

Pestotnik, R., S. Korpar, H. Chagani, R. Dolenec, P. Krizan, and A. Stanovnik. Silicon photomultipliers as photon detectors for PET. *IEEE Nuclear Science Symposium Conference Record*, Dresden, Germany, 2008, pp. 3123–3127.

Piemonte, C. Development of silicon photomultipliers @ IRST. *FNAL*, presentation at ITC-irst, Trento, October 25, 2006.

Piemonte, C. et al. Timing performance of large area SiPMs coupled to LYSO using dark noise compensation methods. *IEEE Nuclear Science Symposium Conference Record*, Valencia, Spain, 2011, p. 59.

Piemonte, C. et al. Characterization of the first FBK high-density cell silicon photomultiplier technology. *IEEE Trans. Electron Devices*, 2013;60:2567–2573.

Popovic, K. et al. Development and characterization of a round hand-held silicon photomultiplier based gamma camera for intraoperative imaging. *IEEE Trans. Nucl. Sci.*, 2014;61:1084.

Powolny, F. et al. Time-based readout of a silicon photomultiplier (SiPM) for time of flight positron emission tomography (TOF-PET). *IEEE Trans. Nucl. Sci.*, 2011;58:597–604.

Raylman, R. R. et al. Simultaneous MRI and PET imaging of a rat brain. *Phys Med Biol.*, 2006;51:6371–6379.

Renker, D. Geiger-mode avalanche photodiodes, history, properties and problems. *Nucl. Instrum. Methods Phys. Res. A*, 2006;567:48–56.

Renker, D. New developments on photosensors for particle physics. *Nucl. Instrum. Methods Phys. Res.*, 2009;A598:207–212.

Renker, D. and E. Lorenz. Advances in solid state photon detectors. *JINST*, 2009;4:P04004.

Roncali, E. and S. R. Cheery. Application of silicon photomultipliers to positron emission tomography. *Ann. Biomed. Eng.*, 2011;39:1358–1377.

Ronzhin, A. et al. Tests of timing properties of silicon photomultipliers. *Nucl. Instrum. Methods Phys. Res.*, 2010;A616:38–44.

Schaart, D. R. et al. First experiments with LaBr$_3$:Ce crystals coupled directly to silicon photomultipliers for PET applications. *IEEE Nuclear Science Symposium Conference Record*, Dresden, Germany, 2008a, pp. 3991–3994.

Schaart, D. R. et al. SiPM-array based PET detectors with depth-of-interaction correction. *IEEE Nuclear Science Symposium Conference Record*, Dresden, Germany, 2008b, pp. 3581–3585.

Schaart, D. R. et al. A novel, SiPM-array based, monolithic scintillator detector for PET. *Phys. Med. Biol.*, 2009;54:3501–3512.

Schaart, D. R. et al. LaBr(3):Ce and SiPMs for time-of-flight PET: Achieving 100 ps coincidence resolving time. *Phys. Med. Biol.*, 2010;55:179–189.

Schaart, D. R., H. T. van Dam, G. J. van der Lei, and S. Seifert. The digital SiPM: Initial evaluation of a new photosensor for time-of-flight PET. *IEEE Nuclear Science Symposium and Medical Imaging Conference*, Valencia, Spain, October 2011, pp. 23–29.

Schneider, F., K. Shimazoe, I. S. Schweiger, K. Kamada, H. Takahashi, and S. Ziegler. A PET prototype based on digital SiPMs and GAGG scintillators. *J. Nucl. Med.*, 2013;54:429.

Seifert, S. et al. A comprehensive model to predict the timing resolution of SiPM-based scintillation detectors: Theory and experimental validation. *IEEE Trans. Nucl. Sci.*, 2012;59:190–204.

Spinelli, A. and A. L. Lacaita. Physics and numerical simulation of single photon avalanche diodes. *IEEE Trans. Electron Devices*, 1997;44:1931–1943.

Szczesniak, T. et al. Time resolution of scintillation detectors based on SiPM in comparison to photomultipliers. *Nuclear Science Symposium Conference Record (NSS/MIC)*, Knoxville, TN, 2010.

Szczesniak, T., M. Moszynski, M. Grodzicka, M. Szawłowski, D. Wolski, and J. Baszak. Time jitter of silicon photomultipliers. *IEEE NSS-MIC Conference Records on CD*, Anaheim, CA, 2012, p. K2467.

Szczesniak, T., M. Moszynski, L. Swiderski, A. Nassalski, P. Lavoute, and M. Kapusta. Fast photomultipliers for TOF PET. *IEEE Trans. Nucl. Sci.*, 2009;56:173–181.

Vacheret, A. et al. Characterization and simulation of the response of multi-pixel photon counters to low light levels. *Nucl. Instrum. Methods A*, 2011;656:69–83.

van Dam, H. T. et al. A comprehensive model of the response of silicon photomultipliers. *IEEE Trans. Nucl. Sci.*, 2010;57:2254–2266.

Vinke, R. et al. Optimizing the timing resolution of SiPM sensors for use in TOF-PET detectors. *Nucl. Instrum. Methods Phys. A*, 2009;610:188.

Wang, Y. et al. Long design and performance evaluation of a compact, large-area PET detector module based on silicon photomultipliers. *Nucl. Instrum. Methods Phys. Res. A*, 2012;670:49–54.

Yamamoto, S. et al. Development of a Si-PM based high-resolution PET system for small animals. *Phys. Med. Biol.*, 2010;55:5817–5831.

Yamamoto, S., S. Takamatsu, H. Murayama, and K. Minato. A block detector for a multi-slice depth of interaction MR compatible PET. *IEEE Trans Nucl Sci.*, 2005;52:33–37.

Yeom, J.Y., R. Vinke, and C.S. Levin. Optimizing timing performance of silicon photomultiplier-based scintillation detectors. *Phys. Med. Biol.*, 2013;58:1207.

Yokoyama, M. et al. Performance of multi-pixel photon counters for the T2K near detectors. *Nucl. Instrum. Methods*, 2010;A622:567–573.

6 Imaging Technologies and Potential Clinical Applications of Photon-Counting X-Ray Computed Tomography

Katsuyuki Taguchi

CONTENTS

6.1 Imaging Technologies 149
    6.1.1 Overall Strategy 150
    6.1.2 X-Ray Beam-Shaping Filters 150
    6.1.3 Calibration and Compensation Methods 152
    6.1.4 PCD Models 154
    6.1.5 Image Reconstruction 157
        6.1.5.1 Interior Problem 157
        6.1.5.2 Spectral Data 159
6.2 Potential Benefits and Clinical Applications 160
    6.2.1 Improved Contrast-to-Noise Ratio and Contrast of CT Images 161
    6.2.2 Dose Reductions of X-Ray Radiation and Contrast Agents 161
    6.2.3 Improved Spatial Resolution 161
    6.2.4 Beam-Hardening Artifacts 162
    6.2.5 Quantitative CT and X-Ray Imaging 162
    6.2.6 Accurate K-Edge Imaging 162
    6.2.7 Simultaneous Multiagent Imaging 164
    6.2.8 Molecular CT with Nanoparticle Contrast Agents and Personalized Medicine 164
References 165

6.1 IMAGING TECHNOLOGIES

We outline the overall strategies, the current status, and our perspective on the imaging technologies that will be necessary to enable photon-counting detector computed tomography (PCD-CT) systems.

6.1.1 Overall Strategy

When an X-ray photon hits a PCD, it generates a pulse. If the PCD detection system is not fast enough, consecutive pulses generated by quasicoincident photons will be integrated and will produce only one count recorded at the wrong energy. This is called *pulse pileup*. And with the loss of counts, the recorded spectrum will be distorted. One can decrease pulse pileups by making PCD pixels smaller and faster. A smaller PCD will receive fewer photons than will a larger PCD at the same X-ray intensity, resulting in fewer coincidences.

A distorted spectrum is also caused by the spectral response effect (SRE), which includes the depth of the interaction effect and splitting energy due to charge sharing, K-escape, and Compton scattering. The SRE occurs even with very weak X-ray beams and is thus potentially more problematic than pulse pileups (which occur only near the object surface where X-rays are intense). The SRE cannot be ignored, because we need to use energy/spectral information to allow a lower dose and many new clinical applications we discuss later. A PCD with no spectral information and geometrical efficiency of 80% would allow for a dose reduction of 30%–40% (depending on the quality of the *current* energy-integrating detectors [EIDs]), but this is not sufficient to reach desirable low-dose levels. Furthermore, ignoring SRE and using the uncorrected output of energy windows would result in shading artifacts and biases in images. Thus, the SRE must be compensated for. One can decrease the SRE by making PCD pixels larger (to avoid splitting energy) and slower (to integrate all of the split energies within each pixel).

Notice that pulse pileup and the SRE have opposite solutions; thus, no PCD can address both of the problems simultaneously. It is desirable to develop imaging technologies that could compensate for the SRE and pulse pileups during the image reconstruction process, similar to how attenuation and scatter are compensated for in single-photon emission computed tomography and positron emission tomography. In addition to improving the detector technologies, we believe it is necessary to advance and integrate imaging methods in the following four areas to make PCD-CT systems viable for imaging:

1. X-ray beam-shaping filters to optimize the intensity and spectrum of X-rays
2. Calibration and compensation methods for the degradation effects of PCDs
3. Models of the PCD's degradation effects
4. Image reconstruction to provide accurate images from PCD data

6.1.2 X-ray Beam-Shaping Filters

New X-ray beam-shaping filters are needed for optimizing the X-ray flux and patient dose. Such a filter may consist of two components, a stationary part and a dynamic part, or a dynamic part only. With the two-component design, the stationary part *shapes* the intensity and spectrum of the X-ray beam across the entire field of view, and the dynamic part specifically shapes the X-ray beam near the edge of the object being imaged.

The stationary part is similar to a conventional attenuating filter used in CT systems, which is often called a bow tie filter because it is thin in the middle and thick at each end. The purpose of the shaping filter is to equalize the X-ray intensity at the detector and to reduce the dose to the patient periphery. It is essential to decrease the intensity of X-rays that go through near or outside the edges of objects for PCDs, because the unattenuated (or less attenuated) X-ray flux would otherwise be very intense. Further, for PCDs, the spectrum incident on the object needs to be shaped to maximize the spectral information acquired from the object.

A single stationary filter alone would not be sufficient, because the fan angles in projections that correspond to the object's edge change as the gantry rotates around the object and different portions of the object are scanned. It will be required to have additional filtrations or collimations that dynamically track the edge for each projection. With such dynamic tracking filters, the maximum count rate requirement for the PCD could be reduced significantly. For example, the count rate of the unattenuated X-ray beam with 120 kVp may be 10^9 cps/mm^2, while it will be reduced to 10^8 cps/mm^2 for the X-ray beam exiting the stationary bow tie filter and further reduced to 10^5–10^8 cps/mm^2 with a dynamic bow tie filter and the presence of the object (Figure 6.1).

Dynamic filters with no stationary components have already been studied. One design splits the stationary bow tie filter into two parts in the middle of the fan beam, and each part moves independently along the fan angles to adjust the intensity of the X-ray beams. Another design has a set of triangular wedges and each wedge moves independently longitudinally.[2] A third design has a hollow ellipse, which rotates in the direction opposite to the gantry rotation.[3]

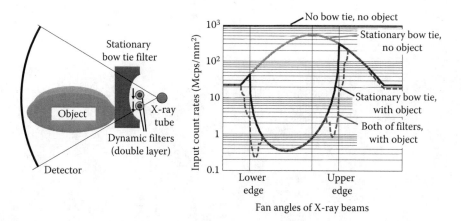

FIGURE 6.1 (See color insert.) Calculated true count rates in the lateral view of an elliptic water phantom when it is 5 cm off-center. Dynamic and stationary bow tie filters (left) decrease the count rates near the edges of the object (right, red curve), compared to the results without the dynamic filters (blue curve). (Figures are from Taguchi, K. and Iwanczyk, J.S., *Med. Phys.*, 40, 100901, 2013.)

6.1.3 Calibration and Compensation Methods

There are two philosophically different approaches to deal with distorted spectral data: corrections and compensation. Corrections attempt to undo the distortion process, while compensation offsets the effect. Before discussing the two approaches, let us first define the terminology using Figure 6.2. Suppose that a forward imaging process to obtain an ideal X-ray spectrum y through an entire object x can be expressed as $h: y = h(x, a)$, where a is the initial X-ray intensity and spectrum exiting the bow tie filters. (Bold letters indicate tensors.) The spectrum y is then skewed to y' by PCD degradation factors g, that is, $y' = g(y)$, which is then recorded as counts within N energy windows, that is, $z = f(y')$. Note that the spectra y and y' can be described reasonably well by counts within narrow energy windows, for example, 1 keV, or by using, for example, 5–10 parameters.

One may be interested in *correcting* SRE and pileups, that is, to estimate y from PCD data, z,[4] and then reconstruct image x from y. The full energy spectrum y may be described by 5–10 parameters; however, estimating so many parameters from, for example, 4-thresholded PCD data z, is an ill-posed problem. Different spectra y may produce the same set of counts, z. And complex cross talk caused by SRE and pileups would make it even worse. We do not think it would work effectively and robustly. Nonetheless, a few approaches have been proposed.[5,6] They work well if,

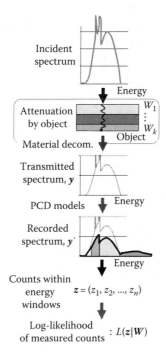

FIGURE 6.2 The model of the forward imaging process used in maximum likelihood methods to compensate for various spectral degradation factors. (Figures are slightly modified from Taguchi, K. and Iwanczyk, J.S., *Med. Phys.*, 40, 100901, 2013.)

and only if, assumptions implicitly used as constraints, for example, the object consists only of water, are correct.

We are interested in *compensating* SRE and pileups, that is, to estimate x from z by iteratively solving the forward process $z = f(g(h(x)))$. In this chapter, we call this algorithm *PIECE*, which stands for *p*hysics-modeled *i*terative reconstruction for *e*nergy-sensitive photon-*c*ounting detector. PIECE incorporates a PCD model of SRE and pulse pileups and estimates either the imaged object or the sinogram using a maximum likelihood approach. The PCD degradation factors will be compensated for during the estimation process.[7–10] This can be formed as a well-posed problem; the method is depicted in Figure 6.2.

First, calibration is performed before the scan to obtain the X-ray intensity and the spectrum exiting the bow tie filters a for each sinogram pixel. Parameters for SRE and pulse pileups of PCDs will also be obtained. Thus, if we know the spectrum incident projected onto PCDs, y, we can calculate the recorded spectrum, y', by using the SRE and pulse pileup model. The expected counts of energy windows, z, can be calculated by integrating y' over the corresponding energy range: $z = f(y')$. Now, from a to z, the only missing link is how to obtain y from a, and we use *material decomposition* to connect the link.

Let us now explain material decomposition[11]. The energy-dependent attenuation of the object at each pixel, $x(E)$, can be accurately modeled by a linear combination of two or three basis functions of energy, $f_k(E)$, and their coefficients, w_k, as $x(E) = \sum_k w_k f_k(E)$, where k is the index of the basis functions. Note that the attenuation model is exact if the number of basis functions is equal to or larger than the sum of the number of physics phenomena and the number of heavy elements inside the patient. It is exact regardless of the number of biological tissue types (e.g., muscle, fat, blood, skin, ligament, tendon, bone). Two predominant physics phenomena, Compton (or incoherent) scattering and photoelectric absorption, are sufficient to model the X-ray interactions with materials within the energy range of diagnostic X-ray. Rayleigh (or coherent) scattering occurs only in low energies and typically accounts for less than 5% of the diagnostic X-ray spectrum range. Pair production requires a photon energy of at least 1.02 MeV and plays no role in diagnostic imaging. Heavy elements include those used as contrast agents (e.g., iodine, gadolinium, or bismuth) and those in medical devices such as implants, stents, and bolts.

Now, let us return to the discussion of PIECE. Line integrals of basis function images w through the object, W, can be calculated for sinogram pixel: $W = \int w dr$. Then, using Beer's law, a and W, the transmitted X-ray spectrum y can be calculated. The entire forward imaging chain is now linked, and the only unknown information is the object we are imaging (or the thicknesses of basis functions), w_k or W_k. These will be estimated using a *maximum likelihood* approach that we outline later.

We model the noisy PCD data $\hat{z}$ using a multivariate normal distribution, which takes into account the SRE and pulse pileups: $\hat{z} \sim Normal(z', \Sigma)$, where z' is the expected value and Σ is the covariance matrix. Both y' and Σ are joint functions of basis functions, W. And W are jointly estimated by maximizing the multivariate normal log-likelihood, that is, $\ln p(\hat{z}|W)$. This process will work robustly and stably, as it is an overdetermined well-posed problem: the number of measurements (e.g., 4 with 4 energy windows) is larger than the number of unknowns (e.g., the thicknesses of 2 or 3 basis functions).

Multivariate normal distribution is more appropriate than the Poisson distribution for modeling nonzero covariance of PCD data because the data are *not* Poisson distributed even without pulse pileups. Poisson assumes data are independent and not correlated. Thus, the Poisson noise model-based methods, which ignore the correlation of PCD data, would result in greater image noise.

We have recently developed a rudimentary version of PIECE, PIECE-1, which only models the intrapixel, energetic cross talk between energy windows.[12] We performed Monte Carlo simulations at high count rates to evaluate the bias and noise standard deviation of the estimation. The results (Figure 6.3a) show that PIECE-1 had very little bias and noise despite having very low detection efficiency (DE) (only 1%–16%) when the water thickness was less than 10 cm. The method without the pileup model (green) had large biases when the water was thinner. The noise was significantly smaller when the covariance of multiple energy windows was used (circle), demonstrating the advantage of using covariance of data. Synthesized abdominal patient data were scanned by PCD-CT at 400 mAs (Figure 6.3b). A significant cupping artifact toward the edge can be observed in the image reconstructed with the model that does not include pulse pileup (Figure 6.3B); this is attributable not to the beam-hardening effect but to pulse pileup effects. In contrast, the image reconstructed with PIECE-1 (Figure 6.3C) shows no such artifacts.

6.1.4 PCD Models

The key to a successful PCD compensation is an accurate model of PCD degradation factors, g. It is logically possible, although it would not be practical, to perform PCD compensation successfully without any model. If the PCD is stable over a long period of time, one can acquire an extensive amount of calibration data to relate every possible x to PCD data $z = f(g(h(x)))$ with every possible combination of conditions (e.g., tube current, tube voltage, materials, and thicknesses of bow tie filters). This approach would not be practical, however, because the number of required calibration datasets is very large and PCD data may change by at least a few percentage points over time. It may be more reasonable to take an approach that is similar to the one implemented with EID-CT systems: an extensive calibration procedure performed less frequently (e.g., semiannually) and a quick calibration procedure employed every day, from which parameters necessary for PCD models are estimated and used to monitor the temporal change of PCD data for quality control. Both the model and the extensive calibration data acquired previously will be used to generate pseudo calibration data, which would be acquired if an extensive procedure was performed frequently.

We model the SRE and pulse pileup separately. The integrated phenomena of the two factors are modeled by cascading models of attenuation, SRE, and pulse

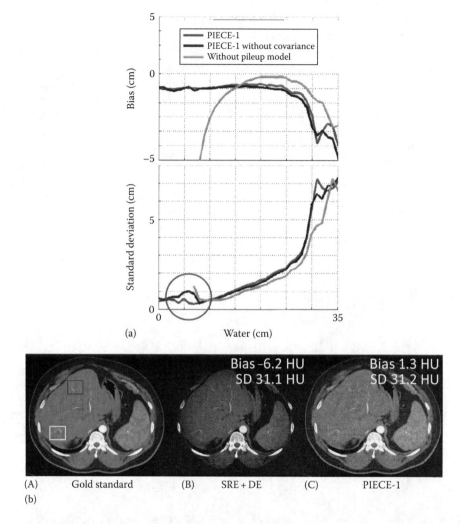

FIGURE 6.3 (See color insert.) (a) Both bias and noise of the estimated water thicknesses were improved by PIECE-1. (b) Images of (A) gold standard, (B) reconstructed compensating for the spectral response effect and detection efficiency (but ignoring pileups), and (C) reconstructed by PIECE-1. *Conditions*: A count rate exiting from the tube of 10^9 cps/mm^2; 1 cm water flat filter; detector dead time of 20 ns; pixel size of 0.5 × 0.5 mm^2; 4 energy thresholds set at 20, 50, 80, and 110 keV; and photopeak ratio of ~0.5 for SRE. The attenuators were 0–40 cm water and 0–5 cm bone.

pileup.[13–15] Next, we discuss examples both of the pulse pileup and SRE models and the cascaded model.

The SRE can be integrated and described as a single spectral response function (SRF), which can be modeled based on measurements using radioisotopes or synchrotron radiation at a very low count rate.[16] Considering the stochastic nature of the SRE, $SRF(E, E_0)$ models the probability density distribution of the recorded energy E, given the true photon energy E_0.[7,8] A small number of input energies E_0 can be used to measure *SRF*, and they will be interpolated to estimate *SRF* at desirable energies. When a polychromatic X-ray spectrum $S(E_0)$ is projected onto the PCD, the recorded spectrum can be calculated by the integration of the $SRF(E, E_0)$ weighted by $S(E_0)$ over E_0. Note that this process is usually *not* a convolution because $SRF(E, E_0)$ changes over E_0 (thus, the SRF is shift variant). An example of the true and recorded spectra is shown in Figure 6.4. It can be seen that the SRE of the PCD blurs the spectrum and increases counts, especially at low energies.

The spectrum distortion caused by pulse pileup is most difficult to model because it is a very complex phenomenon. But it is necessary to model, because the output depends on the input count rates and spectra and, thus, depends on the object to be imaged. Simple models such as linear corrections or self-convolution[17] are not accurate for modeling complex mechanisms of distortion. Various pulse pileup models have been developed,[18–22] and we have developed a model[21,22] that satisfies the accuracy, efficiency, and ability to handle a large number of coincidence

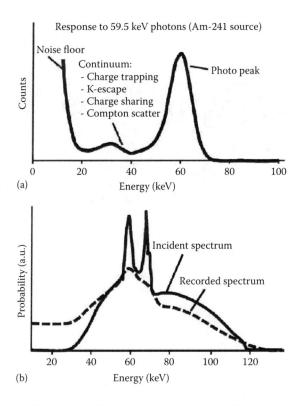

FIGURE 6.4 (a) An illustration of a typical spectrum recorded by a photon-counting detector using Am-241. The spectrum is distorted even at a very low count rate (i.e., the pulse pileup effects are minimal). (b) There is a significant discrepancy between the true and recorded polychromatic X-ray spectra. (Figures are from Taguchi, K. and Iwanczyk, J.S., *Med. Phys.*, 40, 100901, 2013.)

requirements for high input count rates. The pulse pileup model accounts for the (bipolar) shape of the pulse, the distribution function of time intervals between random events, and the transmitted spectrum as the probability density function. The model showed excellent agreement with Monte Carlo simulation[21] and with PCD data.[22] The coefficients of variation (i.e., the root mean square difference divided by the mean of measurements) were as small as 5.3%–10.0% for dead-time loss up to 50% in a Monte Carlo simulation[21] and 7.2% with dead-time loss of 46% in a PCD experiment.[22]

The cascaded model of attenuation, SRE, and pulse pileups start with the spectrum incident onto the object. Using Beer's law to model the attenuation inside the object, the spectrum exiting from the object and incident onto PCD is calculated. The spectrum is then used as an input to SRF and the intermediate spectrum that results from SRE is calculated. Finally, the intermediate spectrum is used as an input to the pulse pileup model, which provides the expected recorded PCD data.

The cascaded model showed excellent agreement with the PCD data (Figure 6.5). The weighted coefficient of variation (i.e., the root mean square difference weighted by the standard deviation of measurements, divided by the mean of measurements) averaged over all channels was as small as 1.5%–6.7% for dead-time losses of (or DE reductions of) 1.1%–55.2% with PMMA. In contrast, models that lack the pulse pileup model or both pulse pileup and SRE resulted in much larger coefficient of variation values: 1.7%–36.3% without the pileup model and 8.3%–67.5% with neither pileup nor SRE models.[15]

6.1.5 Image Reconstruction

The fourth area for advancing and integrating imaging methods is to adapt advanced image reconstruction methods for photon-counting CT data for the interior problem and spectral data.

6.1.5.1 Interior Problem

Even with the earlier discussed PCD compensation schemes, photon-counting data may be inaccurate, especially for X-rays that go through the edge of the object or just outside the object when the object is off-center. Reconstructing images from such inaccurate data will result in undesirable artifacts. From the algorithmic point of view, this is a unique, softly posed interior problem. The detector size defines the physical data truncation range. However, for acceptable data quality, only a subset of all detector channels may be used for reconstruction, for example, because the count rates were high in the periphery. The usable range depends on the PCD compensation method and can be decided retrospectively for PCD-CT. Insight into this unique problem can be gained by studying the trade-off between acceptable data quality and image fidelity using simulation and phantom studies.

There are two approaches to addressing the interior problem: (1) to estimate unmeasured data and *detruncate* the projection data and follow that with a standard image reconstruction method or (2) to reconstruct (quasi-)exact images only

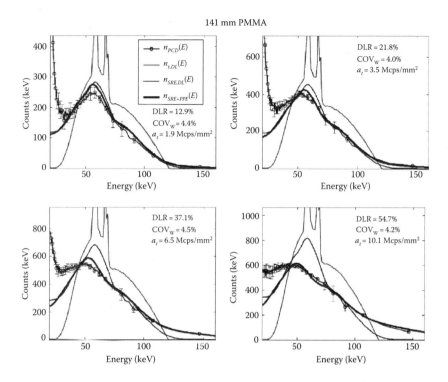

FIGURE 6.5 (See color insert.) The spectrum recorded by a photon-counting detector (PCD), $n_{PCD}(E)$, was severely distorted by the spectral response effect (SRE) and pulse pile-ups, and there are significant discrepancies from the spectrum predicted by a linear model (i.e., the true spectrum linearly scaled by the dead-time loss ratio [DLR]), $n_{t,DL}(E)$. The spectrum predicted by the SRE model and scaled by the DLR, $n_{SRE,DL}(E)$, had better agreement with $n_{PCD}(E)$ than did $n_{t,DL}(E)$; however, the deviation increases with increasing DLR. The fully cascaded PCD model proposed in Ref. [15] accurately estimated the recorded spectrum over a wide range of DLRs (or count rates). a_t is the count rate incident onto the detector. (Figures are from Cammin, J. et al., *Med. Phys.*, 41, 041905, 2014.)

from the truncated measured data. Studying these methods for the softly posed interior problem is certainly of interest.

For the first approach, various detruncation methods have been proposed, which include empirical approaches aimed to decrease an abrupt change between the estimated and measured data[23,24] or more mathematically rigorous approaches that use consistency conditions.[25,26] The use of prior conventional CT images for photon-counting data has recently been proposed.[27]

Regarding the second approach, two important algorithms have recently been developed to solve the interior problem. First, when a small region located inside the region of interest is known, the region-of-interest image can be reconstructed exactly using a differentiated backprojection framework.[29–32] Second, if the region of interest is piecewise constant, an exact image can be reconstructed using the total variation minimization algorithm without other *a priori* knowledge.[33,34]

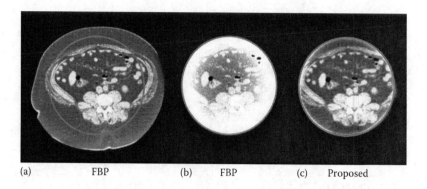

(a) FBP   (b) FBP   (c) Proposed

FIGURE 6.6 (See color insert.) Reconstructed images (a) without or (b, c) with truncation outside the yellow circle, using (a, b) filtered backprojection or (c) the proposed sequential method. The image reconstructed by the proposed method showed very little bias throughout the region of interest except near the edge of the region of interest, while the image appeared very similar to that reconstructed without truncation. (Images are from Taguchi, K. et al., *Med. Phys.*, 38, 1307, 2011.)

Clinical CT data satisfy neither of the requirements; however, it was demonstrated that quasi-exact region-of-interest images can be reconstructed even from noisy clinical CT projections by sequentially using filtered backprojection, total variation minimization, and differentiated backprojection (Figure 6.6).[28] Pixel values of a tiny flat region obtained by total variation minimization were used as *a priori* information during differentiated backprojection.

6.1.5.2 Spectral Data

Spectral data that become available with PCDs provide room to investigate and develop new methods for improved contrast-to-noise ratio, material decomposition, and statistical reconstruction. A study[35] showed that weighting energy-window data by a factor of E^{-3}, where E is the effective energy of the window, improved the contrast-to-noise ratio of images (see Section 6.2.1 for various study results), and other weighting schemes have also been investigated.[36,37] The practical value of these methods in the presence of energetic cross talk between energy windows, however, is not clear. A portion of signals obtained at lower-energy windows come from higher-energy photons, and photons that are supposed to be detected at lower-energy windows may be counted by a higher-energy window. Without appropriate handling of energetic cross talk, energy-window-weighting approaches may enhance artifacts and biases. An application of local, highly constrained backprojection reconstruction (HYPR-LR) broke free from the trade-off between the contrast and the noise of monoenergetic images,[38] although a challenge of this approach is how to handle the energetic cross talk.

Recently, a new class of image reconstruction methods has been developed for PCD-CT.[39] The method, JE-MAP for joint estimation maximum *a posteriori*, jointly reconstructs basis function images and tissue maps of the object directly from PCD data, making full use of the rich information that spectral PCDs acquire. Using the knowledge if the pixels are at an organ boundary or inside an organ, JE-MAP decreased image noise effectively while maintaining the sharpness of organ boundaries and heterogeneous patterns inside organs.

In addition, there are several representational schemes for PCD images such as monoenergetic CT images, material-specific (e.g., iodine) density maps, effective atomic number maps, and electron density maps. Different types of images may be optimally obtained by using different algorithms. Integrating three steps—material decomposition, image reconstruction, and final output calculation—into a single step may improve the accuracy and precision of images.

6.2 POTENTIAL BENEFITS AND CLINICAL APPLICATIONS

Here, we outline the clinical merits and applications of PCD-CT from improved and evolutionary versions of what is currently available to innovative and revolutionary new ones. We have performed a simulation study to demonstrate some of the merits for coronary CT angiography that we discuss later (Figure 6.7). The scan conditions were as follows: 120 kVp; tube current modulation up to 667 mA for lateral direction, down to 200 mA for the AP direction; aluminum bow tie filter with thicknesses of 5–30 mm; focus to center, 600 mm; focus to detector, 1100 mm; 1892 channels for field of view of Ø500 mm for PCD-CT and 946 channels for EID-CT; and 2560 projections per rotation. Images shown in Figure 6.7 were reconstructed while compensating for the spectral distortion

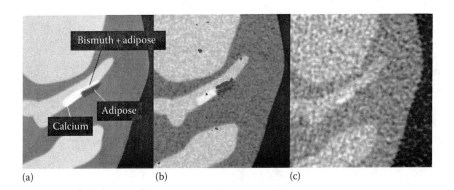

(a) (b) (c)

FIGURE 6.7 (See color insert.) (a) A computer simulated XCAT phantom image with bismuth at the surface of fatty atherosclerosis in a coronary artery. (b, c) Reconstructed images of the phantom scanned at the equivalent dose using a photon-counting detector computed tomography (PCD-CT) (b) and an energy-integrating detector computed tomography (EID-CT) (c). Densities of bismuth are shown in red in (b). The PCD image has a better contrast-to-noise ratio and appears sharper than the EID image. This is also an example of K-edge, molecular, and simultaneous multiagent imaging. (Images are from Cammin, J. et al., Spectral response compensation for photon counting clinical X-ray CT and application to coronary vulnerable plaque detection, in: Noo, F., ed., *Proceedings of the Second International Meeting on Image Formation in X-Ray Computed Tomography*, Salt Lake City, UT, 2012, pp. 186–189.)

due to SRE using a penalized maximum likelihood approach[9] and filtered backprojection for PCD-CT and filtered backprojection only for EID-CT.

6.2.1 Improved Contrast-to-Noise Ratio and Contrast of CT Images

The image quality metrics of CT will improve with PCDs and appropriate algorithms, and material decomposition will allow for reconstructing monoenergetic images at desirable energies. These improvements are significant for any applications but are particularly important for molecular imaging, since weaker signals can be detected.

One point of caution is that, as shown in Ref. [41], results strongly depend on the conditions under which the studies are conducted. Close attention must be paid to factors such as the choice of objects and lesions, the degree of spectral distortion of PCDs, and the algorithms employed for compensation and image reconstruction. One simulation study showed that with optimal energy weighting, the contrast-to-noise ratios of PCD-CT images were better than those of EID-CT images by 15%–57% depending on the materials.[36] When the spectral distortion caused by SRE was incorporated and compensated for, contrast-to-noise ratios of PCD-CT images were improved from EID-CT by 1.4%–11.6% in one study[8] and by 40%–63% in another study.[9] An experimental study using PCD-CT and clinical dual-energy CT, with contrast-to-noise ratio of oil and water, resulted in 57%–96% improvement.[42] Another experimental study showed that the contrast-to-noise ratio of iodine solution against water increased by up to 20%.[43]

6.2.2 Dose Reductions of X-ray Radiation and Contrast Agents

PCD-CT has the potential to improve the contrast-to-noise ratio of contrast-enhanced lesions at a given dose by as much as 30% or more. Expecting such an improvement, one could decrease the amount of contrast agent or radiation dose while maintaining the contrast-to-noise ratio of the lesion at the current level. The contrast dose reduction will be preferable for patients with renal function issues, while the radiation dose reduction will decrease a risk of cancers in general. Using the linear method shown in Appendix B of Ref. [1], the contrast dose might be reduced by 23% or the radiation dose by 41%. The amount of actual dose reduction achieved may be smaller than these values in practice, however, because the PCD-CT system and image reconstruction methods may be nonlinear.

6.2.3 Improved Spatial Resolution

In order to handle the high count rates required for clinical CT, the pixel size of PCDs will likely be smaller than that of EIDs: 0.2–0.5 mm for PCDs in contrast to 1.0–1.4 mm for EIDs. Each scintillator pixel of EIDs is surrounded by light reflectors that physically and optically separate pixels. The reflectors prevent light cross talk between adjacent pixels and direct scintillation lights to be collected by a photo diode underneath the scintillator pixel. The reflectors do not detect X-rays though, and thus, they decrease the geometrical efficiency of EIDs. The thickness of the reflectors is constant regardless of the pixel size; thus, the geometrical efficiency of EIDs decreases with a decrease in pixel size. This is the reason why the pixel size of EIDs cannot be as small as those of PCDs.

The intrinsic spatial resolution of PCD-CT images defined by the Nyquist frequency of the sampling condition will thus be superior to that of EID-CT images. Reconstructed images may become sharper and more accurate due to decreased partial volume effects from small structures such as calcium plaques, although it will come with increased noise.

6.2.4 Beam-Hardening Artifacts

CT vendors have developed beam-hardening correction methods for water and bone.[44] However, beam-hardening artifacts with contrast agents remain a problem for cardiac images.[45,46] PCD-CT will address this problem and improve images where soft plaque, calcium/bone, and contrast-enhanced lumen are present.

6.2.5 Quantitative CT and X-ray Imaging

Current CT pixel values are not as quantitative as one may think. They are measured in Hounsfield units, which are linearly related to the linear attenuation coefficients of X-rays at some energy. However, it is not clear which energy it is. The effective energies of the transmitted X-ray spectrum vary greatly during a scan, depending on factors such as fan/cone angles due to effects of bow tie filters, the attenuation of the object, and projection angles. Thus, the effective energy for an image pixel cannot be calculated. Pixel values of the same tissue vary as the effective energy varies. PCDs can make CT images quantitative using well-defined energies. The physical properties of each image pixel can be accurately modeled using the concept of material decomposition and reconstruction from PCD data. The concentration of contrast agents at regions of interest can then be quantified, which will benefit applications such as cardiac perfusion CT. One problem of current perfusion CT is that it is necessary to subtract a baseline image from target images at different phases to calculate the enhancement due to the injection of the contrast agent. The subtraction will increase noise and misregistration due to motion results in inaccurate time–density curves, and thus, perfusion measurements such as blood flow may be noisy and inaccurate. In addition, the calculated enhancement may change from scan to scan, because the pixel values of CT images are not quantitative. Measuring the concentrations of the contrast agent in target images without subtracting the baseline image, enabled by the quantitative PCD-based CT imaging methods, will improve the accuracy of perfusion CT and other applications.

6.2.6 Accurate K-Edge Imaging

Dual-energy CT provides only two measurements with different energies[11]; however, it is desirable to have three or more measurements for K-edge imaging for contrast-enhanced CT exams[47,48] and for corrections of various data quality degradation factors, as discussed earlier. A third basis function is necessary to model the attenuation curves of contrast agents with high atomic numbers (e.g., iodine, gadolinium, barium) because the curves are discontinuous due to their material-specific K-shell binding energies (Figure 6.8). Using the material decomposition with a third basis function for an atom used in the contrast agent of interest will make it possible to quantify the spatial distribution of contrast agents on a pixel basis. This is called K-edge CT imaging, which will enable quantitative imaging of the contrast agent.

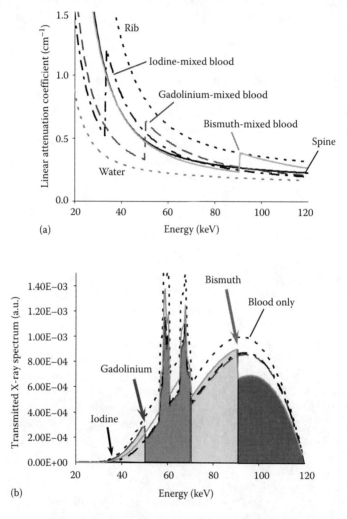

FIGURE 6.8 (See color insert.) (a) Energy-dependent linear attenuation coefficients of various materials. Contrasts between different materials are greater at lower energies in general. Four materials, spine, 0.49% w/w iodine-mixed blood, 0.26% w/w gadolinium-mixed blood, and 0.28% w/w bismuth-mixed blood, result in the same pixel value with the current energy-integrating detector computed tomography (EID-CT), although they have distinctly different attenuation curves. (b) Transmitted spectra with 25 cm water and 5 cm blood without or with one of the three contrast agents. The K-edges of gadolinium and bismuth are clearly seen. (Figures are from Taguchi, K. and Iwanczyk, J.S., *Med. Phys.*, 40, 100901, 2013.)

6.2.7 Simultaneous Multiagent Imaging

Simultaneous multiagent imaging[7] for different functionalities may become possible. Large biological variations between animals and patients make it difficult to interpret measured quantities of agents. By injecting two agents simultaneously, one with target receptors and labeled by one element and the other without receptors and labeled by another element, and imaging both simultaneously, the agent without receptors can be used as a control.[49] This will solve interpretation problems.

There are two blood supplies to the liver: hepatic artery and portal vein. Primary cancers receive 80% of their blood supply from the hepatic artery, while the liver parenchyma receives 80% from the portal vein. Therefore, the liver is usually scanned at two different phases, one at the hepatic arterial phase and the other at the portal venous phase, which are separated by ~50 s. Patients are instructed to breath between the scans, which leads to misregistration between the two images. If two different contrast agents are administered, one early for the portal venous phase and the other later for the hepatic arterial phase, and the patient is scanned once, the single multiagent image may present the distribution of two blood supplies.

6.2.8 Molecular CT with Nanoparticle Contrast Agents and Personalized Medicine

A new type of contrast agents may enable molecular CT imaging.[50–52] Nanoparticles of various sizes and functions are labeled by atoms for CT imaging. The so-called blood pool agents, which consist of large particles with a particle nanometer size of a few hundred (blood pool contrast agents),[53,54] stay in the system longer than 24 h because they are not filtered out by the kidneys. Such large particles can carry more receptors and will increase chances of interaction with target sites and, thus, enhance target-specific therapy and imaging. For example, $\alpha_v\beta_3$-targeted nanoparticles[55–58] have been used to detect, characterize, and treat angiogenesis. Labeling particles for X-ray CT is achieved by attaching atoms with high atomic numbers (e.g., bismuth), which are preferable because signal-to-noise ratios are higher than iodine due to the following reasons: (1) they attenuate more photons with the same particle concentrations than those with lower numbers; and (2) there are more X-ray photons near the K-edges (Figure 6.8).

There are many challenges to this development including toxicity, stability, and clearance for safety, uniformity of particle size for functionality, and particle concentration or uptake for functionality and signal detection. Significant investment from pharmaceutical companies is needed for manufacturing high-quality agents, which will be challenging in an environment where there are both limited market (due to its specific target) and rigorous regulatory hurdles to overcome. Nonetheless, nanomedicine research aligns well with the NIH's goal of personalized medicine, and solutions may be, and should be, found. PCD-CT will be ideal for these biomedical applications and will play a vital role in advancing nanomedicine research.

REFERENCES

1. K. Taguchi, J.S. Iwanczyk, Vision 20/20: Single photon counting X-ray detectors in medical imaging, *Medical Physics* **40**, 100901 (2013).
2. S.S. Hsieh, N.J. Pelc, The feasibility of a piecewise-linear dynamic bowtie filter, *Medical Physics* **40**, 031910 (2013).
3. E. Roessl, R. Proksa, Dynamic beam-shaper for high flux photon-counting computed tomography, in *Workshop on Medical Applications of Spectroscopic X-ray Detectors* (CERN, Geneva, Switzerland, 2013).
4. S. Miyajima, K. Imagawa, M. Matsumoto, CdZnTe detector in diagnostic X-ray spectroscopy, *Medical Physics* **29**, 1421–1429 (2002).
5. S. Kappler, S. Hoelzer, E. Kraft, K. Stierstorfer, T.G. Flohr, Quantum-counting CT in the regime of count-rate paralysis: Introduction of the pile-up trigger method, in *Proceedings of SPIE 7661, Medical Imaging 2011: Physics of Medical Imaging*, Vol. 7661, N.J. Pelc, R.M. Nishikawa, B.R. Whiting, eds. (SPIE, Orlando, FL, 2011), p. 79610T.
6. E. Kraft, F. Glasser, S. Kappler, D. Niederloehner, P. Villard, Experimental evaluation of the pile-up trigger method in a revised quantum-counting CT detector, in *Proceedings of SPIE 8313, Medical Imaging 2012: Physics of Medical Imaging*, Vol. 8313, N.J. Pelc, R.M. Nishikawa, B.R. Whiting, eds. (San Diego, CA, 2012), p. 83134A.
7. J.P. Schlomka, E. Roessl, R. Dorscheid, S. Dill, G. Martens, T. Istel, C. Baumer et al., Experimental feasibility of multi-energy photon-counting K-edge imaging in pre-clinical computed tomography, *Physics in Medicine and Biology* **53**, 4031–4047 (2008).
8. E. Roessl, B. Brendel, K. Engel, J. Schlomka, A. Thran, R. Proksa, Sensitivity of photon-counting based K-edge imaging in X-ray computed tomography, *IEEE Transactions on Medical Imaging* **30**, 1678–1690 (2011).
9. S. Srivastava, J. Cammin, G.S.K. Fung, B.M.W. Tsui, K. Taguchi, Spectral response compensation for photon-counting clinical X-ray CT using sinogram restoration, in N.J. Pelc, R.M. Nishikawa, and B.R. Whiting, eds. *SPIE Medical Imaging 2012: Physics of Medical Imaging*, Vol. 8315 (SPIE, San Diego, CA, 2012), pp. 8315–8337.
10. S. Srivastava, K. Taguchi, Sinogram restoration algorithm for photon counting clinical X-ray CT with pulse pileup compensation, in *Proceedings of the First International Meeting on Image Formation in X-ray Computed Tomography*, F. Noo, ed. (Salt Lake City, UT, 2010), pp. 5–9.
11. R.E. Alvarez, A. Macovski, Energy-selective reconstructions in X-ray computerised tomography, *Physics in Medicine and Biology* **21**, 733–744 (1976).
12. K. Taguchi, K. Nakada, K. Amaya, Compensation for spectral distortions due to spectral response and pulse pileup effects for photon counting CT, in *IEEE Nuclear Science Symposium and Medical Imaging Conference* (IEEE, Seoul, Korea, 2013), pp. M20–M22.
13. J. Cammin, J.S. Iwanczyk, K. Taguchi, Spectral/photon-counting computed tomography, in *Emerging Imaging Technologies in Medicine*, M.A. Anastasio, P.J.L. Riviere, eds. (Taylor & Francis Group, Boca Raton, FL, 2012), pp. 23–39.
14. J. Cammin, J. Xu, W.C. Barber, J.S. Iwanczyk, N.E. Hartsough, K. Taguchi, Modeling photon-counting detectors for X-ray CT: Spectral response and pulse pileup effects and evaluation using real data, in *Proceedings of the SPIE 8668, Medical Imaging 2013: Physics of Medical Imaging*, R.M. Nishikawa, B.R. Whiting, eds. (SPIE, Lake Buena Vista, FL, 2013), p. 86680R.
15. J. Cammin, J. Xu, W.C. Barber, J.S. Iwanczyk, N.E. Hartsough, K. Taguchi, A cascaded model of spectral distortions due to spectral response effects and pulse pileup effects in a photon-counting X-ray detector for CT, *Medical Physics* **41**, 041905 (2014).

16. H. Ding, S. Molloi, Image-based spectral distortion correction for photon-counting X-ray detectors, *Medical Physics* **39**, 1864–1876 (2012).
17. R. Guenzler, V. Schuele, G. Seeliger, M. Weiser, K. Boeringer, S. Kalbitzer, J. Kemmer, A multisegment annular Si-detector system for RBS analysis, *Nuclear Instruments and Methods in Physics Research Section B: Beam Interactions with Materials and Atoms* **35**, 522–529 (1988).
18. L. Wielopolski, R.P. Gardner, Prediction of the pulse-height spectral distribution caused by the peak pile-up effect, *Nuclear Instruments and Methods in Physics Research* **133**, 303–309 (1976).
19. R.P. Gardner, L. Wielopolski, A generalized method for correcting pulse-height spectra for the peak pileup effect due to double sum pulses, *Nuclear Instruments and Methods in Physics Research Section A: Accelerators, Spectrometers, Detectors and Associated Equipment* **140**, 289–296 (1977).
20. N.P. Barradas, M.A. Reis, Accurate calculation of pileup effects in PIXE spectra from first principles, *X-ray Spectrometry* **35**, 232–237 (2006).
21. K. Taguchi, E.C. Frey, X. Wang, J.S. Iwanczyk, W.C. Barber, An analytical model of the effects of pulse pileup on the energy spectrum recorded by energy resolved photon counting X-ray detectors, *Medical Physics* **37**, 3957–3969 (2010).
22. K. Taguchi, M. Zhang, E.C. Frey, X. Wang, J.S. Iwanczyk, E. Nygard, N.E. Hartsough, B.M.W. Tsui, W.C. Barber, Modeling the performance of a photon counting X-ray detector for CT: Energy response and pulse pileup effects, *Medical Physics* **38**, 1089–1102 (2011).
23. B. Ohnesorge, T. Flohr, K. Schwarz, J.P. Heiken, K.T. Bae, Efficient correction for CT image artifacts caused by objects extending outside the scan field of view, *Medical Physics* **27**, 39–46 (2000).
24. A.A. Zamyatin, S. Nakanishi, Extension of the reconstruction field of view and truncation correction using sinogram decomposition, *Medical Physics* **34**, 1593–1604 (2007).
25. J. Hsieh, E. Chao, J. Thibault, B. Grekowicz, A. Horst, S. McOlash, T.J. Myers, A novel reconstruction algorithm to extend the CT scan field-of-view, *Medical Physics* **31**, 2385–2391 (2004).
26. J. Xu, K. Taguchi, B.M.W. Tsui, Statistical projection completion in X-ray CT using consistency conditions, *IEEE Transactions on Medical Imaging* **29**, 1528–1540 (2010).
27. T.G. Schmidt, F. Pektas, Region-of-interest material decomposition from truncated energy-resolved CT, *Medical Physics* **38**, 5657–5666 (2011).
28. K. Taguchi, J. Xu, S. Srivastava, B.M.W. Tsui, J. Cammin, Q. Tang, Interior region-of-interest reconstruction using a small, nearly piecewise constant subregion, *Medical Physics* **38**, 1307–1312 (2011).
29. M. Courdurier, F. Noo, M. Defrise, H. Kudo, Solving the interior problem of computed tomography using a priori knowledge, *Inverse Problems* **24**, 065001 (2008).
30. H. Kudo, M. Courdurier, F. Noo, M. Defrise, Tiny a priori knowledge solves the interior problem in computed tomography, *Physics in Medicine and Biology* **53**, 2207–2231 (2008).
31. H. Yu, Y. Ye, G. Wang, Interior reconstruction using the truncated Hilbert transform via singular value decomposition, *Journal of X-ray Science and Technology* **16**, 243–251 (2008).
32. G. Wang, H. Yu, Y. Ye, A scheme for multisource interior tomography, *Medical Physics* **36**, 3575–3581 (2009).
33. H. Yu, G. Wang, Compressed sensing based interior tomography, *Physics in Medicine and Biology* **54**, 2791–2805 (2009).
34. H. Yu, J. Yang, M. Jiang, G. Wang, Supplemental analysis on compressed sensing based interior tomography, *Physics in Medicine and Biology* **54**, N425–N432 (2009).

35. P.M. Shikhaliev, Beam hardening artefacts in computed tomography with photon counting, charge integrating and energy weighting detectors: A simulation study, *Physics in Medicine and Biology* **50**, 5813–5827 (2005).
36. T.G. Schmidt, Optimal "image-based" weighting for energy-resolved CT, *Medical Physics* **36**, 3018–3027 (2009).
37. T.G. Schmidt, CT energy weighting in the presence of scatter and limited energy resolution, *Medical Physics* **37**, 1056–1067 (2010).
38. S. Leng, L. Yu, J. Wang, J.G. Fletcher, C.A. Mistretta, C.H. McCollough, Noise reduction in spectral CT: Reducing dose and breaking the trade-off between image noise and energy bin selection, *Medical Physics* **38**, 4946–4957 (2011).
39. K. Nakada, K. Taguchi, G.S.K. Fung, K. Amaya, Maximum a posteriori reconstruction of CT images using pixel-based latent variable of tissue types, in *The Third International Conference on Image Formation in X-ray Computed Tomography*, F. Noo, ed. (Salt Lake City, UT, 2014), pp. 5–8.
40. J. Cammin, S. Srivastava, G.S.K. Fung, K. Taguchi, Spectral response compensation for photon counting clinical X-ray CT and application to coronary vulnerable plaque detection, in *Proceedings of the Second International Meeting on Image Formation in X-ray Computed Tomography*, F. Noo, ed. (Salt Lake City, UT, 2012), pp. 186–189.
41. M.S. Polad, The upper limits of the SNR in radiography and CT with polyenergetic X-rays, *Physics in Medicine and Biology* **55**, 5317 (2010).
42. J. Cammin, S. Srivastava, Q. Tang, W.C. Barber, J.S. Iwanczyk, N.E. Hartsough, K. Taguchi, Compensation of nonlinear distortions in photon-counting spectral CT: Deadtime loss, spectral response, and beam hardening effects, in *SPIE Medical Imaging 2012: Physics of Medical Imaging*, Vol. 8315 (SPIE, San Diego, CA, 2012), pp. 8313–8366.
43. S. Kappler, E. Kraft, B. Kreisler, F. Schoeck, T.G. Flohr, Imaging performance of a hybrid research prototype CT scanner with small-pixel counting detector, in *Workshop on Medical Applications of Spectroscopic X-ray Detectors* (CERN, Geneva, Switzerland, 2013).
44. J. Hsieh, R.C. Molthen, C.A. Dawson, R.H. Johnson, An iterative approach to the beam hardening correction in cone beam CT, *Medical Physics* **27**, 23–29 (2000).
45. A. So, J. Hsieh, J.-Y. Li, T.-Y. Lee, Beam hardening correction in CT myocardial perfusion measurement, *Physics in Medicine and Biology* **54**, 3031–3050 (2009).
46. P. Stenner, B. Schmidt, T. Allmendinger, T. Flohr, M. Kachelriess, Dynamic iterative beam hardening correction (DIBHC) in myocardial perfusion imaging using contrast-enhanced computed tomography, *Investigative Radiology* **45**, 314–323 (2010).
47. S. Feuerlein, E. Roessl, R. Proksa, G. Martens, O. Klass, M. Jeltsch, V. Rasche, H.-J. Brambs, M.H.K. Hoffmann, J.-P. Schlomka, Multienergy photon-counting K-edge Imaging: Potential for improved luminal depiction in vascular imaging, *Radiology* **249**, 1010–1016 (2008).
48. E. Roessl, R. Proksa, K-edge imaging in X-ray computed tomography using multi-bin photon counting detectors, *Physics in Medicine and Biology* **52**, 4679–4696 (2007).
49. Y. Li, V.R. Sheth, G. Liu, M.D. Pagel, A self-calibrating PARACEST MRI contrast agent that detects esterase enzyme activity, *Contrast Media & Molecular Imaging* **6**, 219–228 (2011).
50. F.A. Jaffer, R. Weissleder, Seeing within: Molecular imaging of the cardiovascular system, *Circulation Research* **94**, 433–445 (2004).
51. D. Pan, S.D. Caruthers, G. Hu, A. Senpan, M.J. Scott, P.J. Gaffney, S.A. Wickline, G.M. Lanza, Ligand-directed nanobialys as theranostic agent for drug delivery and manganese-based magnetic resonance imaging of vascular targets, *Journal of the American Chemical Society* **130**, 9186–9187 (2008).

52. A.H. Schmieder, P.M. Winter, S.D. Caruthers, T.D. Harris, T.A. Williams, J.S. Allen, E.K. Lacy et al., Molecular MR imaging of melanoma angiogenesis with 3-targeted paramagnetic nanoparticles, *Magnetic Resonance in Medicine* **53**, 621–627 (2005).
53. F. Hyafil, J.-C. Cornily, J.E. Feig, R. Gordon, E. Vucic, V. Amirbekian, E.A. Fisher, V. Fuster, L.J. Feldman, Z.A. Fayad, Noninvasive detection of macrophages using a nanoparticulate contrast agent for computed tomography, *Nature Medicine* **13**, 636–641 (2007).
54. O. Rabin, J. Manuel Perez, J. Grimm, G. Wojtkiewicz, R. Weissleder, An X-ray computed tomography imaging agent based on long-circulating bismuth sulphide nanoparticles, *Nature Materials* **5**, 118–122 (2006).
55. P.M. Winter, H.P. Shukla, S.D. Caruthers, M.J. Scott, R.W. Fuhrhop, J.D. Robertson, P.J. Gaffney, S.A. Wickline, G.M. Lanza, Molecular imaging of human thrombus with computed tomography, *Academic Radiology* **12**, S9–S13 (2005).
56. S.A. Anderson, R.K. Rader, W.F. Westlin, C. Null, D. Jackson, G.M. Lanza, S.A. Wickline, J.J. Kotyk, Magnetic resonance contrast enhancement of neovasculature with alpha-v beta-3-targeted nanoparticles, *Magnetic Resonance in Medicine* **44**, 433–439 (2000).
57. S. Flacke, S. Fischer, M.J. Scott, R.J. Fuhrhop, J.S. Allen, M. McLean, P. Winter et al. Novel MRI contrast agent for molecular imaging of fibrin: Implications for detecting vulnerable plaques, *Circulation* **104**, 1280–1285 (2001).
58. P.M. Winter, S.D. Caruthers, A. Kassner, T.D. Harris, L.K. Chinen, J.S. Allen, E.K. Lacy et al., Molecular imaging of angiogenesis in nascent Vx-2 rabbit tumors using a novel {alpha}{nu}{beta}3-targeted nanoparticle and 1.5 Tesla magnetic resonance imaging, *Cancer Research* **63**, 5838–5843 (2003).

7 Photon-Counting Detectors and Clinical Applications in Medical CT Imaging

Ira Blevis and Reuven Levinson

CONTENTS

7.1 Photon-Counting Detectors for Medical Imaging 169
    7.1.1 Direct Conversion 169
    7.1.2 Material Properties 170
    7.1.3 Candidate Materials 172
    7.1.4 Detector Fabrication and Signal Formation 172
    7.1.5 Prospects for CT Application 176
7.2 Requirements for Photon-Counting Detectors in Medical CT Imaging 177
    7.2.1 Introduction 177
    7.2.2 Image Quality (Spatial and Contrast Resolution) 179
    7.2.3 Dose Efficiency 180
    7.2.4 Examination Time 181
    7.2.5 Rings (Bands) 184
    7.2.6 Multienergy Imaging 185
7.3 Clinical Applications with Photon-Counting Detectors 185
    7.3.1 K-Edge Imaging 185
    7.3.2 Improved Spatial Resolution 188
    7.3.3 Dose Reduction 190
References 191

7.1 PHOTON-COUNTING DETECTORS FOR MEDICAL IMAGING

7.1.1 DIRECT CONVERSION

Direct conversion radiation detectors offer new capabilities for medical CT imaging over indirect conversion detectors currently in use. The capabilities include energy resolution, variable energy weighting, noise reduction to the quantum statistical limit, as well as increased spatial resolution. These advantages in turn enable new applications such as material decomposition as well as improved image quality in low-dose screening applications and high-spatial-resolution imaging. The advantages stem from basic physics limits on the fidelity and entropy of cascaded energy conversion processes [CUN99]. In direct conversion detectors, absorbed X-ray imaging photons generate individually detectable and measurable electric current pulses without the production and detection of light photons as an intermediary step. Since the signals are immediately available on the back of the detectors, they are sensed with close fitting miniature electronics that significantly reduces sources of noise and expensive infrastructure overhead. The technique, as it occurs in medical imaging, is called photon counting (PC).

7.1.2 Material Properties

Direct conversion materials are large band gap (BG) semiconductors with high mobilities, μ, and lifetimes, τ, for both electrons, e^-, and holes, h^+. Blocks of material are coated with metal electrodes on facing sides to which a "bias" voltage (V) is applied producing an electric field throughout the interior, bulk, of the material. Because BG ≥ 1.5 eV, there are very few free e^-s or h^+s at room temperature and thus negligible "dark" currents—Id. The low Id allows signal sensing and readout using miniature ASIC electronics, usually without the need for bulky capacitive coupling.

In CT, the incident photon energy is in the range of 30–140 keV. Such photons are absorbed in detector materials by the photoelectric effect, the Compton effect, or a small cascade of both resulting in one or more energetic free "photo" electrons. The photoelectrons travel through the semiconductor by multiple scattering, producing primary ionization by atomic scattering and secondary ionization from the energetic primaries, finally dissipating their energy into a cloud of electron–hole pairs (ehp). The amount of charge produced is accurately determined by the ionization energy in the semiconductor material that is 4.5 eV in CZT in accordance with Klein relation (BG ~ 1.6 eV). Thus, a 100 keV X-ray photon produces 3.5 fC of e^-'s and $h+$'s.

The initial size of the cloud is a few microns across (Figure 7.1), but the electrons and holes separate from each other in the applied electric field into two clouds traveling to opposite electrodes. Once separated, the two charge balls each undergo self-repulsive Coulomb explosions and inflate an order of magnitude in size (all charges separating simultaneously as in a Hubble expansion) during the transit time to the collecting electrodes. This final size is important because it determines limits to the spatial and energy resolution of the detection process.

The flux of the charge clouds arriving at the electrodes constitutes a measureable photo current; at sufficient V, the current becomes independent of V and limited by the incident radiation flux. With sufficiently high $\mu\tau_e$, the time structure of the electron current is composed of current pulses and allows the detection and counting of individual photons by external electronics, even from deep within thick detectors. At low flux, such as in nuclear medicine, PC eliminates the step of the detection chain where UV light is produced by scintillator detectors and then sensed with the low efficiency of photomultipliers; enough UV photons are detected to ensure sensitivity to even a single photon, but few enough are detected so as to limit the energy resolution of the technique (this limitation is termed a "secondary quantum sink"). With sufficiently high $\mu\tau_h$, the detector response is stable and proportionate even at much higher fluxes and is useful also in medical CT imaging.

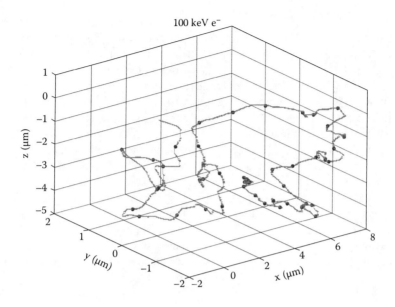

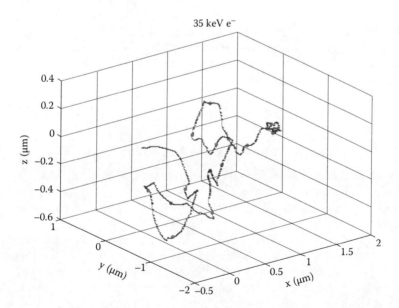

FIGURE 7.1 Photoelectron trajectory and initial ionization in CZT using GEANT. A gray dot marks cumulative 2 keV E_{loss}. The dots cluster at the end of the trajectory. The overall E_{loss} is ~5 eV/nm. The initial size of the cloud of ionization is less than 10 μm across for a 100 keV photo electrons and 2 μm for a 35 keV photoelectron.

7.1.3 CANDIDATE MATERIALS

The main materials being considered for PC include $Cd_{1-x}(Zn_x)Te$, denoted CZT for $0 < x < 0.2$, CdTe, GaAs, Si. Each has its advantages and challenges. CZT has higher BG and lower leakage current than CdTe but is newer, and growth techniques for the high crystallinity needed to achieve high $\mu\tau$'s may be less mature. The low BG of CdTe has led to the use of techniques such as doping and Schottky blocking contacts to limit the Id, but a slowly changing response or instability that may be associated with the use of these techniques has been reported in the literature (e.g., [COL99,HAG95]). Figure 7.2a shows that the mean depth of absorption versus energy in the clinical range is approximately 1–2 mm and that only 3–5 mm of CZT is sufficient to absorb most of the radiation even at the higher energies.

GaAs research is also being reported [VEA14], but since its stopping power is low and detectors must be substantially thicker, it requires even better carrier properties, which at present are still inferior to the alternatives. Figure 7.2b shows that 8–14 mm of GaAs are needed to approach the stopping power of CZT. This provides a significant challenge to crystal growers to achieve sufficiently good electronic transport properties in such thick crystals.

Finally, Si is also considered because of its maturity and availability in the industrial marketplace and its achievably good charge transport properties. The challenge with Si is to overcome the extremely low stopping power with significant material thickness while maintaining production costs. Figure 7.2c shows that its low atomic number and low density require 35–60 mm thicknesses to approach the stopping power of CZT for medical imaging radiation other than the low energies used for mammography. Then, the stopping mechanism and the charge diffusion (in a monolithic detectors structure) would result in large charge balls and reduced spatial resolution. Thus, we can project that CZT is the best candidate to provide a stable, high-performance CT detector.

7.1.4 DETECTOR FABRICATION AND SIGNAL FORMATION

Blocks of semiconductor, in particular CZT, are made into radiation detectors by depositing and configuring thin metal electrodes on a pair of facing surfaces. The electrodes are biased with voltage to give electric fields in the bulk material up to and exceeding 100 V/mm.

Since h⁺s are the slower carriers (in CZT) by 20–50×, impeded by scattering from both shallow and deep traps [JAM95], detectors are configured with radiation incident through the cathode to give the h⁺s the shortest path to exit from the bulk. The e⁻s travel the longer distance to the anodes inducing the larger signal, according to the Shockley–Ramo theorem [HE00]. Ideally, the complete transit of a single charge, an electron to the anode or a hole to the cathode, or a combination that adds up to the same, produces a single charge sensed in the external circuit (anode to cathode) through the power supply and amplifier. Then, the external signal is proportional to the conversion charge and in turn to the incident photon energy.

Traps are unoccupied localized carrier states with energy levels in the gap between the band states. The energy differences with the band edges determine if they are

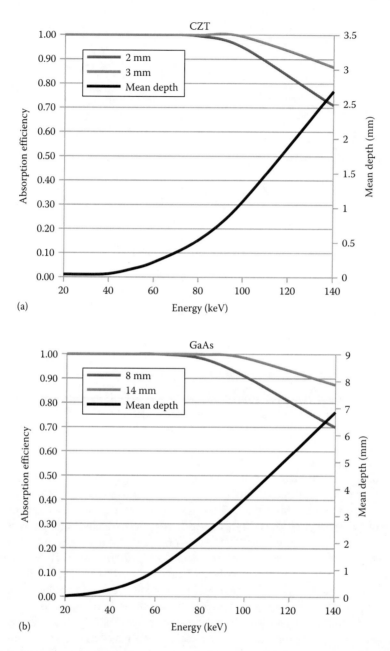

FIGURE 7.2 Absorption efficiency and mean absorption depth versus energy in the clinical range for CZT, GaAs, and Si showing the large variation in the required material thickness for high absorption as needed in medical CT. (*Continued*)

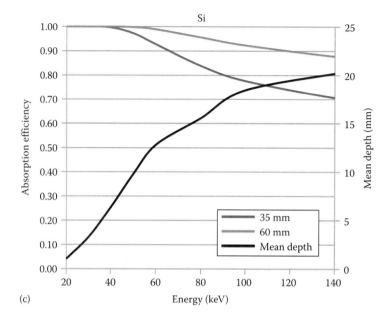

(c)

FIGURE 7.2 (*Continued*) Absorption efficiency and mean absorption depth versus energy in the clinical range for CZT, GaAs, and Si showing the large variation in the required material thickness for high absorption as needed in medical CT.

shallow or deep for either electrons or holes, respectively. If carriers encounter shallow traps and bind to them, they are still released quickly by thermal fluctuations and simply appear to have low mobility μ. If carriers encounter deep traps and bind to them, they are not released within the rise time of the signal in the external electronics and will appear to have low lifetime τ.

The pixilation of the anodes in a monolithic detector therefore has two critical effects. The first is to provide localization of the incident radiation to produce images. The second is to counteract the limitations of slow or trapped carriers in the detection of the energy of the incident photons. In this respect, the pixelated anode effectively divides the bulk to a far region (near the cathode) for X-ray absorption and a near region (near to the anodes and of order of size of the pixel pitch) for signal formation. Then, the charge induction from an absorption event in the CZT bulk, initially spread out on a large number of equidistant anode pads, concentrates geometrically onto a single "hit" anode pad as the charge reaches the "near region" under the propulsion of the applied electric field.

This design gives a number of essential advantages: (1) The opposite polarity induction of slow or trapped holes *on the hit anode* is reduced by the number of equidistant anodes that can shield the hole charge. (2) The signal rise on the induced anode is delayed and importantly shortened (compared to a nonpixelated anode) giving larger amplifier pulses with less-integrated shot noise. In this way, the measured signals become proportional to the incident photon ionization and thereby the incident photon energy, independent of the absorption depth. These two effects together are called the "small pixel effect."

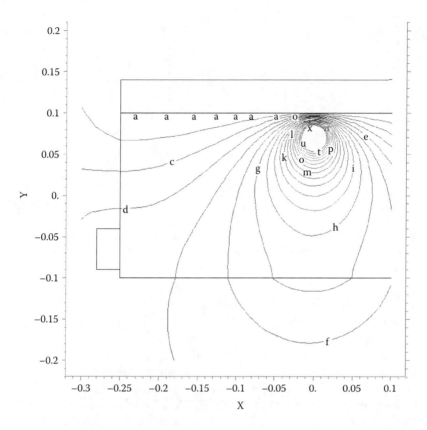

FIGURE 7.3 (See color insert.) Equipotential contours surrounding a test charge as well as the induced potentials on a continuous cathode on top and on a pixelated anode on bottom. The cathode sits at a single potential, whereas the various anodes sit at the local potential caused by the test charge.

Figure 7.3 shows the equipotential contours surrounding a test charge as well as the induced potentials on a continuous cathode on top and on a pixelated anode on bottom [BLE06]. The cathode sits at a single potential, whereas the various anodes sit at the local potential caused by the test charge. Attaching amplifiers with virtual grounds allow measurable charge to flow to the anode pads to bring the local potential to 0. As the test charge is driven toward the anode plane by the applied electric field, the current out of the destination anode will rise first slowly and then quickly as the $1/r^2$ Coulomb field of the test charge moves. This response is sometimes mapped out with respect to the anode and termed a "weighting potential." The nondestination anodes are returned to their initial charge state after the test charge has arrived at the destination anode and caused a signal to appear on the amplifier output.

Figure 7.4 shows the consequent energy resolution measured with: 40 × 40 × 5 mm CZT; 120 V/mm, 2.5 mm pixels, 3 mm source distance, 200 ns peaking time, flux ~ 1/mm/s [BLE08,BOU10].

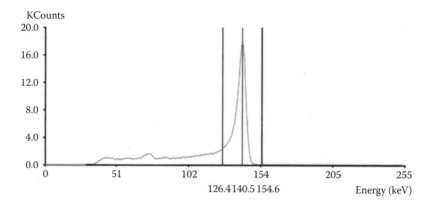

FIGURE 7.4 (See color insert.) CZT energy response spectrum for a ^{99m}Tc flood source. There is a moderate spectrum "tail." The E resolution is 4.8% FWHM. The vertical lines indicate a ±10% energy window for nuclear medicine.

7.1.5 Prospects for CT Application

In the CT application, the flux can exceed 10^8/mm^2/s. This high flux introduces a number of challenges that are currently being investigated. At such rates, the buildup of slow-moving hole charge, called polarization, has been observed to change and reduce the internal electric field of the CZT [BLE05]. Figure 7.5 shows how the spectral peak shifts to lower energy as the parameter μ_{eff} (mobility/flux) decreases in a model that recalculates the bulk electric field based on steady-state flux and steady-state hole mobility [BLEV08]. Since CT systems count photons above a given energy threshold, the count rate plummets above a critical flux determined by the

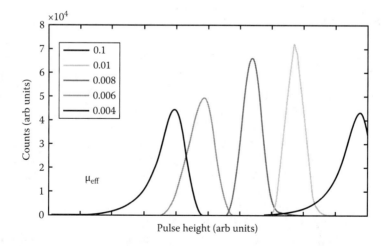

FIGURE 7.5 Spectral peak shifts to lower energy as the parameter μ_{eff} (mobility/flux) decreases in a model that recalculates the bulk electric field based on the steady-state flux and positive charge dynamic equilibrium.

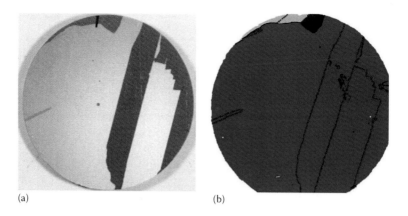

FIGURE 1.7 (a) Cross section of a typical 50 mm diameter CdZnTe ingot and (b) the matching orientation map measured by EBSD.

FIGURE 2.2 Commercial a-Se semiconductor direct X-ray detector for full-field digital mammography. (From http://www.anrad.com/products-direct-xray-detectors.htm, Analogic, Direct Conversion X-ray Detectors, accessed August 12, 2013.)

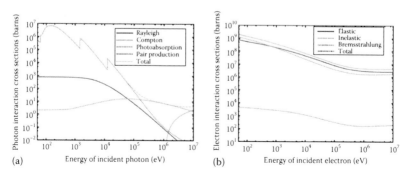

FIGURE 2.4 (a) PENELOPE photon interaction cross sections in selenium from 100 eV to 10 MeV. (b) PENELOPE electron interaction cross sections in selenium from 100 eV to 10 MeV. *Note*: 1 barn = 10^{-24} cm^2. (Reprinted from Fang Y. et al., Monte Carlo simulation of amorphous selenium imaging detectors, *Proc of SPIE*, 7622, 762214, 2010. With permission.)

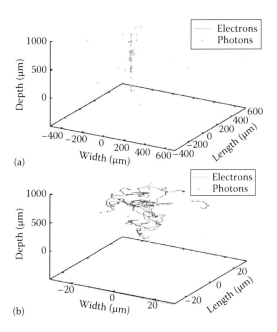

FIGURE 2.6 (a) Particle track of 100 keV incident photons (100 histories) in selenium. (b) Close-up of (a). (Reprinted from Fang Y. et al., Monte Carlo simulation of amorphous selenium imaging detectors, *Proc of SPIE*, 7622, 762214, 2010. With permission.)

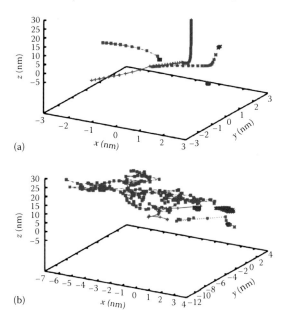

FIGURE 2.9 (a) Sample transport simulation track of three electron–hole pairs in electric field taking into account drift. (b) Sample transport simulation track of three electron–hole pairs in electric field taking into account drift and diffusion. (Blue and red dots represent hole and electron tracks, respectively.)

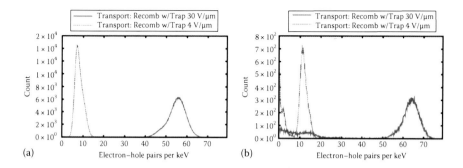

FIGURE 2.10 (a) Results of the detailed spatiotemporal Monte Carlo simulations with ARTEMIS. Plots of the pulse-height spectra, for transport with 4 and 30 V/μm applied electric field with recombination only and with recombination and trapping for 12 keV monoenergetic incident photon energies. (b) For 100 keV.

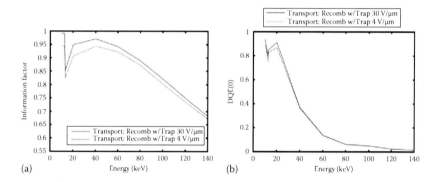

FIGURE 2.11 (a) Simulated information factor as a function of incident photon energy. (b) Simulated DQE at zero spatial frequency as a function of incident photon energy. (Reprinted from Fang Y. et al., *Med. Phys.*, 39, 308, 2012. With permission.)

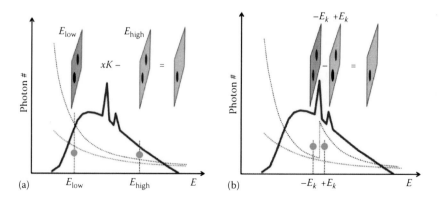

FIGURE 3.2 Schematics of the material decomposition using a photon-counting X-ray/CT system: material decomposition by dual-energy subtraction (a) and by K-edge subtraction (b).

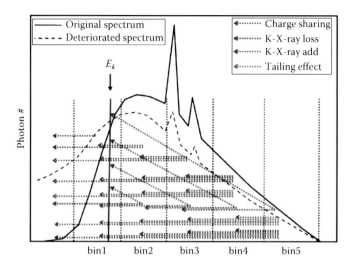

FIGURE 3.14 Schematics that show moving photon counts from one to another energy bin due to the different mechanisms. The original spectrum of the X-ray is deteriorated in unpredictable manner.

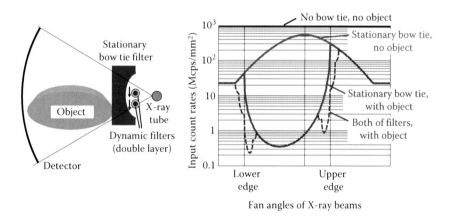

FIGURE 6.1 Calculated true count rates in the lateral view of an elliptic water phantom when it is 5 cm off-center. Dynamic and stationary bow tie filters (left) decrease the count rates near the edges of the object (right, red curve), compared to the results without the dynamic filters (blue curve). (Figures are from Taguchi, K. and Iwanczyk, J.S., *Med. Phys.*, 40, 100901, 2013.)

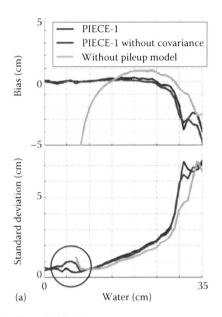

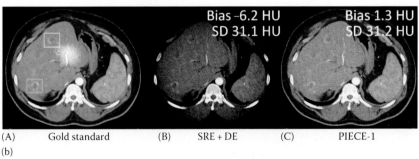

FIGURE 6.3 (a) Both bias and noise of the estimated water thicknesses were improved by PIECE-1. (b) Images of (A) gold standard, (B) reconstructed compensating for the spectral response effect and detection efficiency (but ignoring pileups), and (C) reconstructed by PIECE-1. *Conditions*: A count rate exiting from the tube of 10^9 cps/mm^2; 1 cm water flat filter; detector dead time of 20 ns; pixel size of 0.5×0.5 mm^2; 4 energy thresholds set at 20, 50, 80, and 110 keV; and photopeak ratio of ~0.5 for SRE. The attenuators were 0–40 cm water and 0–5 cm bone.

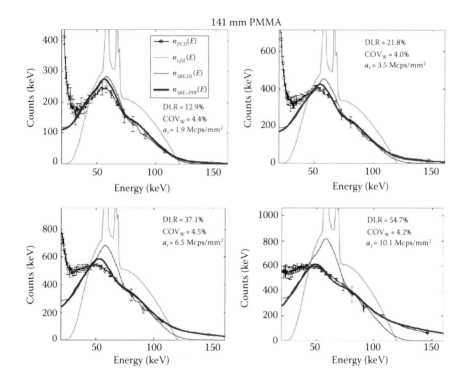

FIGURE 6.5 The spectrum recorded by a photon-counting detector (PCD), $n_{PCD}(E)$, was severely distorted by the spectral response effect (SRE) and pulse pileups, and there are significant discrepancies from the spectrum predicted by a linear model (i.e., the true spectrum linearly scaled by the dead-time loss ratio [DLR]), $n_{t,DL}(E)$. The spectrum predicted by the SRE model and scaled by the DLR, $n_{SRE,DL}(E)$, had better agreement with $n_{PCD}(E)$ than did $n_{t,DL}(E)$; however, the deviation increases with increasing DLR. The fully cascaded PCD model proposed in Ref. [15] accurately estimated the recorded spectrum over a wide range of DLRs (or count rates). a_t is the count rate incident onto the detector. (Figures are from Cammin, J. et al., *Med. Phys.*, 41, 041905, 2014.)

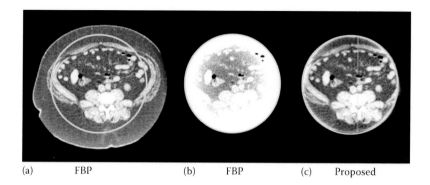

(a)   FBP   (b)   FBP   (c)   Proposed

FIGURE 6.6 Reconstructed images (a) without or (b, c) with truncation outside the yellow circle, using (a, b) filtered backprojection or (c) the proposed sequential method. The image reconstructed by the proposed method showed very little bias throughout the region of interest except near the edge of the region of interest, while the image appeared very similar to that reconstructed without truncation. (Images are from Taguchi, K. et al., *Med. Phys.*, 38, 1307, 2011.)

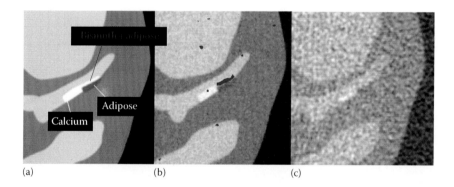

(a)   (b)   (c)

FIGURE 6.7 (a) A computer simulated XCAT phantom image with bismuth at the surface of fatty atherosclerosis in a coronary artery. (b, c) Reconstructed images of the phantom scanned at the equivalent dose using a photon-counting detector computed tomography (PCD-CT) (b) and an energy-integrating detector computed tomography (EID-CT) (c). Densities of bismuth are shown in red in (b). The PCD image has a better contrast-to-noise ratio and appears sharper than the EID image. This is also an example of K-edge, molecular, and simultaneous multiagent imaging. (Images are from Cammin, J. et al., Spectral response compensation for photon counting clinical X-ray CT and application to coronary vulnerable plaque detection, in: Noo, F., ed., *Proceedings of the Second International Meeting on Image Formation in X-Ray Computed Tomography*, Salt Lake City, UT, 2012, pp. 186–189.)

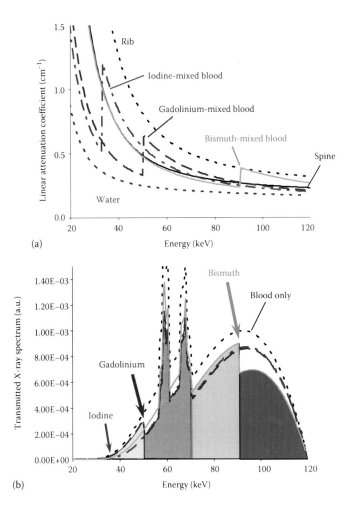

FIGURE 6.8 (a) Energy-dependent linear attenuation coefficients of various materials. Contrasts between different materials are greater at lower energies in general. Four materials, spine, 0.49% w/w iodine-mixed blood, 0.26% w/w gadolinium-mixed blood, and 0.28% w/w bismuth-mixed blood, result in the same pixel value with the current energy-integrating detector computed tomography (EID-CT), although they have distinctly different attenuation curves. (b) Transmitted spectra with 25 cm water and 5 cm blood without or with one of the three contrast agents. The K-edges of gadolinium and bismuth are clearly seen. (Figures are from Taguchi, K. and Iwanczyk, J.S., *Med. Phys.*, 40, 100901, 2013.)

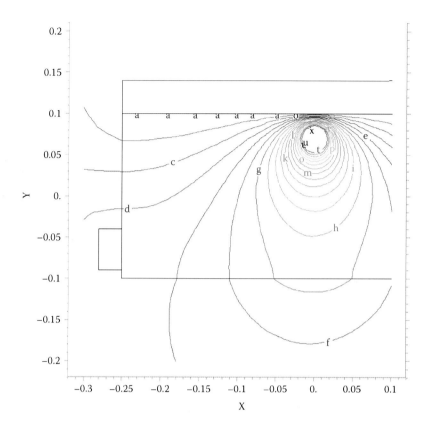

FIGURE 7.3 Equipotential contours surrounding a test charge as well as the induced potentials on a continuous cathode on top and on a pixelated anode on bottom. The cathode sits at a single potential, whereas the various anodes sit at the local potential caused by the test charge.

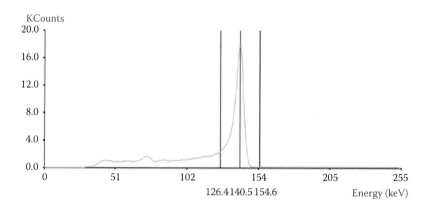

FIGURE 7.4 CZT energy response spectrum for a ^{99m}Tc flood source. There is a moderate spectrum "tail." The E resolution is 4.8% FWHM. The vertical lines indicate a ±10% energy window for nuclear medicine.

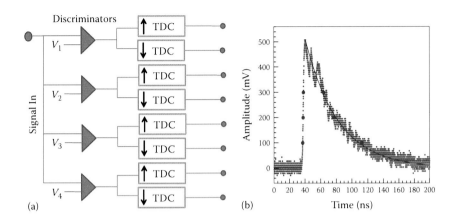

FIGURE 8.8 (a) The MVT method samples a pulse by determining the time it crosses a number of voltage thresholds, using discriminators and TDCs. (b) A sample pulse (blue curve), the MVT samples (red dots), and the fitted pulse from samples (red curve).

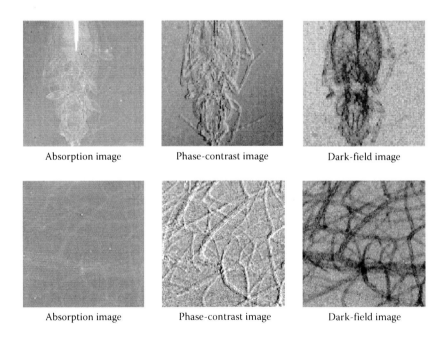

FIGURE 10.5 Reconstructed absorption, differential phase-contrast, and dark-field images of a bug (top) and lichen (bottom).

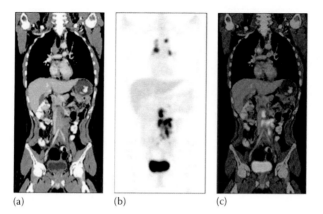

FIGURE 11.1 (a) CT scan of full body, (b) PET scan of full body, and (c) PET + CT scan fused images. There is an obvious improvement of the image quality when PET and CT scan are fused together. (Image taken from: www.dh.org/about-pet-ct.)

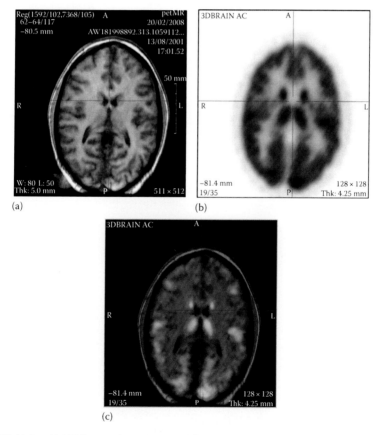

FIGURE 11.2 (a) MRI scan of a brain, (b) PET scan of the same brain, and (c) PET + MRI fused images. There is an obvious improvement of the image quality when PET and MRI scans are fused together. (Image taken from: en.wikipedia.org/wiki/PET-MRI.)

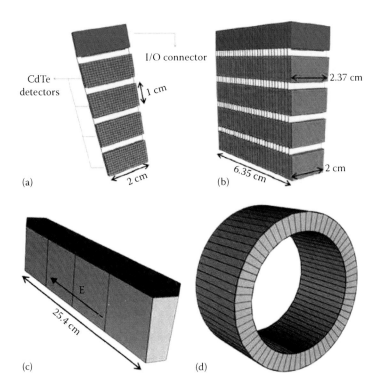

FIGURE 11.4 Detailed depiction of the design of the VIP scanner. (a) One single VIP unit module, consisting of a single layer with CdTe pixel detectors mounted on a thinned readout ASIC (not visible) and both together mounted on thin kapton PCB. (b) A stack of such layers to form a module block. The number of layers per stack is flexible. (c) How a set of module blocks form a sector of the PET scanner. (d) How the full scanner can be made seamless by using the trapezoidal shape of the VIP unit module. (Reprinted from Mikhaylova, E. et al., Simulation of pseudo-clinical conditions and image quality evaluation of PET scanner based on pixelated CdTe detector, *IEEE NSS MIC Conf. Rec.*, 2716–2722, 2011. Copyright 2011 IEEE with permission.)

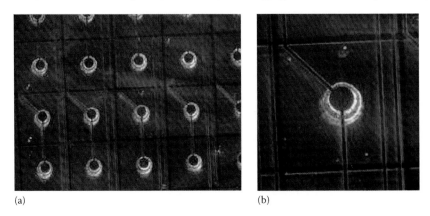

FIGURE 11.7 (a) shows the pixel CdTe detector bonded to glass substrate. The pixel pitch is 1 mm and the gap between neighboring electrodes is 50 µm. In the pixel center one can see the 250 µm BiSn solder ball between the detector and the glass substrate. (b) shows a more detailed picture of the pixel CdTe detector and the bump bonding connection.

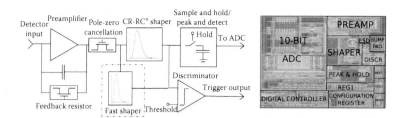

FIGURE 11.8 Schematic depiction and layout of the VIP-PIX pixel cell. (Reprinted from Macias-Montero, J.-G., Sarraj, M., Chmeissani, M. et al., A 2D 4 × 4 channel readout ASIC for pixelated CdTe detectors for medical imaging applications, *IEEE NSS and MIC Conf. Rec.*, 2013. Copyright 2013 IEEE with permission.)

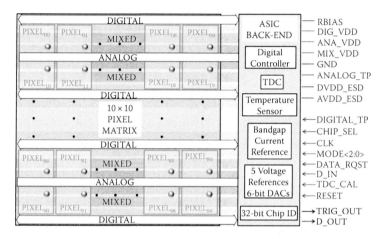

FIGURE 11.9 Depiction of the entire VIP-PIX chip pixel array. (Reprinted from Macias-Montero, J.-G., Sarraj, M., Chmeissani, M. et al., A 2D 4 × 4 channel readout ASIC for pixelated CdTe detectors for medical imaging applications, *IEEE NSS and MIC Conf. Rec.*, 2013. Copyright 2013 IEEE with permission.)

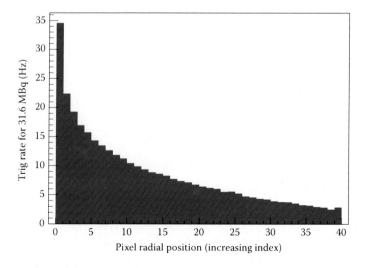

FIGURE 11.17 The trigger rate per single channel as a function of the channel radial position from, position 0 (the closest to the FOV center) to position 39 (the farthest from the FOV center).

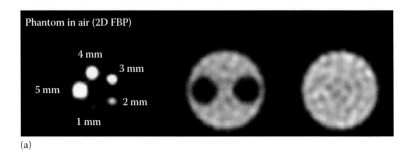

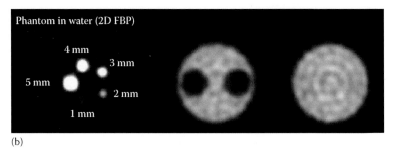

FIGURE 11.19 Reconstructed, images of the three sections of the NEMA NU 4-2008 image quality phantom for the standard procedure with the phantom in air (a) and for the modified scenario with the phantom in water (b). (Reprinted from Mikhaylova, E., De Lorenzo, G., Chmeissani, M. et al., Simulation of the expected performance of a seamless scanner for brain PET based on highly pixelated CdTe detectors, *IEEE Trans. Med. Imaging*, 2014. Copyright 2014 IEEE with permission.

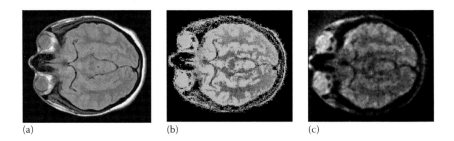

FIGURE 11.23 (a) Example of a DICOM file. (b) A slice of the simulated 3-D brain phantom corresponding to the DICOM file on the left. (c) The same slice after the whole brain reconstruction. (Reprinted from Mikhaylova, E. et al., Simulation of pseudo-clinical conditions and image quality evaluation of PET scanner based on pixelated CdTe detector, *IEEE NSS MIC Conf. Rec.*, 2716-2722, 2011. Copyright 2011 IEEE with permission.)

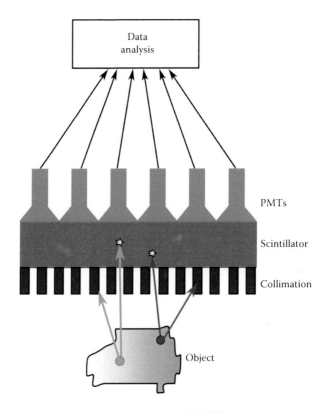

FIGURE 11.28 Schematic depiction of an Anger SPECT camera.

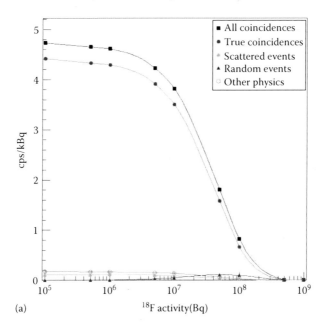

FIGURE 11.33 Sensitivity of the Compton gamma camera as a function of source activity for different isotopes: (a) ^{18}F. *(Continued)*

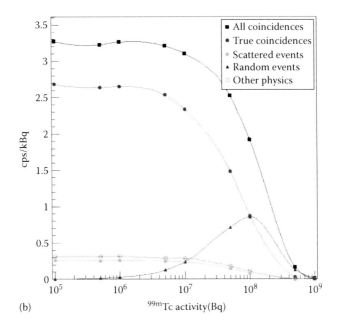

FIGURE 11.33 (*Continued*) Sensitivity of the Compton gamma camera as a function of source activity for different isotopes: (b) ^{99m}Tc.

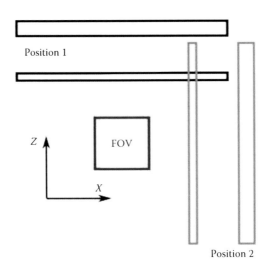

FIGURE 11.34 Depiction of two orthogonal Compton cameras to obtain a good PSF in all direction. (From (Kolstein M. et al., *J. Instrum.*, 9, 2014), licensed under CC BY.)

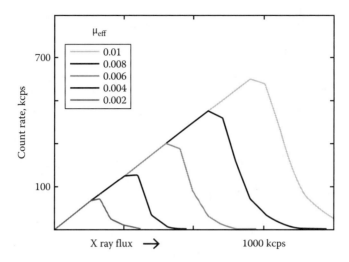

FIGURE 7.6 CT systems count photons above a given energy threshold causing the count rate to plummet above a critical flux determined by the parameter μ_{eff}.

parameter μ_{eff} as shown in Figure 7.6. Note that (1) the count rate curves were found to match experimental data very well and (2) the buildup of charge and the effects on the pulse height spectrum and the count rate curves are all called polarization in different contexts.

Different strategies to counteract polarization by detrapping the holes and helping them to exit the bulk have been devised. These strategies may include heating, illuminating, higher electric fields, or decreased thickness. Small pixels, of order of the bulk thickness or smaller, have also been used since they maintain the signal strength somewhat as the strength of the electric field diminishes.

Small pixels also overcome a strong limitation from electronic counting channels that cannot keep up with high counting rates and show pileup saturation. The pileup saturation resembles closely the CZT polarization in Figure 7.6 but can be distinguished by properly designed experiments. A significant drawback of small pixels is the loss of spectral information as the pixel size approaches the charge cloud size described earlier. Figure 7.7 shows the degradation of the measured spectrum from a CT X-ray source as the pixel size is reduced. The prominent peak at 60 keV is the tungsten characteristic from the X-ray tube anode. Since spectral information is key to the added value of PC, this strategy must be evaluated further.

7.2 REQUIREMENTS FOR PHOTON-COUNTING DETECTORS IN MEDICAL CT IMAGING

7.2.1 Introduction

CT scanners have evolved to provide submillimeter resolution tomographic images of complete body sections in a few seconds. This remarkable performance enables a myriad of imaging applications for CT and has produced many years of double-digit

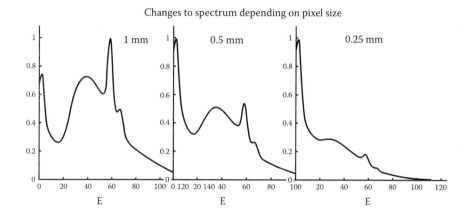

FIGURE 7.7 The measured spectrum from a CT X-ray source in a CZT detector degrades as the pixel size is reduced from 1 mm to 0.25 mm. The prominent peak at 60 keV is the tungsten characteristic from the X-ray tube anode.

growth in the number of CT scans. Indicative of the benefits of this performance is the rapid growth of CT in emergency room medicine. Trauma, stroke, and emergency cardiac imaging extensively use CT, and there are dedicated emergency department CT scanners in many hospitals.

The CT imaging performance has improved due to a number of the technology advances. A major technology breakthrough was the introduction of multirow detector systems. Over a period of 10 years (1998–2008), the number of detector rows grew from 1 to 64, and current (circa 2014) introductions feature over 300 rows. The multirow detectors enable both short scan times (~5 s; less than a single breath hold) and submillimeter isotropic spatial resolution.

Metrics for X-ray medical imaging system performance are image quality (spatial and contrast resolution), radiation dose efficiency, and exam time. An additional metric, specific to CT imaging (now dominated by third-generation (rotate–rotate) gantries, is image ring/band artifacts. The detectors influence all the system performance metrics and the detectors, alone, are the cause of the image ring artifacts. The dimensions of the detector pixel surface area, typically on the order of 1 mm × 1 mm, are a main factor in determining the spatial resolution. The quantum efficiency (QE) of the detector, typically greater than 90%, is the main factor determining the dose efficiency. The response of the detector to high flux levels, measured (in PC detectors) by the pulse pileup, is the main factor in determining the scan time. The potential instability of the detector output creates ring/band artifacts. In the next section, the PC detector design and performance will be derived from the imaging requirements of CT scanning.

The ability of PC detectors to provide a new detection method for CT is dependent on their ability to equal (and surpass) the performance of the current scintillator–photodiode (PD) detectors in conventional single-energy imaging and also enable detector-based multienergy imaging, including, as not yet shown in the clinical setting, K-edge imaging (Table 7.1).

TABLE 7.1
The Following Table Shows the Imaging Performance Metrics for Current Premium CT System and the Detector Parameters/Functionality that Influence Each Metric

	Image Quality (Spatial Resolution)	Radiation Dose Efficiency (Image Noise)	Examination Time (360° Gantry Rotation Time)
Premium CT system (Philips iCT)	16.0	0.27	0.5
Units	Line-pairs/cm @ cutoff (specified for high-resolution scan mode)	% (noise/signal) with specified phantom (typ. 20 cm water) and technique factor (typ. 300 mAs)	Seconds (for routine body scanning)
Detector parameters	• Pixel surface area • Crosstalk	• Photon attenuation • Swank factor • Dead space bet. pixels	• Electronics dead time • Charge mobility

7.2.2 Image Quality (Spatial and Contrast Resolution)

For X-ray imaging systems, the contrast resolution overlaps with the dose efficiency metric: the base requirement is to utilize ("count") every x-photon transmitted through the patient's body. In this section, the photon utilization will be measured with the dose efficiency metric, and the image quality metric will be focused on spatial resolution.

CT scanners have the highest spatial resolution of all the medical tomographic imaging methods; only projection X-ray imaging has higher spatial than CT. Dedicated high-resolution scan modes (orthopedic and middle ear examinations) achieve image pixels sizes less than 0.5 mm; normal scanning modes produce image pixels sizes is in the range of 0.8 mm.

Spatial resolution in CT scanners is a function of the X-ray focal spot size, the detector pixel size, and the "imaging geometry" (i.e., magnification factor). X-ray focal spot sizes are on the order of 0.5–1.5 mm. The magnification factors are on the order of 1.5–2. In order to optimize the spatial resolution (for the X-ray focal spot size and magnification factor of commercial medical CT systems), the detector pixel dimensions should be in the range of 1 mm, as is the case for current CT detectors.

To increase the spatial resolution of a CT system (i.e., reduce the image pixel size), all three factors—X-ray focal spot size, magnification factor, and detector pixel size—must be adjusted accordingly. Changing a single factor will have limited impact on the system resolution. Reduced-size detector pixels to achieve improved spatial resolution also require reduction in X-ray focal spot sizes and an appropriate magnification factor.

The "effective size" of the detector pixel dimensions is influenced by crosstalk in the detectors. Crosstalk increases the effective size of the detector pixels and reduces the spatial resolution. For scintillator–PD detectors, crosstalk results mainly from optical effects (light generated in the scintillator crosses over to a neighboring pixel and is absorbed in a neighbor PD). In direct conversion detectors, crosstalk is the result of photoelectrons traveling outside of the sensor pixel and creating an electrical signal in neighboring pixels. Results from measurements (performed at Philips Haifa) with small PC pixels and a 120 kVp beam show an increase of 15% in counts (with energy level, 20 keV) in the next row of pixels due to crosstalk effects.

The sensors in PC detectors are "blocks" of semiconductor material. The detector pixel size is determined by the pixelated anode structure of the sensor, which enables very small pixel dimensions (to below 0.1 mm). Detector pixel dimensions of 0.225 mm have been reported by Siemens on a prototype PC CT system [WHI1]. The focus of the small pixel size has been the improved count rate performance (see section on scan time), and the Siemens group has not reported on any improvements in the image spatial resolution.

Direct conversion detectors have the potential to deliver significant improvement in the spatial resolution of medical CT systems, more than a factor of 2. Realization of the improvement requires also reduction to the focal spot size and increase in the image reconstruction and display matrices. Both elements are within the engineering realm of today's X-ray and computer technology.

7.2.3 Dose Efficiency

The QE of the detectors is the major influence on the dose efficiency. For high QE, the detectors must absorb "nearly all" the energy of the incident X-ray photons. In practice, the linear attenuation coefficient (μ) of the detector material and the detector thickness (t) are used to estimate the QE (= $100\% * (1 - \exp(-\mu * t))$). Scintillator–PD detectors in medical CT systems have QE about 98%. For typical CT X-ray beam (120 kVp) and 2 mm thickness of CZT, the QE is above 95%.

The dose efficiency of the detectors is also influenced by the "dead space" in the detector: %area of dead space ~ %reduction in dose efficiency. Due to the pixilated structure of the sensor, typically there is dead space between the pixels. For scintillator–PD detectors, the dead space is the result of the reflecting walls (used to optically isolate the crystals) of the scintillation crystal. For direct conversion detectors with pixelated anodes, the electric field in the sensor will determine the charge collection efficiency at each point and the existence of any dead spaces. The electric field is a function of the anode geometry, which should be designed to minimize dead space. It appears that there is a potential for negligible dead space and improved QE with direct conversion detectors versus scintillator–PD detectors.

The dead space is also influenced by the antiscatter grids used in CT. Multislice detectors imaging require an antiscatter grid. The grids are positioned immediately in front of the detector array, and the grid walls shadow the detectors. Typically, the grids are positioned such that the grid walls shadow the pixel border regions. If there is an existing dead space in the detectors at the border region (i.e., reflecting walls in scintillator–PD detectors), then the increase in dead space (from the shadow of the grid wall) may be zero or negligible. For direct conversion detectors with small border region dead space, the shadow area of the antiscatter grid will reduce the QE. For antiscatter grid walls of 0.1 mm with and grid wall pitch of 1 mm, the dead space and reduction in dose efficiency in both cases is 20%. For submillimeter detector pixels (mentioned earlier), the current grid configurations with 1 mm cell pitch and 0.1 mm grid walls will provide the same antiscatter functionality, and the reduction in dose efficiency for the submillimeter detector pixels remains 20%.

The Swank factor is a measure of the reduction in the detective quantum efficiency (DQE) resulting from the statistics of the X-ray photon absorption in the sensor and the signal formation (transformation of X-ray energy to electronic charge) in the sensor. For X-ray scintillation detectors, reports have shown a 20% reduction in DQE resulting from the scintillation crystal transformation of X-ray energy to visible [GIN1]. Direct conversion detectors do not have an intermediate light production step in the signal generation, and it has been speculated that there is a potential for up to 20% improvement in dose efficiency of photon counting versus scintillator–PD detectors resulting from Swank factor in the signal formation.

All (room temperature) electronic detectors have noise ("detector" [electronic] noise). The electronic noise adds in quadrature with the photon shot noise. In most clinical CT scanning scenarios, the detector noise is small compared to the photon shot noise and has negligible influence on the dose efficiency. However, with current interest in dose reduction in CT and the use of lower photon flux levels (and the increase in the relative photon shot noise levels), the dose efficiency is more sensitive to the detector noise. PC detectors have less detector noise than PD detectors. With the pulse-counting technique, there is the potential for "zero-electronic" noise, and indeed, this has been demonstrated in a prototype CT system (see [LEV1]). For low-flux/low-dose/screening applications, PC detectors have the potential for improved dose efficiency versus scintillator–PD detectors.

7.2.4 Examination Time

Required examination time in medical imaging is a function of image artifacts from patient motion, both voluntary and involuntary. The examination time must be "short enough" to freeze the patient motion and eliminate motion artifacts. The beating heart (60 bps) is the fastest involuntary patient motion and cardiac scans require the shortest examination times. Trauma scanning also requires short times. High tube currents and therefore high photon flux are used with short examination times to maintain adequate image quality. Current premium CT systems feature routine subsecond rotation times with 500 mA and higher tube currents. High flux rates are problematic for photon-counting detectors, as they have reduced dose efficiency at high flux rates. Input count rate (ICR) levels produced by CT systems can reach up to 1000 Mcps/mm^2, for "in-air" irradiation conditions.

The major factor on the dose efficiency reduction at high ICR is the PC electronics, paralyzable and nonparalyzable configurations. The limited count rate capabilities of the counting electronics produces pulse pileup and reduces the DQE at high flux levels (Figure 7.8). The DQE is dependent on tau, dead time of the counting electronics. For a given tau, the nonparalyzable configuration has larger DQE than the paralyzable configuration.

Direct conversion sensors (e.g., CdTe, CZT) are known to have memory/lag or even catastrophic breakdown effects at high ICR. These phenomena have an influence on detector DQE performance at high ICR. Currently, this problem has been overcome;

$$DQE = (\partial(OCR)/\partial(ICR) * (ICR/\sqrt{(VAR)_{OCR}})/(ICR)/(\sqrt{VAR_{ICR}}))^2$$

OCR: Output count rate
ICR: Input count rate
$(VAR)_{OCR}$: Variance of OCR
$(VAR)_{ICR}$: Variance of ICR
DQE: Detective quantum efficiency
τ(tau): Dead time (counting electronics)

Nonparalyzable: $DQE = 1/(1 + ICR * \tau)$

Paralyzable: $DQE = (1 - ICR * \tau)^2 * \exp(-ICR * \tau)/1 - 2 * ICR * \tau * \exp(-ICR * \tau)$

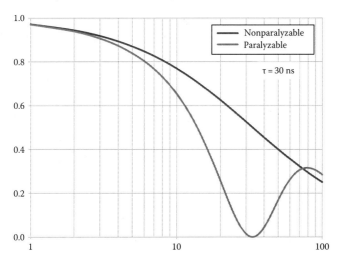

FIGURE 7.8 Plot of DQE versus ICR for dead time of 30 ns.

however, the DQE of direct conversion sensors, as a function of the ICR, has not appeared in the reviewed literature (according to the best knowledge of the authors).

The photon shot noise, propagated through the convolution algorithm in the image reconstruction, sets the noise level in the CT image. The noise at each point in the CT image is proportional to the reciprocal of the photon flux (ICR) summed over all the projection angles (see [BAR1]). For PC detectors, with flux-dependent DQE, the value 1/(DQE*ICR) at the periphery must remain smaller than the value 1/(DQE*ICR) at the center point.

In medical CT imaging of the human body ("quasi-elliptical" shapes), the largest image noise values occur in the center region of the body, as the center points have the largest beam path lengths and smallest photon flux values. The peripheral body points have higher flux levels resulting from smaller beam path lengths (versus the center points). In order to preserve the image quality (i.e., image noise level in the periphery does not exceed the image noise in the center), the reciprocal of the product (DQE*ICR) for the peripheral points must remain smaller than the (DQE*ICR) values for the center point. In current clinical practice, the ICR values for center points are less than 5 Mcps/mm^2 (Levinson, private communication). Assuming that DQE is the maximum level (= 1) at (5 Mcps/mm^2), then the condition DQE(ICR) ≥ 5/ICR

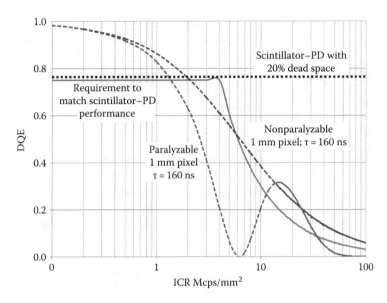

FIGURE 7.9 DQE for both paralyzable and nonparalyzable detector with a dead time (tau) of 160 ns.

for ICR > 5 maintains an image noise level, throughout the body, below the image noise in the center region, for all ICR levels.

Serendipitously, the flux dependence of DQE of PC detectors has a good match to this requirement. Figure 7.9 shows an example of the DQE for both paralyzable and nonparalyzable detector with a dead time (tau) of 160 ns. The requirement curve shows the DQE performance of scintillator–PD detector with 20% DQE reduction (due to dead space) and scan protocols with a maximum ICR of 5 Mcps/mm² at the center pixels. The overlaid curves show that even with a tau of 160 ns, the nonparalyzable configuration does not suffer significant image quality degradation (i.e., increased image noise) with high ICR. The paralyzable configuration has lower DQE value and periodic zero values and therefore requires smaller tau values to achieve equivalent DQE performance of the nonparalyzable configuration.

The analysis of the behavior of pulse-counting electronics shows that the reduced DQE at high ICR does not degrade image quality performance for suitable tau and paralyzable/nonparalyzable counting configurations. A detailed description of the behavior of the direct conversion sensors at high flux rates has not been formalized as with pulse-counting electronics. Indeed, the feasibility of utilizing PC detectors with clinical CT scanning is critically dependent on the high-flux behavior of the sensors.

All commercial medical CT systems have bowtie filters, which modify the X-ray beam profile and reduce ICR values to peripheral detectors. The BTFs, originally designed to reduce the dynamic range of ICR for energy integrating detectors, also provide patient radiation dose reduction. For PC detectors, the bowtie filters provide "much needed" reduction in peripheral ICR. Current commercial BTFs all have static configurations. Maximal ICR reduction is achieved with dynamic bowtie filters: dynamically changing X-ray beam profiles matching the patient attenuation profile from each projection angle. The ideal dynamic bowtie filter would reduce the ICR, to all peripheral detectors, to the ICR of the center region detectors, that has a maximum of 5 Mcps/mm^2. Practically, dBTFs configuration can limit the ICR to approximately 100 Mcps/mm^2 for current clinical protocols and reduce the "in-air" ICR by a factor of 10.

7.2.5 Rings (Bands)

Third-generation (rotate–rotate) scanning geometry creates high sensitivity to ring artifacts. The ring artifact results from nonstable detector behavior for adjacent (x-dimension) detector pixels. The "ring sensitivity" is a strong function of the distance of the detector pixel from the center of the detector array: there is high sensitivity for central pixels (close to the center of the detector array) and low sensitivity for edge (distal to the center of the detector array) pixels. Typically, for clinical scan scenarios, 0.1% change in the detector output, during a single-gantry rotation, produces a visible ring artifact in the center region of the image, while a 1% change produces a visible ring artifact in the edge region of the image.

Ring artifacts can be removed with software algorithms. There is a long history of ring-removal techniques since the introduction of third-generation CT systems, and they are an integral component in all medical CT systems. The software techniques are limited in the magnitude of the ring artifacts that they can remove, and therefore the detectors are required to have sufficient stability to keep the ring level within the action level of the ring-removal techniques. Due to the ring sensitivity as a function of the distance from the center of the detector array, the stability requirements also vary a function of the distance from the center of the array: the highest stability is required in the center region of the detector array.

The introduction of ASIC electronics and "silicon PD arrays" in detector systems has resulted in improved uniformity of detector response and favorably impacted the ring problem. Adjacent detector pixels, from a single ASIC and PD array, have better uniformity than detector pixels from single electronic/PD pixels. In the ASIC/PD array configuration, the ring problem is transformed to a band problem, with band borders corresponding to the edge pixels of adjacent ASIC/PD.

PC detectors have the same stability requirements as the current scintillator–PD detectors. No doubt, the stability requirements of third-generation CT systems are challenging for any detector configuration. The use of ASIC pulse-counting electronics and "blocks" of sensors in the PC detectors should have a positive effect on adjacent pixel uniformity. The sensor uniformity and, most important, the stability of the sensor output are a critical performance criteria to eliminate the ring/band artifacts. The output of semiconductor detectors is temperature dependent, and temperature control is a critical element in achieving a stable output. First indications of successful control of the ring artifact level can be seen in the phantom images from the latest article on the Siemens photo-type scanner [YU], which conforms to the requirements of clinical CT images in use today.

7.2.6 MULTIENERGY IMAGING

PC detectors are ideally suited to dual-energy and K-edge imaging. The ability, not only to count but also measure the energy of each detected photon, enables a detector-based multienergy CT system. The imaging performance of multienergy CT systems is a function of the so-called "energy separation" of the acquired data, that is, the measured difference in energy levels of the acquired data.

The energy discrimination performance of PC detectors is a function of the shape of the charge pulse (height and temporal duration) produced by the sensors and the pulse response of the counting electronics. A number of effects are known to degrade energy performance, for example, ballistic effect, spectral distortion from pileup, and crosstalk. The effect of different mechanisms on the energy response of PC detector has been calculated and measured [BOC1]. The influence of the detector pixel size on the energy performance is particularly interesting; on the one hand, a smaller pixel size reduces the pulse rate to counting electronics and reduces the spectral distortion from pileup; on the other hand, crosstalk increases with the smaller pixel size. The optimal pixel size for medical CT systems, most likely less than 1 mm^2, has yet to be determined.

Dual-energy imaging requires two energy bins. K-edge imaging requires three energy bins. Alvarez has proposed that additional energy bins, beyond two or three, improves the imaging performance [ALV1]. The number of energy bins is a property of the electronics; configurations up to four energy bins have been reported.

The competitive imaging performance of photon-counting-detector-based dual-energy systems versus tube-based dual-energy systems is of interest to the CT imaging community and has spurred interest on the part of CT vendors for the development of PC detector systems and PC based medical CT systems (see clinical section for reference to multienergy, preclinical PC imaging).

7.3 CLINICAL APPLICATIONS WITH PHOTON-COUNTING DETECTORS

CT scanning has evolved to provide imaging solutions for a wide range of diagnostic tasks. From emergency room stroke imaging to adrenal scanning for endocrinologic evaluation, CT finds use in almost every internal medicine and surgical specialty. The helical, multislice detectors technologies, introduced at the end of the 1990s, spurred annual double-digit growth in the medical CT scanning reaching over 80 million annual scans in the United States.

Recently, dual-energy CT imaging has been introduced by the major medical CT vendors. Dual-energy CT has spurred development of new CT clinical applications, further extending the range of CT in the diagnostic arena.

What can PC detectors offer to the big business of medical CT imaging?

7.3.1 K-EDGE IMAGING

Due to the intrinsic energy measurement capability of PC detectors, it is natural to explore the benefits and improvements for multienergy CT imaging with PC detectors. This question has been highlighted by the commercial introduction of dual-energy systems. The first commercial dual-energy systems were based on novel tube technologies (dual source and kVp-switching), and more recently, a detector-based dual-energy system has been introduced. As a note, these configurations are practically limited to two energy levels, as additional energy levels entail hardware modifications beyond the scope of the currently available technology.

PC detectors are built with energy sensitivity and energy measurement capabilities, and indeed, the history of PC detectors is filled with applications with energy resolution well beyond the requirements of a simple "two-bin" energy configuration as required for dual-energy imaging. Albeit these applications are at much lower flux than employed in medical CT and the energy performance is degraded as a function of flux, the lure of multienergy CT imaging is a powerful motivation for PC detectors.

The subject first appeared in the literature with the article of Roessl and Proksa [ROE1], who provided the theoretical background and simulation results for clinical CT imaging with three or more energy bins. The authors introduced a new concept of K-edge imaging: producing tomographic CT images displaying the K-edge attenuation coefficient. The images have a "positron emission tomography (PET)-like" material specificity, that is, tomographic images that display only a single material. The "specificity" is achieved by exploiting the K-edge attenuation of various materials. While the K-edge of human tissue is outside the energy range of medical CT scanners, there are many materials with high atomic number Z that have K-edge attenuation peaks in the medical X-range, and serendipitously, such a material (gadolinium; K-edge is 50.3 keV) is already in widespread clinical use as a contrast agent. The authors show a simulation of a cardiac scan with gadolinium contrast in which complete specificity is achieved and remarkably good SNR with gadolinium contrast concentrations typically used in clinical imaging (Figure 7.10). Figure 7.10d shows the K-edge gadolinium image with excellent specificity and sensitivity.

The use of the K-edge image can be enhanced by overlaying the K-edge image on the conventional CT image. This is the same technique employed with PET-CT; however, in the case of the K-edge CT/conventional CT overlay, the two images are produced with the exact same projection data, and there is exact spatial registration of the two images, which is a significant advantage versus other hybrid (composite) imaging techniques, PET-CT and single-photon emission computed tomography (SPECT)-CT, that require two separate scanning acquisitions and suffer from patient motion and image misregistration in the overlay process.

The imaging specificity of the K-edge technique suggests its use with contrast agents that have tissue specificity, that is, the targeting capability of the contrast agent. The combination of contrast material with a known target tissue (i.e., tumor cells) displayed in material-specific image (K-edge) should produce an image with high specificity for diagnostic tasks (i.e., cancer detection). To date, CT imaging is practiced with blood pool contrast agent, without any targeting functionality. The clinical uses for blood pool agents are varied: angiography, tissue characterization with perfusion, and differential parenchymal enhancement. However, there is no direct targeting with blood pool agents.

As is implemented with PET and SPECT applications, the exploitation of the K-edge technique requires a targeted contrast medium. The requirement of the contrast material is a K-edge in the diagnostic medical X-ray range (50–120 keV). The

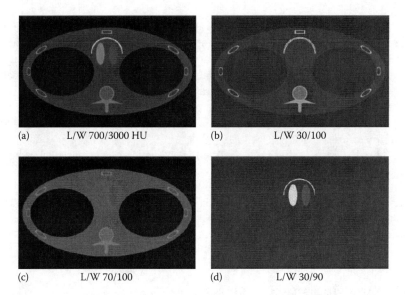

(a) L/W 700/3000 HU (b) L/W 30/100
(c) L/W 70/100 (d) L/W 30/90

FIGURE 7.10 Conventional polychromatic reconstruction of the phantom (a) and reconstructions after spectral decomposition corresponding to a photo-effect basis image (b), a compton-effect basis image (c) and a gadolinium basis image (d). Level and window values are given in Hounsfield units (a) or refer to the minimum and maximum in the images (b)–(d).

candidate materials are typically metals with a high atomic number and high physical density. Materials that have been mentioned in the literature are gadolinium, tantalum, bismuth, tungsten, and gold ([RAB1, CLA1, BON1]). These materials also have high X-ray attenuation, and therefore the contrast media also produces high contrast in conventional, single-energy CT images.

The process of combining a high atomic number material to a targeting entity is the subject of research in many chemical and pharmaceutical laboratories. Metal nanoparticles, with the required high atomic number component, have become a significant area of investigation. Among the metal nanoparticles, gold nanoparticles (GNPs) have attracted special attention. Other than the high atomic number of gold, GNPs have unique optical properties that make them appealing also for optical imaging and for photothermal therapy [SPE1]. The wide applications of GNPs and their potential for clinical implementation have led to substantial research regarding their in vivo chemical stability, pharmacokinetics, biodistribution, and biotoxicity. The well-known biosafety of gold since the 1950s, along with the high degree of flexibility in terms of particles' size, shape, and functional groups for coating and targeting, provides the GNPs with potential to become important CT contrast agents. Tantalum is another option for CT contrast agent due to its bioinertness and relative nontoxicity. In addition, tantalum is much cheaper in comparison to gold.

The K-edge imaging technique has an additional unique imaging capability in differentiating between different contrast media in a single scan acquisition. For example, the simultaneous application of a blood pool contrast agent and a targeted agent can deliver both an angiographic map with a detailed picture of stenotic structures. This capability was reported in a preclinical mice study imaging vulnerable plaque. Gold high-density lipoprotein nanoparticles (Au-HDL) were the targeted contrast agent. The targeted nanoparticles induced CT contrast enhancement specifically in macrophage-rich, rupture-prone plaques, while no significant enhancement was observed for stable plaques that are not rich in macrophages. In addition, iodine-based blood pool contrast agent provided a map of the arteries. The resultant image enables differentiation of the Au-HDL and the iodine-based contrast for a full picture of the rupture-prone (vulnerable) plaques with their location in the arterial tree.

Another K-edge imaging study reported visualization of intravascular pathologic epitopes with fibrin-targeted bismuth nanoparticles in rabbit models of atherosclerosis. Indeed, it is possible with K-edge imaging to have a simultaneous tracking of two differently labeled cell populations and a study of their interactions or the interaction of cells with specific receptors while each cell population retains its own unique label.

The clinical impact of K-edge imaging will await the regulatory approval of the K-edge contrast agents.

7.3.2 Improved Spatial Resolution

PC detectors are direct conversion detectors with no light-scintillating crystal. The detectors are made from sensor "chunks," and the detector pixels are created by anode structures attached to the sensor surface. This technology enables a packing advantage of the PC detectors versus the scintillator crystal–PD detectors, which require reflector walls to isolate (and prevent crosstalk) between the detector pixels in the scintillation crystal. As the detector pixel size is decreased, the relative area of the reflector walls increases, and the dead space of the sensor increases. Practically, the reflector walls are on the order of 0.1 mm, resulting in 20% dead space at 1 mm pitch and increasing to 40% dead space at 0.5 mm pitch. The PC detectors do not have any "interpixel" dead space issues. So, this is an opportunity for improved spatial resolution versus scintillator–PD detector, without the price of reduced geometric efficiency.

The interpixel dead space of scintillator–PD detector is used for the placement of the antiscatter grid walls. The grid walls overlap the interpixel dead space. This is current practice with 1×1 mm detector pixels. For submillimeter detector pixels, there is no need for grid walls with submillimeter pitch. Therefore, for 0.5 mm detector size with 1 mm grid pitch, there is a 20% gain in geometric efficiency for PC detectors versus scintillator–PD detectors.

The spatial resolution of the CT system is dependent also on the X-ray focal spot size and the imaging geometry (magnification factor). In order to maximize an improvement in the spatial resolution, for a constant magnification factor, the focal spot size must also be reduced versus focal spots sizes currently in use.

Finally, the noise in the tomographic images increases as the cube of the linear spatial resolution. It is doubtful that radiologists will accept an eightfold increase in noise for a two fold improvement in spatial resolution. However, with the recent introduction of reconstruction-based noise reduction techniques, there may be room for a gain in spatial resolution with acceptable increase in image noise.

The premium clinical CT market currently features $\sim 1 \times 1$ mm detector pixel sizes delivering $\sim 0.6 \times 0.6$ mm image pixel resolution. For most clinical brain and body examinations, the displayed pixel resolution is on the order of 0.8 mm with slice thicknesses varying from high resolution (0.6 mm) to standard (2–3 mm) and low resolution (5 mm).

A 50% reduction in detector pixel size to 0.5 × 0.5 mm with a concurrent reduction in the X-ray focal spot size (from 0.6 × 0.6 mm to 0.3 × 0.3 mm) and an increase in the display matrix (from 512^2 to 1024^2) can improve the displayed image resolution limit to 0.3 × 0.3 mm. Does this improved spatial resolution bring additional diagnostic capabilities to clinical CT imaging?

An interesting example is pancreatic imaging. The table shows the timeline of the technology advances and the resulting improvement in organ visualization.

	Scan Technique	Resolution	Scan Time	Dose	What Do We See?
1990	Single slice step and shoot	1.2 mm × 1.2 mm @SW = 5 mm	5 min	10 mSv	Pancreas – YES Main duct – NO Pancreas Lesions - NO Communicating* - NO
2010	64 slice helical	0.8 mm × 0.8 mm @SW = 3 mm	5 s	15 mSv	Pancreas – YES Main duct – YES Pancreas Lesions - YES Communicating* - NO
2020	128 slice helical photon-counting iterative recon	0.4 mm × 0.4 mm @SW = 1 mm	2 s	20 mSv	Pancreas – YES Main duct – YES Pancreas Lesions - YES Communicating* - YES

Communicating*: pancreatic lesion communicating with the main duct.

The introduction of multislice detector helical scanning "opened up" the internal anatomy of the pancreas: the main duct, cystic, and malignant lesions became visible in CT. New contrast injection techniques were developed to optimize image quality and lesion conspicuity. Current diagnostic dilemmas include characterization of pancreatic cysts/lesions, one of the factors being a communicating duct between a lesion and the main duct of pancreas. Currently, the communicating ducts are not routinely visible on CT scans. In the clinical arena, different modalities bring varied imaging capabilities to each diagnostic task. Currently, magnetic resonance cholangiopancreatography provides excellent images of the biliary and pancreatic ductal systems. The need and utility for improved imaging capability, in each modality, must be evaluated in a landscape of competing imaging modalities. Improved spatial resolution in CT can bring added diagnostic utility to pancreatic evaluation and propel CT to the leading modality for this organ's evaluation.

Evaluation of lung nodules has developed using high-resolution CT imaging. At the end of 2013, the USTFS issued a recommendation of CT lung screening for high-risk population groups (heavy smokers), and in November, 2014, the Centers for Medicare and Medicaid Services approved reimbursement for CT lung screening for heavy smokers. The nodule evaluation is based on a measurement of the nodule size and tracking the nodule size over time. Automated methods to measure the nodule volume are available from all major CT vendors. Nodules of interest are in the range of 5 mm diameter. A 1 mm uncertainty in the diameter measure of a 5 mm nodule introduces a 50% uncertainty in the volume measure. The accuracy of the diameter measure is a strong function of the "edge detection" of the volume calculation, and the edge detection is strongly dependent on the spatial resolution of the scanner system. Prof. David Yankelevitz has noted that "...higher resolution will allow for the improvement in our ability to measure accurately. The most obvious area where this is of benefit is in the volumetric measurements of pulmonary nodules (and ultimately any volumetric measurement). Higher resolution provides for better measurements with less error and therefore the ability to determine (nodule) change in shorter time intervals and with greater confidence." Improved spatial resolution in the CT scanning system will be a positive impact on the accuracy and efficiency of lung nodule evaluation and early lung cancer detection.

7.3.3 Dose Reduction

Since the publication of the NCRP Report No. 160 in 2006 and the identification of CT imaging as the largest contributor from all the X-ray imaging techniques to the ionizing radiation burden of the U.S. population, there has been a concerted effort in the radiology community to decrease the radiation dose from CT imaging examinations.

As reviewed in the detector section, there are properties of PC that indicate a potential for dose reduction versus the scintillator–PD detectors. However, the dose reduction shown in the last few years from the new reconstruction methods (iterative methods) is significantly greater than dose reduction values projected for PC. Breakthroughs in dose reduction for most CT scanning applications are, at this moment, in the hands of the mathematicians and computer scientists.

For lung screening applications, the situation is different. Ten years ago, low-dose CT was introduced to lung examinations for nodule detection. A fivefold reduction in dose was achieved (tube technique was reduced from 250 to 50 mAs) as radiologists proved that a lower mAs rating was equally effective as the standard mAs technique for detection and evaluation of lung nodules (the validation was done without any of the new iterative reconstruction techniques). With the 50 mAs technique, the scintillator–PD detectors are operating at photon flux signal levels close to electronic noise signal levels. Any further reduction in the flux levels would result in acquired data dominated by the detector electronic noise.

The iterative dose reduction has not been successful, to date, in reducing the image noise and consequently enabling a dose reduction for scanning techniques with both the photonic and electronic noise components. The PC detectors have been shown to have zero-electronic noise. Therefore, further dose reduction for lung screening and possibly other CT screening techniques (virtual colonoscopy) may be possible with PC detectors.

REFERENCES

[ALV1] R.E. Alvarez, *Med. Phys.*, 40, 111909, 2013.

[BAR1] H.H. Barrett and W. Swindell, *Radiological Imaging*, Academic Press, New York, 1981, p. 532.

[BLE06] I.M. Blevis, Analysis of action of guard ring on monolithic CZT, *IEEE Room Temperature Semiconductor Detectors*, San Diego, CA, 2006.

[BLE05] I.M. Blevis, US7514692, Method and apparatus for reducing polarization within an imaging device.

[BLE08] I.M. Blevis, L. Volokh, L. Tsukerman, J.W. Hugg, F.P. Jansen, J.P. Bouhnik, CZT gamma camera with pinhole collimator: Spectral measurements, *IEEE Medical Imaging Conference*, Dresden, Germany, 2008.

[BLEV08] I.M. Blevis, J.P. Bouhnik, A Cohen measurements of dark current in CZT with variable flux, *IEEE Room Temperature Semiconductor Detectors*, Dresden, Germany, 2008.

[BOC1] M. Bocher, I.M. Blevis, L. Tsukerman, Y. Shrem G. Kovalski, L. Volokh, *Eur. J. Nucl. Med. Mol. Imaging*, 37, 1887–1902, 2010.

[BON1] Bonitatibus et al., Preclinical assessment of a zwitterionic tantalum oxide nanoparticle X-ray contrast agent, *ACS Nano*, 6(8), 6650–6658, 2012.

[BOU10] M. Bocher, I.M. Blevis, L. Tsukerman, Y. Shrem, G. Kovalski, L. Volokh, A fast cardiac gamma camera with dynamic SPECT capabilities: Design, system validation and future potential, *Eur. J. Nucl. Med.*, 37(10), 1887–1902, 2010.

[COL99] A. Cola, I. Farella, The polarization mechanism in CdTe Schottky detectors, *Appl. Phys. Lett.*, 94, 102113, 2009.

[CUN99] I.A. Cunningham, R. Shaw, Signal-to-noise optimization of medical imaging systems, *J. Opt. Soc. Am. A*, 16(3), 621–632, March 1999.

[GIN1] A. Ginzburg and C.E. Dick, *Med. Phys.*, 20(4), 1013, 1993.

[HAG95] M. Hage-Ali, P. Siffert, CdTe nuclear detectors and applications. In *Semiconductors for Room Temperature Nuclear Detector Applications*, Schlesinger, T.E., James, R.B., eds. Academic Press, San Diego, CA, 1995, Vol. 43, pp. 291–331.

[HE00] Z. He, Review of the Shockley–Ramo theorem and its application in semiconductor gamma-ray detectors, *Nucl. Instrum. Methods Phys. Res. A*, 463, 250–267, 2001.

[JAM95] R.B. James, T.E. Schelsinger, J.C. Lund, M. Scheiber, $Cd_{1-x}Zn_xTe$ spectrometers for gamma and X-ray applications. In *Semiconductors for Room Temperature Nuclear Detector Applications*, Schlesinger, T.E., James, R.B., eds. Academic Press, San Diego, CA, 1995, Vol. 43, pp. 336–378.

[KAP] S. Kappler, Photon counting CT at elevated X-ray tube currents: Contrast stability, image noise and multi-energy performance, *Proc. SPIE*, 9033, 90331C, 2014.

[LEV1] R. Levinson, *Clinical Use of Photon Counting Detectors in CT*, SPIE Medical Imaging Conference, Lake Buena Vista, FL, February 7–12, 2009.

[RAB1] O. Rabin et al., An X-ray computed tomography imaging agent based on long-circulating bismuth sulphide nanoparticles, Vol. 5, *Nat. Mater.*, 5, 118–122, 2006.

[ROE1] E. Roessl and R. Proksa, *Phys. Med. Biol.*, 52, 4679–4696, 2007.

[SPE1] Sperling et al., Biological applications of gold nanoparticles, *Chem. Soc. Rev.*, 37, 1896–1908, 2008.

[VEA14] M.C. Veale et al., Investigating the suitability of GaAs:Cr material for high flux X-ray imaging, *International Workshop on Radiation Imaging Detectors*, Trieste, Italy, June 2014.

[YU] Z. Yu, Initial results from a prototype whole-body photon-counting computed tomography system, *Proc. SPIE*, 9412, 94120W, 2015.

8 Radiation Detection in SPECT and PET

Chin-Tu Chen and Chien-Min Kao

CONTENTS

8.1 Introduction 194
    8.1.1 Historic Perspectives 194
    8.1.2 Early Developments of SPECT 195
    8.1.3 Early Developments of PET 196
8.2 Overview 197
    8.2.1 Overview of SPECT 197
    8.2.2 Overview of PET 198
8.3 Current Technologies in Routine Uses 199
    8.3.1 Scintillation Detectors and Pulse-Height Analysis 199
    8.3.2 SPECT 200
        8.3.2.1 Gamma Cameras 200
        8.3.2.2 Anger Position Logic 201
        8.3.2.3 Collimators 202
    8.3.3 PET 202
        8.3.3.1 Block PET Detector Modules 202
        8.3.3.2 Coincidence Detection 204
        8.3.3.3 Factors Affecting Spatial Resolution 205
        8.3.3.4 Considerations for the Scintillator 208
8.4 Recent Advances 209
    8.4.1 SPECT 209
        8.4.1.1 Scintillators 209
        8.4.1.2 Semiconductor Detectors 210
        8.4.1.3 Photodetectors 212
    8.4.2 PET 213
        8.4.2.1 DOI Detectors 213
        8.4.2.2 TOF Detectors 215
        8.4.2.3 Silicon Photomultipliers 216
        8.4.2.4 Novel Readout Methods 216
        8.4.2.5 Waveform Sampling 217
8.5 Future Trends and Summary 219
    8.5.1 Modular and Reconfigurable Detectors 219
    8.5.2 Application-Specific Imaging Systems 219

8.5.3	Multimodality Imaging Systems	220
8.5.4	SPECT and PET in Image-Guided Therapy	220
8.5.5	Summary	221
References		221

8.1 INTRODUCTION

The 1903 Nobel Prize in Physics was awarded to Henri Becquerel for his discovery of spontaneous radioactivity (Becquerel 1896) and to Pierre and Marie Curie for their research on radiation phenomena leading to the discovery of radium and polonium (Currie et al. 1898). The 1943 Nobel Prize in Chemistry was awarded to George de Hevesy for his long-time work on the use of isotopes as tracers that he first proposed three decades earlier (de Hevesy 1913). These two monumental achievements in science paved the way to the creation and expansion of the field of nuclear medicine, which has benefitted millions in health and disease.

Nuclear medicine imaging consists of planar scintigraphy (2D gamma-ray projection imaging) and emission computed tomography (ECT, 3D tomographic imaging). ECT includes two modalities: single-photon emission computed tomography (SPECT) and positron emission tomography (PET). This chapter focuses on radiation detection in SPECT and PET, including a brief review of the major historic developments, followed by discussions on current state-of-the-art technologies in routine clinical uses, recent advances, and future trends. Note that other aspects of SPECT and PET, for example, image reconstruction, quantitative imaging, radiotracers, or the clinical and research applications, are not covered in this chapter. Information regarding these topics can be found in many textbooks and review articles (e.g., Cherry et al. 2003/2012; Wernick and Aarsvold 2004).

8.1.1 HISTORIC PERSPECTIVES

Early radiation detection devices included the cloud chamber (Wilson 1911, for which he was awarded the 1927 Nobel Prize in Physics) and Geiger counter; first as the Rutherford–Geiger tube (Rutherford and Geiger 1908), and then later as the Geiger–Muller (G–M) tube (Geiger and Muller 1928). Radiation detectors were employed, fairly early on, in both biological experiments (e.g., de Hevesy 1923) and in vivo human studies (e.g., Blumgart and Weiss 1927; Blumgart and Yens 1927). Rectilinear scanners for routine preclinical and clinical uses were built with calcium tungstate scintillation crystals coupled to the window of photomultiplier tubes (PMTs) (Cassen et al. 1950, 1951), which were employed for imaging of patient's thyroid with the use of I-131 as routine clinical procedures (Allen et al. 1951). A gamma-ray pinhole camera using a thallium-activated sodium iodide, NaI(Tl), scintillation crystal and a photographic plate was developed and used for imaging of metastatic tumor of thyroid cancer (Anger 1952). These devices were 100-fold more sensitive than the G–M counter, making routine clinical imaging applications practically feasible.

The major breakthrough came when gamma camera was developed with a NaI(Tl) scintillation crystal and multiple PMTs in conjunction with an X–Y position-determining electronic circuit, first using a pinhole collimator (Anger 1958) and then later using a parallel-hole collimator (Anger 1964), which practically has been the standard configuration of modern gamma cameras used in both planar scintigraphy and SPECT for more than the past five decades. The significance of these scintillation cameras, often referred to as *Anger* cameras, was their relatively large field of view (FOV) that allows for imaging of entire organs without rectilinear scanning, making substantial improvements in gamma-ray detection efficiency in routine uses and clinical practice.

8.1.2 Early Developments of SPECT

Tomographic gamma-ray imaging later evolved in two parallel paths: transaxial section tomography and longitudinal (focal-plane) tomography, both based on NaI(Tl) scintillation crystals and PMTs arranged in various configurations with different scanning strategies. Even though longitudinal tomography succeeded first commercially as a clinical imaging product, transaxial tomography eventually took over as the standard clinical SPECT systems.

Longitudinal tomography, which is in essence limited-angle tomography, utilized focused collimators to select a particular plane of interest to image while also capturing blurred images from other out-of-focus planes (Crandall and Cassen 1966). The successful commercial *PHO-CON* scanner was based on a longitudinal tomographic design to produce 6 slices by using a single gamma camera and 12 slices with dual-head cameras (Anger 1969). Another longitudinal tomographic design utilized a rotating slant-hole collimator to provide limited-angle sampling (Muehllehner 1970). These techniques have been overshadowed by the transaxial tomography in the last three decades. However, with recent advances in applying the concept of *tomosynthesis* and the related computing algorithms for limited-angle tomography, this type of emission tomography may find its way to return to clinics in the future.

Kuhl et al. pioneered the concept of both longitudinal and transaxial tomographies and developed several generations of scanners from the early 1960s to the mid-1970s using discrete detectors (e.g., Kuhl and Edwards 1963, 1964; Kuhl et al. 1976). In the same time frame, several other multiple-detector-based transaxial tomographic scanners were also reported (e.g., Patton et al. 1969; Bowley et al. 1973) with somewhat different configurations and/or scanning trajectories. Investigation of using Anger cameras in transaxial tomography began in the early 1960s (Harper et al. 1965), with a few incorporating a rotating chair for the patient in conjunction with a stationary camera (e.g., Muehllehner 1971; Budinger and Gullberg 1974).

These early SPECT developments advanced this relatively new field significantly. However, the rotating-camera approach (Jaszczak et al. 1977; Keyes et al. 1977) was the seminal development that led to the broader uses of SPECT in routine clinical practices (Murphy et al. 1978). Multiple-camera (especially dual-head or triple-head) whole-body SPECT systems finally became the norm in nuclear medicine clinics (Jaszczak et al. 1979). Throughout the 1980s, new or improved approaches for developing rotating-camera SPECT systems and the related image reconstruction and processing methods were reported (e.g., Larsson 1980; Tanaka et al. 1984). The development of stationary SPECT systems using more complete-sampling configurations without rotating the patients also made progresses (Genna and Smith 1988; Rogers et al. 1988), especially for specific applications such as cardiac or brain imaging.

8.1.3 Early Developments of PET

Positron-emitting radioisotopes were suggested for use in locating brain tumors in the early 1950s (Wrenn et al. 1951). The first successful positron imaging device, consisting of a pair of NaI(Tl) detectors, was soon designed and built to reveal brain tumor location and size (Brownell and Sweet 1953). The concept of positron camera was also proposed (Anger and Rosental 1959). The first transverse sectional scanner for positron imaging was composed of 32 NaI(Tl) crystals arranged in a circular geometry (Ranhowitz et al. 1962; Robertson and Niell 1962).

It was not until the early 1970s, probably stimulated by the success of x-ray computed tomography (CT), that the development of PET flourished in many different design approaches. PC-I and PC-II used two opposing banks of multiple NaI(Tl) detector arrays (Brownell and Burnham 1973; Brownell et al. 1977), while positron emission transaxial tomography III (PETT-III) employed 48 NaI(Tl) crystals in a hexagonal array (Phelps et al. 1975; Ter-Pogossian et al. 1975). The use of a multiwire proportional chamber (MWPC) was also explored for positron imaging (Lim et al. 1975). A high-resolution single-slice system was constructed with the use of a fixed ring array of 280 closely packed NaI(Tl) crystals (Derenzo et al. 1975, 1977; Budinger et al. 1977). A system using two large-field-of-view (LFOV) Anger gamma cameras was also developed (Muehllehner et al. 1977). PETT-IV and PETT-V, the multislice versions of PETT-III for body and brain imaging, respectively, were built (Ter-Pogossian 1977; Mullani et al. 1978), while ECAT-II was developed by the use of 66 NaI(Tl) detectors (Phelps et al. 1978). Note that except for the MWPC-based positron imaging system, all these PET devices utilized NaI(Tl) crystals.

In 1977, bismuth germinate (BGO) was suggested for use in PET to improve detection efficiency because of its high density (Cho and Farukhi 1977; Derenzo 1977). POSITOME II was the first PET system developed by using BGO crystals (Thompson et al. 1979), which led the PET instrumentation into another era including the development of Neuro-PET (Brooks et al. 1980), Donner 280-BGO-Crystal Tomograph (Derenzo et al. 1981), and ECAT-III (Hoffman et al. 1983). Two other scintillation crystals with relatively fast decay time were also explored for use in PET instrumentation: cesium fluoride (CsF) and barium fluoride (BaF_2). CsF was used in building several PETT-VI (Mullani et al. 1980; Ter-Pogossian et al. 1982) systems for brain imaging. These two fast scintillators also motivated the development of the first-generation time-of-flight (TOF) PET systems including SUPER PETT-I (Mullani et al. 1980a; Ter-Pogossian et al. 1981), TOF PET (Mullani et al. 1982), and LETI TOF PET (Gariod et al. 1982).

More information regarding the historic perspectives and early developments of SPECT and PET can be found in several textbooks (e.g., Wernick and Aarsvold 2004; Cherry et al. 2012) and many review papers (e.g., Nutt 2002; Jaszczak 2006; Muehllehner and Karp 2006; Hutton 2014).

8.2 OVERVIEW

8.2.1 Overview of SPECT

Single-photon imaging employs radiotracers labeled by gamma-ray (single-photon) emitting radioisotopes, introduced into live animals or human objects, to follow in vivo function and physiology associated with the specific radiotracer in use. It includes planar scintigraphy, in which 3D radioactivity distribution within the object under investigation is projected onto 2D planar images, and SPECT, which produces 3D volumetric image data. These two modalities are routine clinical nuclear medicine imaging techniques that are also utilized in many biomedical research investigations. Currently, planar scintigraphy is used only in a very few specific application areas such as thyroid imaging or breast imaging. 3D SPECT, in which projection views of the 3D radioactivity distribution are acquired at various projection angles, constitutes the large majority of the uses of single-photon imaging technology in both clinical and research applications. Therefore, in this chapter, we focus mainly on SPECT.

SPECT is considered a functional or molecular imaging method revealing physiological characteristics and/or molecular pathway and signatures that the specific radiotracer in use is designed to probe, which is in contrast to the anatomical imaging methods that reveal the structural architecture of the organism—including CT and magnetic resonance imaging (MRI).The most commonly used single-photon emitting radioisotope is technetium-99m (Tc-99m), which has a half-life of approximately 6 h and decays by emitting primarily 140 keV gamma rays. Other single-photon radioisotopes used in nuclear medicine clinics include thallium-201 (Tl-201, with a half-life of 73 h and gamma rays of 70–80, 135, and 167 keV), iodine-123 (I-123, 13.2 h, and 159 keV), indium-111 (In-111, 2.8 days and 171 and 245 keV), and gallium-67 (Ga-67, 3.25 days and 93–393 keV). Radiotracers labeled with these and other single-photon emitters have been used routinely in imaging of the brain, heart, bone, lung, liver, prostate, and other organs and tissues.

SPECT imaging detector systems capture gamma rays emitted from the single-photon radiotracers in use and convert the radiation to photonic signals and then, in turn, electrical signals to be processed in order to form the final images of the radioactivity distribution. Since gamma rays emit, in general, isotropically, and they impinge upon scintillator detectors also from all directions, a detected event in the detector without other design considerations would not necessarily correspond to a unique gamma-ray traveling path or direction that can lead to the definition or estimate of the origin of the radiation decay event. In order to better establish the relationship between the detected signals and their original locations of radiation decay, physical collimators with holes or channels only allowing passage of gamma rays traveling in specific directions are usually designed and used. Physical collimation, even though it defines and provides the needed spatial resolution, reduces the system sensitivity significantly because only those photons not stopped by the collimator materials and passing through the holes and open channels would have the chance to be detected; the majority of the gamma rays are absorbed and stopped by the collimator materials, which are usually high Z materials with relatively high attenuation coefficients for absorbing gamma rays.

A typical SPECT imaging system consists of one or more gamma cameras, usually with relatively large FOV. A gantry capable of rotating the gamma cameras around the objects under study, as well as a scanning bed, provides the necessary angular and axial sampling mechanism for the 3D tomographic imaging.

8.2.2 Overview of PET

PET was first introduced in the 1950s as stated earlier. Like SPECT, PET is also regarded as a noninvasive functional and molecular imaging technology for measuring the spatial, or spatiotemporal, distribution of certain positron-emitting compounds inside a live organism to reveal useful information about a certain function or disease of the organ. Because carbon-11 (C-11), nitrogen-13 (N-13), oxygen-15 (O-15), and fluorine-18 (F-18) are all positron-emitting isotopes, one potential advantage of PET is the possibility to label a great many organic molecules relevant to life and life processes for studying a variety of normal or disease biology. Another advantage of PET is its quantification capability due to its superior sensitivity to other functional imaging modalities including SPECT and the tractability of the mathematical problem of reconstructing the unknown radioactivity distribution from the measurements, which is somewhat simpler and more solvable than that of SPECT.

PET was widely used for studying brain functions before functional MRI becomes the method of choice for that task. Currently in the clinic, PET is most often used in conjunction with CT, by employing a multimodality PET/CT system, for cancer imaging with fluorodeoxyglucose (FDG), which is a glucose analog labeled by F-18. A high uptake of FDG can depict abnormal glucose metabolism associated with the increased aerobic glycolysis in cancer cells, and it is possible to detect cancer with PET before there are observable structural changes on CT or MRI images. The utility of PET for evaluating the outcome of cancer treatment has been well documented as well. With the advent of molecular biology and genetics, the usefulness of PET imaging for studying in vivo the molecular basis of cell functions and diseases, as well as gene expression and translation, and for developing therapies that cure disease at the molecular level has also been demonstrated. As a result, the potential role of PET imaging in personalized medicine, in which a disease is treated based on the patient's particular genetic makeup and physiology with quantitative evaluation of treatment outcome to optimize the treatment strategy, has been widely recognized.

As the potential of PET in research, diagnosis, and treatment depends critically on the PET agents, there are substantial activities in developing novel PET agents, including cell- and nanoparticle-based agents, to provide high sensitivity and specificity and/or theranostic functions. Many also believe that PET can significantly improve drug discovery and development. Nowadays, dedicated PET systems for imaging small animals such as rodents are also widely employed in preclinical and translational research.

PET imaging is based on coincidence detection of two 511 keV gamma-ray photons that are generated when a positron released by the PET isotope annihilates with an electron. As these two annihilation photons travel in almost opposite directions, the annihilation shall take place on the line connecting the positions where they are detected, which is often referred to as the line of response (LOR). Conversely, the rate of coincidence detection observed on a particular LOR is proportional to, if ignoring attenuation of the gamma-ray photons by the subject for the present discussion, the rate of annihilations occurring on the LOR, which in turn is proportional to the sum of the radioactivity on the LOR. Therefore, PET measurement provides ray sums of the 3D radioactivity distribution (the distribution of the PET agent in the 3D space) on a set of LORs that are determined by the geometrical configuration of the PET system. Modern PET systems often employ multiple rings of small detectors surrounding the subject to provide a large number of LORs through the subject.

8.3 CURRENT TECHNOLOGIES IN ROUTINE USES

This section provides brief discussions on the main radiation detection technologies currently used for routine SPECT and PET imaging in clinical and research applications. More detailed information can be found in the relevant textbooks and their references (see, e.g., Wernick and Aarsvold 2004; Cherry et al. 2012).

8.3.1 Scintillation Detectors and Pulse-Height Analysis

Currently, most gamma-ray detectors used in either SPECT or PET are based on inorganic scintillators and PMTs, and the principle of their operation is depicted in Figure 8.1. Scintillator is a material that converts the entire energy or a portion of the energy of the gamma-ray photon it interacts with into visible lights. Typically, the emission of these lights increases quickly to reach a peak and then decays exponentially. Some materials can also have multiple decay components. The light photons are detected and converted to electrical charges by a PMT with a high gain of about 10^6, producing a charge pulse reminiscent of the scintillation light pulse. From the charge pulse, the time when it appears is determined by using a constant fraction discriminator (CFD) and time-to-digital converter (TDC). The charge pulse is also integrated by using a shaper and digitized by an analog-to-digital converter (ADC), to obtain the pulse height, which is related to the amount of gamma-ray energy deposited in the scintillator. In commercial systems, these electronic operations are implemented by application-specific integrated circuit (ASIC). The operation of radiation detectors often relies on a proportional relationship between the pulse height and the gamma-ray energy deposited in the detector, although for some scintillation materials, nonproportional responses are sometimes observed. The pulse time is also determined on the summed pulse.

The pulse-height spectrum is a histogram of the detected events (counts) as a function of the detected energy. Ideally, one would expect that the relative counts are proportional to their relative abundance in the decay scheme at the specific gamma-ray

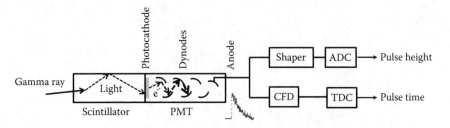

FIGURE 8.1 Detection of gamma-ray photons by using scintillator and PMT.

energy levels of the radioisotope in use, forming a set of sharp lines referred as *photopeaks*, and no counts at any other energy levels. However, since the scintillation detector has only limited *energy resolution*, the observed pulse-height spectrum is usually a blurred version of the idea version of only sharp lines. Considering also gamma rays only deposited part of their energy inside the scintillator via the Compton scattering, there usually is a continuum of detected events of lower energies, which is also subject to the blurring by the limited energy resolution of the detector. Furthermore, including the scattered photons emitted from the object under study that are also detected within the scintillator, the actual measured pulse-height spectrum of a radioisotope inside an object is often a broadband continuum of detected counts with blurred photopeaks corresponding to the gamma-ray energies characteristic to the radioisotope in use.

Those events associated with primary photopeaks contain the most useful information in imaging, while the scattered events are considered less important or even as noise or artifacts in certain circumstances. Therefore, in practice, a *window* setting usually centered around the photopeaks is applied to the pulse-height spectrum to accept the desirable primary photons and also to exclude the unwanted scatter counts.

Evidently, the detection scheme described earlier assumes that there is only one incoming gamma-ray photon interacting with the entire detector within the time frame that the output is taken. If there are multiple interactions by multiple incoming photons, some of these photons can be lost or erroneously registered as a single interaction with incorrect pulse height and pulse time generated with imprecisions. Such events are known as *pileups*. Also, the electronic processing time for a pulse can be considerably longer than the duration of the pulse. Pulses that occur during the processing time of the preceding one will be ignored. This loss of events, called *dead-time loss*, can be described by two mathematical models: nonparalyzable and paralyzable. In the nonparalyzable model, a pulse occurring during the processing time of a preceding pulse is simply ignored and it has no effect on subsequent pulses. In the paralyzable model, a pulse will introduce an addition processing time regardless of whether it is actually counted. The dead-time model for a real imaging system is more complicated, typically containing several components and showing mixed behavior of the paralyzable and nonparalyzable models.

8.3.2 SPECT

8.3.2.1 Gamma Cameras

Modern gamma cameras are, in essence, still very similar to those first proposed in the 1950s by Anger as described earlier. Figure 8.2 illustrates the major components of such Anger camera, which includes a relatively large-area continuous NaI(Tl) crystal, an array of PMTs, the collimator, and electronics for pulse-height analysis, detected event position determination, and other signal processing, which are output and connected to a computer and display.

The thickness of NaI(Tl) used in the typical gamma cameras ranges from approximately 6 to 12.5 mm. While the thicker crystals offer higher system sensitivity, the thinner crystals provide better spatial resolution. Since NaI(Tl) is hygroscopic, the scintillator is usually sealed hermetically in a special housing case. The backside of the NaI(Tl) crystal is attached to the PMT array, often via a light guide layer.

Radiation Detection in SPECT and PET

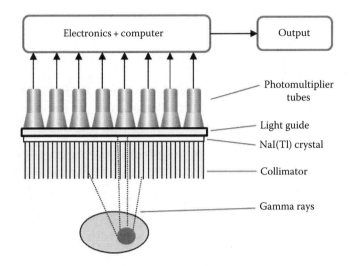

FIGURE 8.2 Major components of an Anger gamma camera.

Both rectangular and circular crystals have been employed in gamma cameras, with the latest ones more in the rectangular configuration. In general, 7 to more than 100 PMTs, mostly round and square in shape, are used to form the PMT array.

One or more gamma cameras are typically mounted on a gantry, which can rotate around the object under imaging to acquire data at various angular positions. Single-head SPECT systems have a simpler and more flexible configuration and scanning trajectory, but with inferior system sensitivity. Dual- and triple-head SPECT systems have more restrictive configuration and scanning geometry, but with marked increase of system sensitivity. Nonrotating, stationary ring-based SPECT systems, with the use of many smaller detector modules, have also been developed and used for dynamic imaging studies such as those involving cardiac functions.

8.3.2.2 Anger Position Logic

One of the key features in the original Anger scintillation camera design was the position logic circuit, which employed a resistive network to *code* the position of each PMT and to provide the *weighted* outputs of each detected event so that both the position and energy deposit of that event could be calculated. This Anger position logic and its associated circuit designs had long been utilized in analog gamma cameras. In modern digital cameras, the outputs of PMTs are digitized and fed into the computer to determine the information relevant to event position and deposited energy.

However, this Anger position logic assumes that the scintillation camera responds to radioactive source completely linearly across the entire face of the detector, which is not the case in reality. For example, the light collection efficiency near the center of a PMT is in general better than that toward its edge; also, when a radioactive source is moving near the edge of the scintillator, the photon detection tends to respond differently compared to the case when the source is within the inner circle toward the center. These nonlinear responses can produce *pincushion* or *barrel* distortions in the image, which need to be corrected for in the signal processing steps.

8.3.2.3 Collimators

As stated earlier, physical collimators are required to define and provide the spatial resolution in gamma-ray and SPECT imaging. They also determine the sensitivity of the SPECT systems since the large majority of the incoming photons are stopped and absorbed by the collimators so that only those that travel along very selected lines of direction defined by the specific collimator design are permitted to reach the scintillator surface to be detected. Various collimators are designed for different considerations of radioisotopes in use and their corresponding gamma-ray energies, as well as for different study objectives. Materials used for constructing the physical collimators often include lead, tungsten, and other high Z materials with significant stopping power of gamma rays.

The commonly used collimators include parallel-hole, diverging, converging, and pinhole collimators. The parallel-hole collimators, most routinely employed in practice, consist of hexagonal or other shape of holes drilled or cast in lead, or formed by lead foils. The lead walls between holes, called septa, are designed with the thickness necessary to stop the photons traveling to the neighboring holes. There is no magnification effect in using a parallel-hole collimator. Diverging collimators have holes diverging from a focal point behind the detector so that a minified image of the object under study is collected. This type of collimators is used in situation when objects of large sizes, often more extended than the dimension of the gamma camera in use, need to be imaged. On the other hand, converging collimators have holes converging from a focal point in front of the detector so that a magnified image of the object is generated. The converging collimators are used to enlarge the target regions so that more details can be revealed. Pinhole collimators can consist of either a single pinhole or multiple pinholes, with the latter usually for providing various angular views in limited-angle tomography and also sometimes for increasing the sensitivity. Each pinhole is an aperture that allows passage of gamma rays within a certain solid angle and produces an inverted image of the object. If the source-to-collimator distance is smaller than the distance from the detector face to the collimator aperture opening, then a magnified image can be generated to achieve higher spatial resolution. Conversely, a minified image of an extended volume of object can be produced.

8.3.3 PET

8.3.3.1 Block PET Detector Modules

Most PET systems are multiring systems. As illustrated in Figure 8.3, a multiring system consists of multiple detector rings and each of which contains a large number of detectors. A state-of-the-art PET system often employs *block detectors*, which, as illustrated in Figure 8.4 as an example, uses a 2 × 2 PMT array to read out a much larger scintillator array, for example, 8 × 8 or 12 × 12. As a result, smaller scintillators can be used to improve the spatial resolution without requiring an increase of the number of PMTs and electronic channels. The scintillator array can be obtained by cutting a large block of scintillator with a pattern of cut depth, which is empirically determined for encoding the position of the scintillator within the block in the

Radiation Detection in SPECT and PET

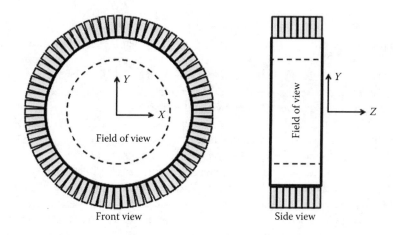

FIGURE 8.3 A PET system typically consists of multiple rings of small detectors.

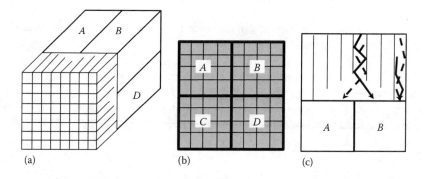

FIGURE 8.4 (a) A PET block detector that uses 2 × 2 PMTs to read 8 × 8 crystals. (b) Top view showing the PMT arrangement. (c) Side view showing crystal position encoding by affecting the light distributions through varying the cut depths.

distribution of light to the PMTs. For example, the cut pattern illustrated in Figure 8.4 will distribute lights originating in the corner crystals to concentrate in one of the PMTs, while distribute lights originating in the central crystals are more widely distributed to all PMTs. Consequently, one can determine the signal-generating crystal—the crystal that interacts with the gamma-ray photon and produces lights—by using the relative pulse heights observed on the PMTs. A more common practice is to calculate the x and y coordinates by

$$x = \frac{(A+B)-(C+D)}{A+B+C+D}, \quad y = \frac{(A+C)-(B+D)}{A+B+C+D}, \tag{8.1}$$

where A, B, C, and D denote the pulse heights obtained by PMTs A, B, C, and D. Based on the x and y values, the signal-generating scintillator is then determined by applying a lookup table that is predetermined during system calibration.

8.3.3.2 Coincidence Detection

In the previous section concerning SPECT, we described the detection of a gamma-ray photon, resulting in a *single event*. In the context of PET, when the detection time of two single events is within a prescribed *coincidence timing window*, they are grouped together to form a *coincidence event*. Generally, only the identifications of the pair of detectors that register the two single events are retained for a coincidence event. In TOF PET systems, the differential time is also stored. Single events that are not in coincidence with any others are discarded. Typically, the coincidence logic is implemented by an electronic board that receives single-event data from a number of detector boards and outputs the coincidence events serially, in chronological order of their detections, to an acquisition workstation. Time tags are also inserted into the list-mode data stream at a regular interval to provide time information. Other special tags, such as tag for respiration, can also be introduced. The list-mode data stream can be stored *as is* to produce *list-mode data*. Or the data can be histogrammed according to the LORs (and also according to the differential-time bin in TOF PET) to obtain *histogrammed-mode data* (also popularly referred to as *sinogram*). For preclinical PET systems, it is also possible to store the single events in list mode and perform coincidence filtering postacquisition on the computer workstation. With this approach, more complicated coincidence logics can be implemented in software. It, however, requires a large bandwidth for data transmission and a large data storage.

The coincidence detection method described earlier can produce three types of coincidence events that are illustrated in Figure 8.5. The first type is *true coincidence*, or true, of which the two gamma-ray photons are associated with the same annihilation and they travel directly from their origination to the detectors. The second type is called *scattered coincidence*, or scatter, of which the two gamma-ray photons are also associated with the same annihilation but at least one of them encounters Compton scattering before it reaches the detector. Because the traveling direction is deflated by scattering, the LOR is mispositioned. Evidently, a larger subject will lead to more scattered coincidences. It also reduces the true coincidences because, in addition to more scattering, more gamma-ray photons are absorbed by the subject through, for example, photoelectric interaction. The third type is called *random coincidence*, or *accidental coincidence* or *random*, of which the gamma-ray photons are associated with two separate annihilations and they are detected as coincidence by chance. For random coincidences, the measured LORs have no

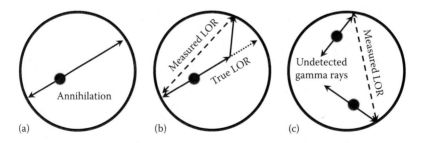

FIGURE 8.5 Three types of coincidence events: (a) true coincidence, (b) scattered coincidence, and (c) random coincidence.

relation to the true LORs. Evidently, more random coincidences are obtained when a larger coincidence timing window is used. Random coincidences can be estimated by delayed-coincidence measurement in which the detection time for a single event is artificially delayed by an amount larger than the coincidence timing window before applying coincidence logic. As a result, the resulting coincidences, called *delayed coincidences*, can only be random coincidences. To distinguish, coincidence events obtained without delay are called *prompt coincidences*. As already discussed, prompt coincidences contain true, scattered, and random coincidences.

A PET system shall detect as many true coincidences as possible while rejecting scattered and random coincidences. For clinical imaging, the amount of scattered coincidences is observed to be substantial. As Compton scattering reduces gamma-ray energy, a standard approach to reduce scattered coincidence is to reject single events having energy below a certain preset threshold, called the lower-level discriminator setting. An upper energy threshold, called the upper-level discriminator setting, is also considered for rejection of pileup events. Such *energy qualification* is performed before coincidence filtering, as illustrated in Figure 8.6.

8.3.3.3 Factors Affecting Spatial Resolution

As elucidated earlier, any departures of the response function from yielding the ideal line integral would lead to data blurring and degrade the image resolution. An important factor that contributes to spatial resolution in PET is, therefore, the crystal size. Consider the coincidence rate obtained by two opposing crystals of width d when a point source is moved along the line halfway the detectors. When the source is outside the common area seen by the detectors, there would be no detection as the two annihilation photons, traveling in opposite directions, cannot be both detected. As the source is moved into this area from the edge to the center, the detection rate increases linearly from zero to reach a peak, reflecting the increase in the detection solid angle of the detectors for the annihilation photons as the source position changes. The resulting response therefore describes, as is illustrated in Figure 8.7a, a triangle having a base equal to d and hence an full-width-at-half-maximum (FWHM) equal to $d/2$.

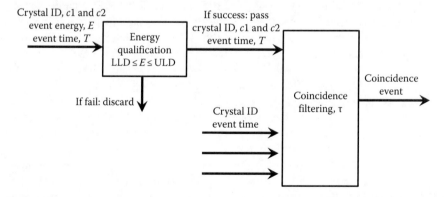

FIGURE 8.6 The signal flow in coincidence detection in PET. τ is the coincidence timing window.

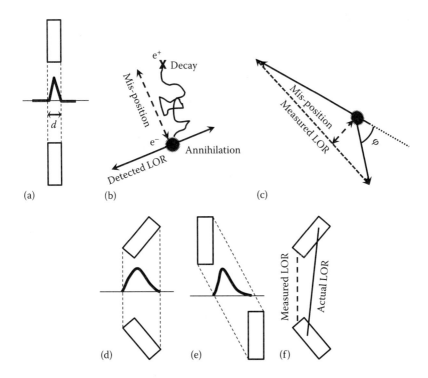

FIGURE 8.7 Spatial resolution in PET is affected by the (a) crystal width; (b) positron range; (c) photon acollinearity, where φ has a zero mean and an FWHM of 0.2°; (d, e) and DOI blurring. DOI blurring for the case involving two crystals on the same ring is shown in (d) and the case involving two crystals in different rings is shown in (e). DOI blurring can also be examined by considering mispositioning of the LOR when DOI is unknown (f).

The second factor is related to the fact that while the goal is to record the distribution of that of the PET tracer, the observed actually is that of annihilation. As illustrated in Figure 8.7b, a positron, after its release and due to its energy, will travel a distance before annihilation. Thus, coincidence events detected by a pair of detectors come from a volume larger than that of annihilation that the pair is responsive to. The distance in the direction normal to that of the annihilation photons is called *positron range*, and this distance depends on the radioisotope in use and the medium within which the positron travels in. For F-18, the positron range in water is about 0.54 mm in FWHM. For Rb-82, it is, however, as large as 6.14 mm.

The third factor, called *photon noncollinearity* and depicted in Figure 8.7c, is due to the fact that the annihilation photons are not always emitted in completely opposite directions because of the small residual momentum of the positron when it reaches the end of its range. The deviation is random; its distribution has a zero mean and an FWHM of about 0.5°. Due to this angular deviation, the measured LOR does not pass through the actual location of annihilation. With respect to an LOR, the distribution of the distance between the LOR and the annihilation photon due to this effect, therefore, has an FWHM equal to $0.0044R$, where R is the radius of the system.

Radiation Detection in SPECT and PET

The fourth factor is specific to block detector of which the signal-generating crystal can be incorrectly identified due to the statistical uncertainty in the quantities given by Equation 8.1. This component depends on the specific design of the block detector and the accuracy of the decoding method.

An empirical rule for the overall spatial resolution, in FWHM, due to these factors is given by

$$r = k\sqrt{\left(\frac{d}{2}\right)^2 + r^2 + (0.0044R)^2 + b^2}, \qquad (8.2)$$

where
r is the positron range
b is the resolution due to block detector design and decoding
$k > 1$ describes the effect of image reconstruction

Among these factors, the predominant factor limiting the spatial resolution is the crystal width. Consequently, the main design approach for increasing PET resolution is to develop detectors by using narrow crystals.

The aforementioned discussion considers the spatial resolution at the center of a PET system. Figure 8.7d and e illustrates situations in a multiring PET system when the annihilation photons enter the crystals at an oblique angle. The sensitivity profile can now be much wider than the width of the crystal if the thickness of the crystal is significantly larger than the width. It also becomes asymmetrical, no longer a triangular shape as shown in Figure 8.7a. Another way to look at the cause of this blurring is to consider mispositioning of the LOR, as illustrated in Figure 8.7f. When a single event is detected, the interaction of the gamma-ray photon with the detector is, by convention, assumed to occur at the center of the front face of the detector. Or it can be assumed to locate at a fixed distance into the front face in accordance with the attenuation length of the scintillator at 511 keV (attenuation length is discussed later). The actual depth into the crystal where the gamma-ray photon deposits energy, called the *depth of interaction* (DOI), is random, however. When it is different from the assumed interaction depth, the measured LOR obtained by connecting the assumed interaction positions of the gamma-ray photons with the detectors is misplaced from the actual LOR. Evidently, this mispositioning is on average greater for longer crystals; the DOI blurring effect, which shall be understood as the blurring additional to the $d/2$ blurring shown in Figure 8.7a, is therefore more pronounced for crystals having a larger thickness/width ratio. This blurring can be removed, or reduced in practice, if the DOI can be measured (see discussion later). It is also reduced when crystals having shorter attenuation lengths are used because mispositioning error of the LOR is smaller on average and the penetration of gamma-ray photons through neighboring scintillators to enter a crystal from the side is also decreased.

The gamma-ray photon can interact with the detector through multiple interactions if the first interaction is Compton scattering. In theory, the first interaction position shall be used for determining the LOR. In practice, individual interactions cannot be resolved. For large crystals, this is not an issue because the multiple interactions are likely to occur inside a single crystal, and as already mentioned, the interaction position is assigned to the front face on the crystal. However, when narrow crystals are used, the interactions can spread over multiple crystals and the identified crystal can be different from the first-interacting crystal. Blurring caused by this *intercrystal scattering* can be an issue when crystal *fingers* are used in attempt to achieve submillimeter resolution.

8.3.3.4 Considerations for the Scintillator

An important consideration in developing a PET system is the selection of the scintillator. Important properties of the scintillator to consider include the attenuation length (at 511 keV), light yield, and decay time. The *attenuation length*, numerically equal to the inverse of the attenuation coefficient, is the distance into the scintillator that the intensity of a gamma-ray beam drops to $1/e$ of the value at surface. It is also the average distance a gamma-ray photon travels in the scintillator before undergoing an interaction. Evidently, a small attenuation length (or a large attenuation coefficient) is preferred so that the annihilation photons can be stopped efficiently by using short scintillators. A material having a high density and/or a high effective Z number will have a short attenuation length. Another related property is the *photofraction*, which is the percentage of events detected by a scintillator that is due to the photoelectric effect. Since energy qualification often keeps only events in the photopeak, scintillator having a large photofraction, which generally is associated with a high effective Z number, is preferred. The *light yield* refers to the amount of scintillation lights generated per gamma-ray energy deposited in the scintillator. It is often expressed in photons/MeV or as a percentage of the light output of NaI. Due to Poisson statistics, a scintillator having a higher light yield will produce a better energy resolution in general. As a result, a tighter energy window around 511 keV can be used in energy qualification to reject scattered events more effectively. For block detectors, the x and y coordinates are calculated from pulse heights; therefore, their statistical precision and the accuracy in identifying the signal-generating scintillator are better when the pulse-height measurement has a better energy resolution. This means that scintillators having a high light yield are needed for developing high-resolution block detectors. Typically, short decay time also means short rise time for the scintillation light pulse. The accuracy in determining the start time of the pulse is to first order proportional to the ratio m/δ, where m is the slope of the rising edge of the pulse and δ is the standard deviation of the noise on the pulse. Therefore, TOF systems, which require superior timing resolution, require the use of scintillators having a short decay time. At the same time, a high light yield is preferred because it reduces the component of δ due to the statistical variation in the number of scintillation lights produced. Using scintillators having a short decay time can also reduce pileups and dead-time loss. Therefore, an ideal scintillator for PET shall have a high density, a high effective Z number, a high light yield, and a short decay time.

Scintillators that have been considered for PET include NaI, BaF_2, BGO, and L(Y)SO. NaI, which is the scintillator of choice for gamma cameras, was used in early PET systems. It has a high light yield to achieve a good energy resolution. But it has a low density and a large attenuation length for annihilation photons. BaF_2 has a fast component (a short decay time) and was successfully used to develop early

TOF PET systems. Unfortunately, the potential gain in the image signal-to-noise ratio (SNR) with the TOF measurement was offset by the low detection efficiency of the material for annihilation photons due to its relatively low density. On the other hand, BGO, having a high density and a high effective Z number, can provide good detection efficiency. As a result, it was the scintillator of choice for developing clinical PET systems in the 1990s. Its two main disadvantages are low light yield and slow response (long decay time). However, for clinical imaging, the resulting energy resolution is acceptable and the spatial resolution of BGO-based block detectors is adequate, but it cannot provide adequate timing accuracy for TOF imaging. L(Y)SO is slightly less dense than BGO but is much brighter and faster. As a result, it is a popular choice for preclinical PET systems that require high resolution. Its energy resolution is better than BGO but has an intrinsic limit of about 9% due to the nonproportionality of L(Y)SO. L(Y)SO-based systems can have a good timing resolution; consequently, they allow the use of a small coincidence timing window for reducing random coincidences. Many TOF PET detectors and systems employ L(Y)SO and the reported coincidence resolution time (CRT) of such systems is in the range of 300–600 ps in FHWM. L(Y)SO has natural radioactivity; the background coincidence rate due to this natural radioactivity can be a hindrance in imaging applications that seek to detect a small amount of, or a small change in, radioactivity. Other scintillators such as G(Y)SO and $LaBr_3$ have also been investigated. A number of other promising scintillators are under further exploration.

8.4 RECENT ADVANCES

In this section, we review a number of recent advances in new technologies of radiation detection in SPECT and PET that, we believe, will impact significantly the future ECT instrumentation development. These relatively new technology advances will lead to new design concepts and imaging systems that may evolve and shape the future clinical and research practices when employing SPECT and PET into a new landscape, which will be further discussed in the next section. Only selected topics are discussed in this section; more complete and detailed information can be found in the relevant textbooks (see, e.g., Knoll 1999; Wernick and Aarsvold 2004; Cherry et al. 2012) and review articles (see, e.g., Madsen 2007; Peterson and Furenlid 2011).

8.4.1 SPECT

8.4.1.1 Scintillators

NaI(Tl) has been the workhorse scintillator widely employed in gamma cameras and SPECT imaging systems, usually in the form of relatively large-area monolithic crystals. *Block detector* designs, similar to those described earlier for PET but consisting of an array of small *pixelated* individual NaI(Tl) crystals of the size of 1–2 mm, have also been developed for gamma-ray and SPECT imaging (Zeniya et al. 2006; Xi et al. 2010). Pixelated CsI(Tl) and CsI(Na) crystals, which are denser and have higher light yield but slower decay time than those of NaI(Tl), have also been used in building prototype gamma cameras (Truman et al. 1994; William et al. 2000). CsI(Tl), in the form of microcolumnar crystal arrays, has been employed in building high-resolution gamma cameras and SPECT imaging systems (Tornai et al. 2001; Nagarkar et al. 2006). YAlO$_3$(Ce), also known as YAP that is denser and has faster decay time but less light yield than those of NaI(Tl), was utilized in building the YAP-(S)PET small-animal scanner that is capable of both SPECT and PET imaging (Del Guerra et al. 2006).

Lanthanum scintillation crystals offer relatively high light output compared to other scintillators (Pani et al. 2006), leading to better energy resolution. Both LaCl$_3$(Ce) and LaBr$_3$(Ce) have been investigated for uses in gamma-ray and SPECT imaging (van Loef et al. 2001; Shah et al. 2001; Alzimami et al. 2008). Since LaBr$_3$(Ce) is denser and has a higher light yield between the two, it has been employed in building gamma-ray cameras and SPECT imaging systems for a variety of applications (Russo et al. 2009; Yamamoto et al. 2010; Roy et al. 2014). More recently, a new scintillator, europium-doped strontium iodide (SrI$_2$(Eu)), has received considerable attention for its potential uses in gamma-ray and SPECT imaging (Cherepy et al. 2008, 2009). SrI$_2$(Eu) offers even more light output than that of LaBr$_3$(Ce) leading to excellent energy resolution and is denser than NaI(Tl). Even though its decay time is slower than that of both NaI(Tl) and LaBr$_3$(Ce), for SPECT imaging, this inferior characteristics is not important. This new scintillation crystal has a great potential to be a candidate to evolve the future landscape of SPECT instrumentation.

More detailed information of advances in scintillation technologies for gamma-ray and SPECT imaging can be found in the relevant review papers (see, e.g., van Eijk 2002; Madsen 2007; Peterson and Furenlid 2011).

8.4.1.2 Semiconductor Detectors

As discussed in several other chapters in this book, semiconductor (solid-state) detectors, such as silicon (Si), germanium (Ge), cadmium telluride (CdTe), and cadmium zinc telluride (CdZnTe, or CZT), have been extensively investigated in the last few decades for applications in medical imaging including those in nuclear medicine. These semiconductor detectors offer excellent energy resolution when compared to the standard NaI(Tl)-PMT approach. The superior energy resolution results from the relatively large number of electron–hole pairs generated per keV of photon energy deposited from those radioisotopes commonly used in nuclear medicine (Knoll 1999; Cherry et al. 2012), leading to relatively low statistical variation in signal response to photon energy. These semiconductor detectors also have the potentials for providing superior spatial resolution. The detector element can be fabricated with very small image pixels, and also the clouds of electrons and holes resulting from photon interactions at the typical radiotracer energies encountered in nuclear medicine are less than a couple of hundred microns after experiencing diffusion and drift in the electric field, leading to a spatial resolution limit on the order of a few hundred microns, which is dependent on the specific photon energy associated with the radioisotope in use. However, factors such as the requirement for cooling especially in the case of Ge and Si, relative costs and availability of the semiconductor detector materials, their temperature-sensitive performance parameters, and the needs for a large number of signal processing channels and rapid computation have prevented from the effective and routine utilization of these semiconductor detector technologies in nuclear medicine, especially in SPECT

(Scheiber and Cambron 1992; Scheiber 1996, 2000; Scheiber and Giakos 2001; Limousin 2003; Sharir et al. 2010). But recent advances in these areas have made future incorporation of their advantage of excellent energy and spatial resolution into routine SPECT uses more feasible.

Even though investigation on the use of Si and Ge for gamma-ray imaging has been explored for several decades, their incorporation into routine applications has been somewhat limited, in part because of the requirement of special cooling in order to avoid excessive thermally generated electronic noise, which could be costly and inconvenient for operation. Several gamma-ray or SPECT imaging systems based on Si detectors have been developed in the last decade, which offered very high spatial resolution, but with relatively small field of view (FOV), for relatively low-photon-energy detection such as that involving I-125 (Peterson et al. 2003, 2009; Choong et al. 2005; Shokouhi et al. 2010). Spatial resolution better than 100 μm can be achieved with specially designed multipinhole collimators. High-purity germanium (HPGe), integrated with advanced electronics and compact mechanical cooling system, has also been exploited for building SPECT imaging system (Johnson et al. 2011). It was demonstrated that better than 1% in both energy resolution and imaging response uniformity can be achieved. Both these Si and Ge technologies have proved to be useful and feasible for building gamma-ray and SPECT imaging systems for specific applications such as small-organ or small-animal imaging. It is also worth noting that HPGe has also recently been employed to build a PET imaging system (Cooper et al. 2009).

CdTe and CZT have also been studied extensively for decades to explore their potential uses in gamma-ray and SPECT imaging, especially because, in contrast to Si and Ge, they can be operated in an ordinary room temperature environment. The broad employment of these semiconductor detectors has also been somewhat limited previously, primarily because of their relatively high costs and scarce availability. In the past decade, multiple gamma cameras and SPECT imaging systems have been designed and built with the use of either CdTe or CZT. For example, MediSPECT, a small-animal imaging system, employed CdTe pixel detectors and a coded aperture mask collimator to image radioisotopes with low (I-125) or medium (Tc-99m) energy and achieved a spatial resolution of approximately 1–2 mm (Accorsi et al. 2007). A single-head MediSPECT system was later used with a 0.4 mm single pinhole collimator to achieve a spatial resolution of 0.2 mm but within an FOV of the size of only 2 mm (Accorsi et al. 2007a). SemiSPECT, another small-animal SPECT imaging system based on an array of 8 CZT detectors, achieved an average spatial resolution of 1.45 mm along each axis within the FOV using pinhole collimators of 0.5 mm diameter and offered an overall system sensitivity of approximately 0.1% depending on the specific window setting (Kim et al. 2009). Clinical gamma cameras and SPECT imaging systems based on CZT detectors have also been successfully utilized in routine practice, especially for cardiac imaging (Esteves et al. 2009; Duvall et al. 2011) and breast imaging (O'Connor et al. 2007; Mitchell et al. 2013). Other room temperature compound semiconductors that potentially can be very promising for medical imaging applications, especially SPECT, include HgI_2, PbI_2, and TlBr, in part because of their relatively high density and strong stopping power when compared to those of CdTe or CZT (Peterson and Furenlid 2011).

It is worth noting that a chapter on using CZT in PET is also included later in this book. More information on solid-state detectors and their uses in gamma-ray and SPECT imaging can be found in the relevant chapters in the textbooks (e.g., Wernick and Aarsvold 2004; Cherry et al. 2012) and selected review articles (e.g., Madsen 2007; Peterson and Furenlid 2011; Peterson and Shokouhi 2012).

8.4.1.3 Photodetectors

In addition to scintillation crystals and semiconductor detectors, advances in photodetector technologies for signal readout have also been a driving force in evolving the gamma-ray and SPECT imaging instrumentation developments. Various conventional PMTs in a round or square shape are employed in the great majority of the current gamma-ray and SPECT imaging systems to convert light signals from the scintillators to become electrical signal outputs. Position-sensitive PMT (PSPMT) technology, of which 2D position information is provided via two sets of wire anodes that are arranged orthogonally to one another, has been widely employed in gamma cameras and SPECT systems to replace the traditional Anger logic circuits in determining the event position (Kume et al. 1986). The PSPMT technology has been integrated with primarily pixelated scintillation crystals including NaI(Tl) (Yasillo et al. 1990; Zeniya et al. 2006), CsI(Na) (Williams et al. 2000), CsI(Tl) (Pani et al. 1999), and $LaBr_3(Ce)$ (Yamamoto et al. 2010).

Photodiodes (PDs), much more compact than PMTs or PSPMTs, can also be used to convert scintillation lights to electrical signals (Choong et al. 2002) in place of the bulky PMTs, especially when compact designs are required. However, PDs cannot amplify the signals like the way PMTs do in order to enhance the signals significantly. Avalanche photodiodes (APDs), which operate at higher reverse-bias voltages in a breakdown mode, can achieve signal amplification since the drifting charges are accelerated so that additional electron–hole pairs are created. But even so, the gain of a typical APD is usually only 20%–30% of that of the traditional PMTs. Large-area tiled APD arrays can be used to replace PMT arrays in conjunction with the use of monolithic scintillators as in traditional gamma camera (Shah et al. 2001). Alternatively, one APD can also be coupled directly to an individual segmented scintillation crystal as the basic detector element for forming the large-area detector array. Position-sensitive APDs (PSAPDs) use charge sharing between additional electrodes on the back surface of the APD to determine event localization with sufficient spatial resolution at a few hundred microns. PSAPDs have been used as the photodetectors in conjunction with CsI(Tl) for use in high-resolution small-animal SPECT systems (Funk et al. 2006). It is usually required to cool the PSAPD-based detectors in order to reduce the dark current for maintaining a reasonable level of the SNR.

Charge-coupled devices (CCDs), unlike the event-driven PMTs and APDs, integrate the events over some integration time before sequentially reading out each pixel and outputting the data in a frame transfer mode. CCDs can also be used as the photodetectors in converting scintillation lights to electrical signals. CCDs have high quantum efficiency with their energy resolution affected by dark current and readout electronic noise, which can be reduced by cooling. Some CCDs also employ electron-multiplying approaches (EMCCDs) to amplify the charge signals serially in order to reduce the readout noise. Since a typical CCD pixel size is about only 20 μm, relatively small compared to the scintillator element (typically CsI(Tl) crystals), demagnification is usually needed in order to extend the detection area and retain the high intrinsic spatial resolution when coupling the scintillator to the CCD. For this demagnification step, both fiber optic tapers (de Vree et al. 2005) and lenses (Nagarkar et al. 2006) have been used to build SPECT imaging systems. To compensate for the light loss by using the lens and fiber optic taper, demagnifier tubes (Meng 2006) and microchannel plate (Miller et al. 2008) have also been used to provide optical gains in the high-resolution SPECT imaging systems using CCDs or EMCCDs. These prototype SPECT imaging devices achieved subhundred microns spatial resolution for I-125 imaging and a few hundred microns resolution for Tc-99m imaging.

The latest technology advances in novel photodetectors for gamma-ray or SPECT imaging are those associated with silicon photomultipliers (SiPMs), which are also discussed in other relevant chapters in this book, as well as in those sections related to PET in this chapter. SiPM has been used in conjunction with EMCCD in building high-resolution SPECT imaging systems to provide prior information of events occurred inside the scintillator CsI(Tl) (Heemskerk et al. 2010). A compact gamma camera using scintillator YSO(Ce) and an array of SiPMs was built and evaluated, demonstrating an achievable spatial resolution of 1.5 mm (Yamamoto et al. 2011). Another recent study showed that SiPMs can be employed successfully in nuclear medicine imaging when coupled with NaI(Tl) for use in SPECT and with LYSO for use in PET, but unsuccessfully when coupled with CsI(Tl) or BGO (Stolin et al. 2014). The use of SiPMs in novel design concepts for building new gamma-ray and SPECT imaging systems in the future is expected to grow and requires substantial research and development efforts to fully explore the potentials of this new technology.

More detailed information regarding photodetectors can be found in the relevant chapters in the textbooks (e.g., Wernick and Aarsvold 2004; Cherry et al. 2012) and review articles (e.g., Madsen 2007; Peterson and Furenlid 2011).

8.4.2 PET

8.4.2.1 DOI Detectors

The spatial resolution of PET imaging has now been improved to reach about 2–4 mm for clinical systems and to approximately 1 mm for preclinical systems. Detectors that can achieve submillimeter resolution have also been reported. The sensitivity of PET, however, remains relatively low: the sensitivity of clinical systems is under 1% (based on NEMA NU-2 standard) and that of most preclinical systems is under 5%. As already discussed, the crystal width shall be less than twice the resolution desired. To achieve high sensitivity, the detectors shall also provide a high detection efficiency. For a dense scintillator such as L(Y)SO, this means a crystal thickness of 2–3 cm for providing a detection efficiency of 83%–93% for 511 keV photons (and 69%–86% for coincidence detection of two 511 keV photons). Therefore, narrow and long crystals are needed for developing high-resolution and high-sensitivity systems but, as discussed earlier, they are particularly prone to DOI blurring. This situation is accentuated when fingerlike crystal elements are used for achieving submillimeter resolution and adequate detection efficiency. Another way to increase the sensitivity of a PET system is to increase its overall solid angle of detection by making it longer and/or reducing its detector-ring diameter. For such systems, in addition to increasing scattered and random coincidences, more events will enter the detectors obliquely and therefore aggravates DOI blurring as well.

Therefore, in the past decade, a substantial effort in PET research is on developing thick detectors that are capable of providing DOI measurement, called the *DOI detectors*. Several DOI technologies have been proposed. The dual-ended readout method is based on the observation that, for a continuous crystal, the amount of lights reaching at its front end relative to that at the rear end is dependent upon the DOI (Moses and Derenzo 1994). In practice, developing the photodetectors and its electronics for the front-end readout is a challenge. Another design idea is to construct a long *crystal* by stacking up shorter segments, or layers, and devise a scheme to determine the signal-generating layer from the read-end measurement. This idea has been employed to extend the conventional block detector to consist of multiple layers; an ingenious scheme of distributing the scintillation lights to the PMTs is devised such that the histograms of the x and y coordinates for different layers are interleaved; therefore, both the layer and the position in the layer of the signal-generating crystal are encoded in these quantities. In a phoswich detector, crystals in different layers have different decay time and the signal-generating layer is determined based on an output that depends on the decay time. These single-ended designs have their own practical challenges: assembling segmented crystals can be difficult especially for fingerlike crystals. Recently, Roncali et al. proposed to cover a portion of a continuous crystal by phosphor that absorbs a fraction of the scintillation lights and reemits them with a delay characteristic of the phosphor. Consequently, the DOI can be encoded in the observed decay time by employing an adequate coating configuration along the length of the scintillator (Roncali et al. 2014).

Using narrower crystals not only poses challenges in detector assembly, it can also significantly increase costs as proportionally more scintillator materials are wasted in cutting the crystals. As a result, there is substantial interest in using monolithic scintillators for developing high-resolution PET detectors. The scintillator is often coupled to a PSPMT for measuring the distribution of lights at the exit surface of the scintillator. An array of small photodetectors such as the APDs can also be used. Ignoring light reflection at the edges of the scintillator, the centroid of the measured light distribution would give the x–y coordinates of the interaction position of the gamma ray with the scintillation and the width of the DOI (as lights are spread more widely when the interaction is far away from the measurement). Therefore, the 3D interaction position can be determined from the light distribution. However, due to light reflection, the positioning accuracy near edges is poorer. This problem can be mitigated by rough surface treatment and use of black paints to absorb lights. More sophisticated positioning algorithms, including maximum-likelihood and nonlinear estimation methods, have also been proposed (Joung et al. 2002; Ling et al. 2007; Li et al. 2010).

Another design approach is to put several photodetectors along the length of the crystal, or an array of photodetectors on a side surface of a scintillator slab (Levin 2012). One can also stack several layers of single-ended readout, non-DOI detectors. These designs are simple in concept but require the use of compact photodetectors such as the APDs and SiPMs. In addition, the electronics for reading these photodetectors need to be minimal. Other DOI technologies, including the use of wavelength-shifting fibers and the hybrids of the aforementioned designs, have also been proposed. Ito et al. have an excellent review on PET DOI detectors (Ito et al. 2011).

8.4.2.2 TOF Detectors

Another recent development in PET is TOF, reinvigorated by the availability of dense and fast scintillators such as L(Y)SO (Moses 2003). Based on using L(Y)SO and PMTs, commercial clinical systems have achieved a CRT on the order of 500–600 ps FWHM (Surti et al. 2007; Jakoby et al. 2011). The scintillator often contributes significantly to the time resolution: for a system that has a CRT of 528 ps FWHM and employs $6.75 \times 6.75 \times 25$ mm^3 crystals, the contribution is estimated to be as much as 326 ps FHWM (Moses and Ullisch 2006). Using LaBr$_3$, Kuhn et al. have reported a CRT of 313 ps FWHM (Kuhl et al. 2006). The timing resolution of LaBr$_3$ has been found to depend substantially on the cesium (Ce) doping. By increasing the Ce concentration to 30%, a CRT better than 100 ps in FWHM was reported (Glodo et al. 2005). Other promising scintillators for TOF PET include LaBr$_3$, LuAG, LuYAP, LaCl$_3$, CeBr$_3$, and LuI$_3$, which are all able to achieve a CRT in the range of 100–400 ps FWHM. Among them, LuI$_3$ is of great interest because its light yield is 2.6 times that of NaI, an energy resolution of 4%, and a CRT of 125 ps FWHM (Moses 2007). Among LSO, LuAG, LuYAP, LaBr$_3$, and LaCl$_3$, LSO is determined to yield the highest value of a figure of merit, taking into consideration the trade-off between detection efficiency and timing resolution (Conti et al. 2009). Recently, TOF detectors based on detecting the Cherenkov light generated in materials such as PbF$_2$ and PWO have also been proposed, and early results indicate that sub-100 ps CRT is feasible (Korpar et al. 2011; Brunner et al. 2014). A limitation of this interesting approach is its poor energy resolution due to the limited number of Cherenkov lights produced.

It is important to recognize that generally the time resolution deteriorates with the length of the crystal. This is due to two factors. First, longer crystals have less light output. Second, the DOI is random and a longer crystal will produce a wider distribution of the DOI and hence a larger spread in the transit time for the scintillation lights to reach the photodetector. By using 2×2 mm^2 LSO, the CRT is measured to degrade from 108 to 176 ps FWHM as the crystal length increases from 3 to 20 mm (Gundacker et al. 2014). On the other hand, results from simulation studies indicated that the cross section of the crystal does not significantly affect the time resolution.

As DOI technologies are already available, the development of DOI detectors capable of TOF measurement (DOI + TOF detectors) is a natural extension, and for such detectors, DOI measurement may be used for improving time resolution. By correcting for the time shift due to DOI, Shibuya et al. improved the CRT of a four-layer detector slightly from 361 to 324 ps FWHM (Shibuya et al. 2008). Spanoudaki and Levin expected the DOI effect to be more significant when the time resolution improves; by conducting simulation studies, they found that rough surface treatment yields strong correlation between the transit time and DOI than polished surface treatment (Spanoudaki and Levin 2011).

8.4.2.3 Silicon Photomultipliers

The introduction of the SiPM is arguably transforming PET instrumentation (Roncali and Cherry 2011). This solid-state photodetector, besides being compact and rugged, has a PMT-like gain and TOF-capable time response, operates at a low voltage, and is insensitive to magnetic fields. As a result, a great many research groups are employing SiPMs to replace PMTs and APDs for developing next-generation PET detectors with improved performance properties or new capabilities. Most SiPM, similar to the PMTs, produce an analog output of which the amplitude is proportional to the number of light photons hitting the photodetector (ignoring the possibility of saturation). Digital SiPM (dSiPM) that generates digital outputs containing the number of light photons detected and the time of detection has also been developed (Haemisch et al. 2012). Based on dSiPM, detectors that have impressive performance are developed (Seifert et al. 2013; van Dam et al. 2013). Due to extra on-chip electronics needed, a relatively large fraction of a dSiPM is not sensitive to light photons, and therefore, the detection efficiency is compromised. Improved designs to address this issue with in-pixel data compression have been proposed (Braga et al. 2011). Due to its fast response, much attention of SiPM-based PET detector development has been on TOF imaging. The principle and response characteristics of SiPM and its applications in TOF PET are covered in another chapter on this topic of this book.

In addition to TOF, SiPMs are also widely considered for developing high-resolution DOI detectors (which can be TOF capable also). Examples include using SiPM arrays to read out pixelated or monolithic scintillator with single-ended readout or dual-ended readout (Scharrt et al. 2009; Delfino et al. 2010; Kang et al. 2010; Llosa et al. 2010; Song et al. 2010; Kwon et al. 2011; Kishimoto et al. 2013; Nishikido et al. 2013; Seifert et al. 2013). It is also possible to build DOI detectors by stacking multiple layers of thin, non-DOI detectors that employ single-ended readout (Herbert et al. 2006; Moeher et al. 2006; Espana et al. 2014). A detector design that encloses all six surfaces of a scintillator block with SiPMs is also reported (Yamaya et al. 2011). Because of their compactness, insensitiveness to magnetic fields, and requiring only a low operating voltage, SiPMs are also exploited for developing SiPM-based detectors for PET/MRI (Schulz et al. 2011; Hong et al. 2012; Thompson et al. 2012; Yoon et al. 2012), endoscopic detectors for pancreas and prostate imaging (Frisch 2013; Garibaldi et al. 2013), and handheld intraoperative imager (Popovic et al. 2014).

8.4.2.4 Novel Readout Methods

As a typical SiPM pixel ranges from 1×1 to 4×4 mm^2 in size, a small $2'' \times 2''$ PET detector module can easily use more than 100 SiPMs. Efficient readout of SiPMs, therefore, poses a considerable challenge. Many groups are addressing this issue with the development of readout ASICs for SiPMs (Deng et al. 2010; Meier et al. 2010; Bagliesi 2011; Powolny 2011; Janecek et al. 2012; Stankova et al. 2012; Castilla et al. 2013; Comerama et al. 2013; Goertzen et al. 2013; Sacco et al. 2013). Complementary to ASIC development, multiplexing methods to significantly reduce the number of readout channels have also been developed or are under investigation. In comparison to one-to-one readout, multiplexing sacrifices the count-rate performance by having multiple SiPMs sharing a readout channel. Generally, the higher the multiplexing ratio, the more reduction in the count-rate performance. A natural choice for the multiplexing method is to do something similar to the conventional block detector: while, in the block detector, the scintillation lights are usually distributed to four PMTs, in this case, a resistive network is devised to distribute the electrical outputs of the SiPMS to four outputs (Siegel et al. 1995; Song et al. 2010; Goertzen et al. 2013). Position decoding is done in the same way as in the block detector. As discussed earlier, there can be decoding errors that degrade the spatial resolution. Another method is to distribute the SiPM outputs in such a way to obtain row and column sums, therefore producing $2N$ outputs for $N \times N$ SiPM arrays (Stratos et al. 2013). In this case, SiPM identification is achieved by identifying the signal-containing row and column, which is easier and more accurate than the previous method. The row and column sums can be further reduced to four by placing resistive chains between them (Wang et al. 2012; Stratos et al. 2013).

The aforementioned methods use resistive networks; as a result, each readout channel is connected to the summed capacitive load, leading to increased pulse rise time and degraded timing accuracy. For fast timing application, preamplifiers can be introduced to individual SiPMs to hide their capacitance from others (Liu and Goertzen 2013). This, however, increases the circuit complexity and costs. The idea of compressed sensing has also been applied to reduce the number of readout channels of a 12×24 SiPM array to 16 (Chinn et al. 2012; Dey et al. 2013). In the strip-line method, multiple SiPMs are connected to a strip line and the signals at the two ends of the strip line are taken (Kim et al. 2010, 2012a). If needed, preamps are placed at the ends of the strip line but no preamp is needed for individual SiPMs. Similar to TOF detection, the position of the signal-generating SiPM on the strip line can be determined based on the differential time the signals arrived at the two ends of the line. The current implementation has 8 SiPMs on a strip line. However, as multiple strip lines can be connected to increase the number of SiPMs on a line, this readout is scalable. Another interesting feature of this method is the capability to separate the scintillator/SiPM from the acquisition electronics. This feature is useful for developing detectors for PET/MRI, endoscoping PET imaging, and zoom-in PET imaging that present severe space constraints and nonideal environments for electronics.

8.4.2.5 Waveform Sampling

Another noteworthy development in PET data acquisition is the development of sampling-based readouts. As discussed earlier, gamma-ray detection is conventionally done by splitting the signal pulse into a fast channel for event-time determination with CFD and TDC and a slow channel for even-energy measurement with a shaper that *slows down* the pulse from a few hundred nanoseconds to tens of microseconds and a flash ADC that samples the shaped pulse at about 40 MHz. Some proposed designs derive both the energy and time information from the pulse samples, sometimes using a higher sampling rate of 50–100 MHz (Streun et al. 2002; Martinez et al. 2004; Ziemons et al. 2005; Fontaine et al. 2006; Olcott et al. 2006; Hu et al. 2009, 2012). These sampling-based methods can achieve a CRT in the range of 2–4 ns FHWM, with a best result of about 0.7 ns FWHM when using 100 MHz sampling and interpolating the initial rise of the pulse using the samples (Hu et al. 2009). To improve the CRT to 200–500 ps FWHM that is needed for TOF PET, it would be difficult as higher-speed flash ADCs are too costly or not available. As sampling speed increases, heat dissipation and power consumption can become significant practical issues also. Recently, inexpensive low-power sampling chips based on switched-capacitor arrays are available. In particular, the DRS4 chip can provide nine channels of 0.7–5 GHz sampling with 1024 sampling cells per channel. This kind of devices is adequate for sampling a short duration at a high rate (e.g., ~200 ns at 5 GHz for DRS4). However, it has a considerable dead time (e.g., ~30 μs to read out 1000 samples for DRS4), but for PET, this issue can be addressed by employing adequate triggering schemes (Ashmanskas et al. 2011). DRS4 samplers have been employed for developing TOF PET detectors (Kim et al. 2010, 2011, 2012a, 2013; Ashmanskas et al. 2011; Ronzhin et al. 2013). When used with LYSO crystals, a CRT in the range of 300–600 ps FWHM is achieved. If the full pulse is sampled, the decay time can also be derived. For LYSO, the sampling rate can be reduced to about 2 GHz (Kim et al. 2012b) and the sampling duration to ~50 ns, without substantially compromising the timing and energy resolutions. For waveform samplers such as DRS4, time calibration is an important consideration when optimized timing is needed (Kim et al. 2014). With waveform samples, digital signal processing methods may be developed to improve timing resolution (Kao et al. 2011) or to remove RF interferences in PET/MRI (Eclov et al. 2014).

An alternative sampling method is the multivoltage threshold (MVT) proposed in Xie et al. (2005) and Kim et al. (2009). This method, as illustrated in Figure 8.8, samples a pulse with respect to several voltage thresholds. Information about the pulse is then obtained by fitting the samples to a mathematical formula describing the pulse shape. The MVT method can always sample the fast rising edge of the pulse if the thresholds are properly defined. This is not the case with DRS4 sampling unless the sampling rate is sufficiently high. Also, while DRS4 generates hundreds of samples per pulse, the MVT method generates only eight samples per pulse when four thresholds are used. As a result, the requirements for onboard processing and communication bandwidth are substantially reduced. In principle, of course, more information about the pulse is obtained by DRS4 sampling. Recently, an FPGA-only implementation of the sampling method is reported (Xi et al. 2013). With this FPGA implementation, readout electronics and systems based on MVT sampling can be rapidly prototyped, and some examples are shown in Figure 8.8. The MVT sampling method has been successfully applied to LYSO coupled to PMT and SiPM (Xie et al. 2013). It is also possible to achieve a CRT of ~300 ps FWHM when applying MVT sampling to an LYSO/SiPM detector.

Radiation Detection in SPECT and PET

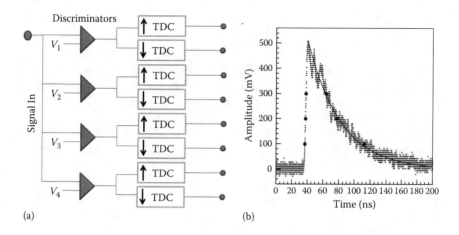

FIGURE 8.8 (See color insert.) (a) The MVT method samples a pulse by determining the time it crosses a number of voltage thresholds, using discriminators and TDCs. (b) A sample pulse (blue curve), the MVT samples (red dots), and the fitted pulse from samples (red curve).

8.5 FUTURE TRENDS AND SUMMARY

Based on the advances made in recent years, which are discussed in the previous section regarding relevant technologies for SPECT and PET, vision for future directions and significant trends of this two molecular imaging modalities are discussed briefly in the following section.

8.5.1 MODULAR AND RECONFIGURABLE DETECTORS

The availability of compact semiconductor detectors such as CdTe and CZT, as well as SiPMs that can replace the bulky PMTs, will enable novel and flexible designs of modular detector components that can be used as basic building blocks to assemble imaging systems very flexibly to meet the specific clinical or research needs. It is also anticipated that these compact detector components can be reconfigured conveniently whenever needed in order to provide the optimal imaging geometry and scanning trajectory to maximize the resulting image quality as well as the optimal clinical or research outcomes. These modular and reconfigurable detectors can facilitate the advances of many application-specific imaging devices discussed as follows.

8.5.2 APPLICATION-SPECIFIC IMAGING SYSTEMS

Most current SPECT and PET systems are designed for use as general-purpose equipment to cover the full range of potential clinical and research needs. Although very useful, these general-purpose imaging systems are usually not optimal for certain specific applications. Modular and reconfigurable detectors made possible by the recent advances of relevant technologies as discussed in the previous section can facilitate the development of application-specific imaging devices that are optimized specially for the targeted applications. Application-specific gamma camera, SPECT, and PET imaging systems have been proposed or developed to provide high sensitivity and resolution for imaging of the breast (Thompson et al. 1994; Doshi et al. 2000, 2001; Levine et al. 2003; Moses 2004; Abreu et al. 2006; Karellas and Vedantham 2008; Raylman et al. 2008; Bowen et al. 2012; Koolen et al. 2012; Moliner et al. 2012; Miyake et al. 2014), prostate (Huber et al. 2001, 2005; Majewski et al. 2011), and brain (Watanabe et al. 2002; Wienhard et al. 2002; Karp et al. 2003; van Velden et al. 2009; Yamamoto et al. 2011; Majewski et al. 2011a).

Another example is special imaging device specifically designed for use under targeted study conditions. For example, typically, animals are imaged under anesthesia. For brain studies, anesthesia, however, alters the brain activity. There are also experiments that require the animal to be awake and perform tasks. High-resolution, wearable *RatCAP* imagers for imaging the brain of an awake mouse has been developed (Schlyer et al. 2007; Vaska et al. 2007; Schulz and Vaska 2012). SiPM-based endoscopic PET detectors and lightweight handheld imaging probes have also been proposed as discussed earlier.

8.5.3 Multimodality Imaging Systems

Dual-modality imaging systems such as PET/CT and SPECT/CT are now considered standard clinical equipment for routine uses. As the aforementioned, there are also active efforts on developing hybrid PET/MRI and SPECT/MRI systems. Most of these efforts are concerned with the development of PET or SPECT insert detectors. Insert detectors for whole-body PET systems have also been proposed for enhancing the resolution (Wu et al. 2008), for improving breast imaging (Matthews et al. 2013), and for zoom-in imaging to enhance the detection of small lesions (Zhou and Qi 2009, 2011).

8.5.4 SPECT and PET in Image-Guided Therapy

Recently, the value of PET to provide range verification and dose monitoring in proton and heavy-ion therapy has been demonstrated (Psheninchnov et al. 2007; Attanasi et al. 2011; Parodi 2012; Aiello et al. 2013; Jan et al. 2013; Zhu and El Fakhri 2013), and in-beam PET systems have been proposed (Enghardt et al. 2004; Attanasi et al. 2008; Parodi et al. 2008; Vecchio et al. 2009; Shakirin et al. 2011; Zhu et al. 2011; Shao et al. 2014; Sportelli et al. 2014). Stationary partial-ring and dual-head configurations are favorable for such systems because they provide access to the patient for the treatment beams (Shakirin et al. 2007; An et al. 2013). Image artifacts that are typically generated with these limited-view systems are removed if they are TOF systems. Similarly, dedicated TOF PET breast systems employing partial rings to provide access for biopsy needles have been proposed (Surti and Karp 2008; Chen et al. 2011). The OpenPET systems are also very promising for supporting image-guided applications (Yamaya et al. 2008; Yoshida et al. 2011, 2013; Tashima et al. 2012).

8.5.5 Summary

The advances in scintillators, semiconductor detectors, photodetectors such as SiPM, and readout electronics in the past decades have led to significant progress in SPECT and PET in terms of performance as well as applications. The advances are likely to finally enable the development of high-resolution, high-efficiency DOI- and TOF-capable detectors. The new detectors are also likely to be much more rugged and versatile. More application-specific SPECT and PET imaging systems, as well as SPECT/MRI and PET/MRI systems, can be expected. The use of SPECT and PET in image-guided therapy can be routine applications soon as well. The development of long-bore systems to greatly increase the sensitivity of SPECT and PET imaging to enable the detection of abnormalities at the earliest stage possible is also of great interest and significance. These advances in SPECT and PET instrumentation however need to be accompanied by parallel advances in image reconstruction methods and imaging tracer probes.

REFERENCES

Abreu, M.C., J.D. Aguiar, F.G. Almeida et al. 2006. Design and evaluation of the clear-PEM scanner for positron emission mammography. *IEEE Trans. Nucl. Sci.* 53:71–77.

Accorsi, R., M. Autiero, L. Celentano et al. 2007a. MediSPECT: Single photon emission computed tomography system for small field of view small animal imaging based on a CdTe hybrid pixel detector. *Nucl. Instrum. Methods A* 571:44–47.

Accorsi, R., A.S. Curion, P. Frallicciardi et al. 2007b. Preliminary evaluation of the tomographic performance of the mediSPECT small animal imaging system. *Nucl. Instrum. Methods A* 571:415–418.

Aiello, M., F. Attanasi, N. Belcari et al. 2013. A dose determination procedure by PET monitoring in proton therapy: Monte Carlo validation. *IEEE Trans. Nucl. Sci.* 60:3298–3304.

Allen, H.C., R.L. Libby, and B. Cassen. 1951. The scintillation counter in clinical studies of human thyroid physiology using I-131. *J. Clin. Endocrinol. Metab.* 11:492–511.

Alzimami, K.S., N.M. Spyrou, and S.A. Sassi. 2008. Investigation of $LaBr_3$:Ce and $LaCl_3$:Ce scintillators for SPECT imaging. In *ISBI 2008, the Fifth IEEE International Symposium on Biomedical Imaging: From Nano to Macro*, May 14–17, 2008.

An, S.J., C.H. Beak, K. Lee et al. 2013. A simulation study of a C-shaped in-beam PET system for dose verification in carbon ion therapy. *Nucl. Instrum. Methods A* 698:37–43.

Anger, H.O. 1952. Use of a gamma-ray pinhole camera for invivo studies. *Nature* 170:200–201.

Anger, H.O. 1958. Scintillation camera. *Rev. Sci. Instrum.* 29:27–33.

Anger, H.O. 1964. Scintillation camera with multichannel collimators. *J. Nucl. Med.* 5:515–531.

Anger, H.O. 1969. Multiplane tomographic gamma-ray scanner. In *Medical Radioisotope Scintigraphy*, International Atomic Energy Agency, Vienna, Austria.

Anger, H.O. and D.J. Rosental. 1959. Scintillation camera and position camera. *Medical Radioisotope Scanning*. IAEA and WHO, Vienna, Austria.

Ashmanskas, W.J., B.C. LeGeyt, F.M. Newcomer et al. 2011. Waveform-sampling electronics for time-of-flight PET scanner. In *The 2011 IEEE Nuclear Science Symposium and Medical Imaging Conference Record (NSS/MIC)*, Valencia, Spain, pp. 3347–3350.

Attanasi, F., N. Belcari, M. Camarda et al. 2008. Preliminary results of an in-beam PET prototype for proton therapy. *Nucl. Instrum. Methods A* 591:296–299.

Attanasi, F., A. Knopf, K. Parodi et al. 2011. Extension and validation of an analytical model for in vivo PET verification of proton therapy—A phantom and clinical study. *Phys. Med. Biol.* 56:5079–5098.

Bagliesi, M.G., C. Avanzini, G. Bigongiari et al. 2011. A custom front-end ASIC for the readout and timing of 64 SiPM photosensors. *Nucl. Phys. B—Proc. Suppl.* 215:344–348.

Becquerel, H. 1896. Sur les radiations invisibles emises par les corps phosphorescents. *Comptes rendus de l'Academie des Sciences, Paris* 122:501–503.

Blumgart, H.L. and S. Weiss. 1927. Studies on the velocity of blood flow II. *J. Clin. Invest.* 4:15–31.

Blumgart, H.L. and O.C. Yens. 1927. Studies on the velocity of blood flow I. The method utilized. *J. Clin. Invest.* 4:1–13.

Bowen, S.L., A. Ferrero, and R.D. Badawi. 2012. Quantification with a dedicated breast PET/CT scanner. *Med. Phys.* 39:2694–2707.

Bowley, A.R., C.G. Taylor, D.A. Causer et al. 1973. Radioisotope scanner for rectilinear, arc, transverse section and longitudinal section scanning: (Ass—Aberdeen Section Scanner). *Br. J. Radiol.* 46:262–271.

Braga, L.H.C., L. Pancheri, L. Gasparini et al. 2011. A CMOS mini-SiPM detector with in-pixel data compression for PET applications. In *The 2011 IEEE Nuclear Science Symposium and Medical Imaging Conference Record (NSS/MIC)*, Valencia, Spain, pp. 548–552.

Brooks, R.A., V.J. Sank, G. Dichiro et al. 1980. Design of a high-resolution positron emission tomograph—Neuro-PET. *J. Comput. Assist. Tomogr.* 4:5–13.

Brownell, G.L. and C.A. Burnham. 1973. MGH positron camera. In *Tomographic Imaging in Nuclear Medicine*, edited by G.S. Freedman. New York: Society of Nuclear Medicine.

Brownell, G.L., C.A. Burnham, D.A. Chesler et al. 1977. Transverse section imaging of radionuclide distributions in heart, lung and brain. In *Reconstruction Tomography in Diagnostic Radiology and Nuclear Medicine*, edited by M.M. Ter-Pogossian, M.E. Phelps, G.L. Brownell, J.R. Cox, D.O. Davis, and R.G. Evens. Baltimore, MA: University Park Press.

Brownell, G.L. and W.H. Sweet. 1953. Localization of brain tumors with positron emitters. *Nucleonics* 11:40–45.

Brunner, S.E., L. Gruber, J. Marton et al. 2014. Studies on the Cherenkov effect for improved time resolution of TOF-PET. *IEEE Trans. Nucl. Sci.* 61:443–447.

Budinger, T.F., S.E. Derenzo, G.T. Gullberg et al. 1977. Emission computer-assisted tomography with single-photon and positron-annihilation photon emitters. *J. Comput. Assist. Tomogr.* 1:131–145.

Budinger, T.F. and G.T. Gullberg. 1974. Three-dimensional reconstruction in nuclear-medicine emission imaging. *IEEE Trans. Nucl. Sci.* 21:2–20.

Cassen, B., L. Curtis, and C.W. Reed. 1950. A sensitive directional gamma-ray detector. *Nucleonics* 6:78–81.

Cassen, B., L. Curtis, C. Reed et al. 1951. Instrumentation for I-131 use in medical studies. *Nucleonics* 9:46–50.

Castilla, J., J.M. Cela, A. Comerma et al. 2013. Evaluation of the FlexToT ASIC on the readout of SiPM matrices and scintillators for PET. *The 2013 IEEE Nuclear Science Symposium and Medical Imaging Conference Record (NSS/MIC)*, Seoul, South Korea, pp. 1–4.

Chen, Y., K. Saha, and S.J. Glick. 2011. Investigating performance of limited angle dedicated breast TOF PET. In *The 2011 IEEE Nuclear Science Sympoium and Medical Imaging Conference Record (NSS/MIC)*, Valencia, Spain, pp. 1346–2349.

Cherepy, N.J., G. Hull, A.D. Drobshoff et al. 2008. Strontium and barium iodide high light yield scintillators. *Appl. Phys. Lett.* 92:083508. http://dx.doi.org/10.1063/1.2885728.

Cherepy, N.J., S.A. Payne, S.J. Asztalos et al. 2009. Scintillators with potential to supersede lanthanum bromide. *IEEE Trans. Nucl. Sci.* 56:873–880.

Cherry, S.R., J.A. Sorenson, and M.E. Phelps. 2012. *Physics in Nuclear Medicine*. Philadelphia, PA: Elsevier/Saunders.

Chinn, G., P.D. Olcott, and C.S. Levin. 2012. Improved compressed sensing multiplexing for PET detector readout. *The 2012 IEEE Nuclear Science Symposium and Medical Imaging Conference Record (NSS/MIC)*, Anaheim, CA, pp. 2472–2474.

Cho, Z.H. and M.R. Farukhi. 1977. Bismuth germanate as a potential scintillation detector in positron cameras. *J. Nucl. Med.* 18:840–844.

Choong, W.S., G.J. Gruber, W.W. Moses et al. 2002. A compact 16-module camera using 64-pixel CsI(Tl)/Si P-I-N photodiode imaging modules. *IEEE Trans. Nucl. Sci.* 49:2228–2235.

Choong, W.S., W.W. Moses, C.S. Tindall et al. 2005. Design for a high-resolution small-animal SPECT system using pixellated Si(Li) detectors for in vivo I-125 imaging. *IEEE Trans. Nucl. Sci.* 52:174–180.

Comerma, A., D. Gascon, L. Garrido et al. 2013. Front end ASIC design for SiPM readout. *JINST* 8:C01048. doi 10.1088/1748-0221/8/01/C01048.

Conti, M., L. Eriksson, H. Rothfuss et al. 2009. Comparison of fast scintillators with TOF PET potential. *IEEE Trans. Nucl. Sci.* 56:926–933.

Cooper, R.J., A.J. Boston, H.C. Boston et al. 2009. Positron emission tomography imaging with the SmartPET system. *Nucl. Instrum. Methods A* 606:523–532.

Crandall, P.H. and B. Cassen. 1966. High speed section scanning of brain. *Arch. Neurol.* 15(2):163.

Curie, P., M. Curie, and G. Bemont. 1898. Sur une nouvelle substance fortement radio-active contenue dans la pechblende. *Comptes Rendus des l'Academie des Sciences, Paris* 127:1215–1217.

de Hevesy, G. 1913. Radioelements as aracers in chemistry and physics. *Chem News* 108:166.

de Vree, G.A., A.H. Westra, I. Moody et al. 2005. Photon-counting gamma camera based on an electron-multiplying CCD. *IEEE Trans. Nucl. Sci.* 52:580–588.

Del Guerra, A., A. Bartoli, N. Belcari et al. 2006. Performance evaluation of the fully engineered YAP-(S)PET scanner for small animal imaging. *IEEE Trans. Nucl. Sci.* 53:1078–1083.

Delfino, E.P., S. Majewski, R.R. Raylman et al. 2010. Towards 1 mm PET resolution using DOI modules based on dual-sided SiPM readout. In *The 2010 IEEE Nuclear Science Symposium and Medical Imaging Conference Record (NSS/MIC)*, pp. 3442–3449.

Deng, Z., A.K. Lan, X. Sun et al. 2010. Development of an 8-channel time based read-out ASIC for PET applications. In *The 2010 IEEE Nuclear Science Symposium and Medical Imaging Conference Record (NSS/MIC)*, pp. 1684–1689.

Derenzo, S.E. 1977. Positron ring cameras for emission-computed tomography. *IEEE Trans. Nucl. Sci.* 24:881–885.

Derenzo, S.E., T.F. Budinger, J.L. Cahoon et al. 1977. High-resolution computed tomography of positron emitters. *IEEE Trans. Nucl. Sci.* 24:544–558.

Derenzo, S.E., T.F. Budinger, R.H. Huesman et al. 1981. Imaging properties of a positron tomograph with 280-Bgo-crystals. *IEEE Trans. Nucl. Sci.* 28:81–89.

Derenzo, S.E., H. Zaklad, and T.F. Budinger. 1975. Analytical study of a high-resolution positron ring detector system for transaxial reconstruction tomography. *J. Nucl. Med.* 16:1166–1173.

Dey, S., E. Myers, T.K. Lewellen et al. 2013. A row-column summing readout architecture for SiPM based PET imaging systems. *The 2013 IEEE Nuclear Science Symposium and Medical Imaging Conference Record (NSS/MIC)*, Seoul, South Korea, pp. 1–5.

Doshi, N.K., Y.P. Shao, R.W. Silverman et al. 2000. Design and evaluation of an LSO PET detector for breast cancer imaging. *Med. Phys.* 27:1535–1543.

Doshi, N.K., R.W. Silverman, Y. Shao et al. 2001. maxPET: A dedicated mammary and axillary region PET imaging system for breast cancer. *IEEE Trans. Nucl. Sci.* 48:811–815.

Duvall, W.L., L.B. Croft, E.S. Ginsberg et al. 2011. Reduced isotope dose and imaging time with a high-efficiency CZT SPECT camera. *J. Nucl. Cardiol.* 18:847–857.

Enghardt, W., P. Crespo, F. Fiedler et al. 2004. Charged hadron tumour therapy monitoring by means of PET. *Nucl. Instrum. Methods A* 525:284–288.

Espana, S., R. Marcinkowski, V. Keereman et al. 2014. DigiPET: Sub-millimeter spatial resolution small-animal PET imaging using thin monolithic scintillators. *Phys. Med. Biol.* 59:3405–3420.

Esteves, F.P., P. Raggi, R.D. Folks et al. 2009. Novel solid-state-detector dedicated cardiac camera for fast myocardial perfusion imaging: Multicenter comparison with standard dual detector cameras. *J. Nucl. Cardiol.* 16:927–934.

Fontaine, R., M.A. Tetrault, F. Belanger et al. 2006. Real time digital signal processing implementation for an APD-based PET scanner with phoswich detectors. *IEEE Trans. Nucl. Sci.* 53:784–788.

Frisch, B. and EndoTOFPET-US Collaboration. 2013. Combining endoscopic ultrasound with time-of-flight PET: The EndoTOFPET-US project. *Nucl. Instrum. Methods A* 732:577–580.

Funk, T., P. Despres, W.C. Barber et al. 2006. A multipinhole small animal SPECT system with submillimeter spatial resolution. *Med. Phys.* 33:1259–1268.

Garibaldi, F., S. Capuani, S. Colilli et al. 2013. TOPEM: A PET-TOF endorectal probe, compatible with MRI for diagnosis and follow up of prostate cancer. *Nucl. Instrum. Methods A* 702:13–15.

Gariod, R., R. Allemand, E. Cormoreche et al. 1982. The LETI positron tomography architecture and time-of-flight improvement. In *Proceedings of Workshop on Time-of-Flight Tomography*, St. Louis, MO. Silver Spring, MD: IEEE Computer Society Press.

Geiger, H. and W. Muller. 1928. Electronic counter for measurement of weak activities. *Naturwissenschaften* 16:617–618.

Genna, S. and A.P. Smith. 1988. The development of ASPECT, an annular single-crystal brain camera for high-efficiency SPECT. *IEEE Trans. Nucl. Sci.* 35:654–658.

Glodo, J., W.W. Moses, W.M. Higgins et al. 2005. Effects of Ce concentration on scintillation properties of $LaBr_3$:Ce. *IEEE Trans. Nucl. Sci.* 52:1805–1808.

Goertzen, A.L., X.Z. Zhang, M.M. McClarty et al. 2013. Design and performance of a resistor multiplexing readout circuit for a SiPM detector. *IEEE Trans. Nucl. Sci.* 60:1541–1549.

Gundacker, S., A. Knapitsch, E. Auffray et al. 2014. Time resolution deterioration with increasing crystal length in a TOF-PET system. *Nucl. Instrum. Methods A* 737:92–100.

Haemisch, Y., T. Frach, C. Degenhardt et al. 2012. Fully digital arrays of silicon photomultipliers (dSiPM)—A scalable alternative to vacuum photomultiplier tubes (PMT). *Phys. Procedia* 37:1546–1560.

Harper, P.V., R.N. Beck, D.E. Charlest et al. 1965. The three dimensional mapping and display of radioisotope distributions. *J. Nucl. Med.* 6:332.

Heemskerk, J.W.T., M.A.N. Korevaar, J. Huizenga et al. 2010. An enhanced high-resolution EMCCD-based gamma camera using SiPM side detection. *Phys. Med. Biol.* 55:6773–6784.

Herbert, D.J., S. Moehrs, N. D'Ascenzo et al. 2007. The silicon photomultiplier for application to high-resolution positron emission tomography. *Nucl. Instrum. Methods A* 573:84–87.

Hoffman, E.J., A.R. Ricci, L.M.A.M. Vanderstee et al. 1983. Ecat-III—Basic design considerations. *IEEE Trans. Nucl. Sci.* 30:729–733.

Hong, S.J., H.G. Kang, G.B. Ko et al. 2012. SiPM-PET with a short optical fiber bundle for simultaneous PET-MR imaging. *Phys. Med. Biol.* 57:3869–3883.

Hu, W., Y. Choi, K. Hong et al. 2012. Free-running ADC- and FPGA-based signal processing method for brain PET using GAPD arrays. *Nucl. Instrum. Methods A* 664:370–375.

Hu, W., Y. Choi, J.H. Jung et al. 2009. A simple and improved digital timing method for positron emission tomography. In *The 2009 IEEE Nuclear Science Symposium and Medical Imaging Conference Record (NSS/MIC)*, Orlando, FL, pp. 3893–3896.

Huber, J.S., W.S. Choong, W.W. Moses et al. 2005. Characterization of a PET camera optimized for prostate imaging. *The 2005 IEEE Nuclear Science Symposium and Medical Imaging Conference Record (NSS/MIC)*, pp. 1556–1559.

Huber, J.S., S.E. Derenzo, J. Qi et al. 2001. Conceptual design of a compact positron tomograph for prostate imaging. *IEEE Trans. Nucl. Sci.* 48:1506–1511.

Hutton, B.F. 2014. The origins of SPECT and SPECT/CT. *Eur. J. Nucl. Med. Mol. Imag.* 41(Suppl. 1):S3–S16.

Ito, M., S.J. Hong, and J.S. Lee. 2011. Positron emission tomography (PET) detectors with depth-of-interaction (DOI) capability. *Biomed. Eng. Lett.* 1:70–81.

Jakoby, B.W., Y. Bercier, M. Conti et al. 2011. Physical and clinical performance of the mCT time-of-flight PET/CT scanner. *Phys. Med. Biol.* 56:2375–2389.

Jan, S., T. Frisson, and D. Sarrut. 2013. GATE simulation of C-12 hadrontherapy treatment combined with a PET imaging system for dose monitoring: A feasibility study. *IEEE Trans. Nucl. Sci.* 60:423–429.

Janecek, M., J.P. Walder, P.J. McVittie et al. 2012. A high-speed multi-channel readout for SSPM arrays. *IEEE Trans. Nucl. Sci.* 59:13–18.

Jaszczak, R.J. 2006. The early years of single photon emission computed tomography (SPECT): An anthology of selected reminiscences. *Phys. Med. Biol.* 51:R99–R115.

Jaszczak, R.J., L.T. Chang, N.A. Stein et al. 1979. Whole-body single-photon emission computed-tomography using dual, large-field-of-view scintillation cameras. *Phys. Med. Biol.* 24:1123–1143.

Jaszczak, R.J., P.H. Murphy, D. Huard et al. 1977. Radionuclide emission computed tomography of head with Tc-99m and a scintillation camera. *J. Nucl. Med.* 18:373–380.

Johnson, L.C., D.L. Campbell, E.L. Hull et al. 2011. Characterization of a high-purity germanium detector for small-animal SPECT. *Phys. Med. Biol.* 56:5877–5888.

Joung, J., R.S. Miyaoka, and T.K. Lewellen. 2002. cMiCE: A high resolution animal PET using continuous LSO with a statistics based positioning scheme. *Nucl. Instrum. Methods A* 489:584–598.

Kang, J., Y. Choi, K.J. Hong et al. 2010. Dual-ended readout PET detector module based on GAPD having large-area microcells. In *The 2010 IEEE Nuclear Science Symposium and Medical Imaging Conference Record (NSS/MIC)*, pp. 3205–3209.

Kao, C.-M., H. Kim, and C.-T. Chen. 2011. Event-time determination by waveform analysis for time-of-flight positron emission tomography. In *The 2011 IEEE Nuclear Science Symposium and Medical Imaging Conference Record (NSS/MIC)*, Valencia, Spain, pp. 3874–3879.

Karellas, A. and S. Vedantham. 2008. Breast cancer imaging: A perspective for the next decade. *Med. Phys.* 35:4878–4897.

Karp, J.S., S. Surti, M.E. Daube-Witherspoon et al. 2003. Performance of a brain PET camera based on anger-logic gadolinium oxyorthosilicate detectors. *J. Nucl. Med.* 44:1340–1349.

Keyes, J.W., N. Orlandea, W.J. Heetderks et al. 1977. Humongotron—Scintillation-camera transaxial tomograph. *J. Nucl. Med.* 18:381–387.

Kim, H., C.-T. Chen, H.-T. Chen et al. 2013. A TOF PET detector development using waveform sampling and strip-line based data acquisition. In *The 2013 IEEE Nuclear Science Symposium and Medical Imaging Conference Record (NSS/MIC)*, Seoul, South Korea, pp. 1–4.

Kim, H., C.-T. Chen, N. Eclov et al. 2014. A new time calibration method for switched-capacitor-array-based waveform samplers. *Nucl. Instrum. Methods A* 767:67–74.

Kim, H., C.-T. Chen, A. Ronzhin et al. 2012a. A silicon photomultiplier signal readout using transmission-line and waveform sampling for positron emission tomography. In *The 2012 IEEE Nuclear Science Symposium and Medical Imaging Conference Record (NSS/MIC)*, Anaheim, CA, pp. 1466–2468.

Kim, H., C.-T. Chen, A. Ronzhin et al. 2012b. A study on the optimal sampling speed of DRS4-based waveform digitizer for time-of-flight positron emission tomography application. In *The 2012 IEEE Nuclear Science Symposium and Medical Imaging Conference Record (NSS/MIC)*, Anaheim, CA, pp. 2469–2471.

Kim, H., H. Frisch, C.-T. Chen et al. 2010. A design of a PET detector using micro-channel plate photomultipliers with transmission-line readout. *Nucl. Instrum. Methods A* 622:628–636.

Kim H., L.R. Furenlid, M.J. Crawford et al. 2006. SemiSPECT: A small-animal SPECT imaging based on eight CZT detector arrays. *Med. Phys.* 3:465–474.

Kim, H., C.-M. Kao, S. Kim et al. 2011. A development of waveform sampling readout board for PET using DRS4. In *The 2011 IEEE Nuclear Science Symposium and Medical Imaging Conference Record (NSS/MIC)*, Valencia, Spain, pp. 2393–2396.

Kim, H., C.-M. Kao, Q. Xie et al. 2009. A multi-threshold sampling method for TOF-PET signal processing. *Nucl. Instrum. Methods A* 602:618–621.

Kishimoto, A., J. Kataoka, T. Kato et al. 2013. Development of a dual-sided readout DOI-PET module using large-area monolithic MPPC-arrays. *IEEE Trans. Nucl. Sci.* 60:38–43.

Knoll, G.F. 1999. *Radiation Detection and Measurement*, 3rd edn. New York: John Wiley & Sons.

Koolen, B.B., W.V. Vogel, M.J. Vrancken et al. 2012. Molecular imaging in breast cancer: From whole-body PET/CT to dedicated breast PET. *J. Oncol.* 2012:438647. doi:10.1155/2012/438647.

Korpar, S., R. Dolenec, P. Krizan et al. 2011. Study of TOF PET using Cherenkov light. *Nucl. Instrum. Methods A* 654:532–538.

Kuhl, D.E. and R.Q. Edwards. 1963. Image separation radioisotope scanning. *Radiology* 80:653–662.

Kuhl, D.E. and R.Q. Edwards. 1964. Cylindrical and section radioisotope scanning of the liver and brain. *Radiology* 83:926–936.

Kuhl, D.E., R.Q. Edwards, A.R. Ricci et al. 1976. Mark-4 system for radionuclide computed tomography of brain. *Radiology* 121:405–413.

Kuhn, A., S. Surti, J.S. Karp et al. 2006. Performance assessment of pixelated LaBr$_3$ detector modules for time-of-flight PET. *IEEE Trans. Nucl. Sci.* 53:1090–1095.

Kume, H., S. Muramatsu, and M. Iida. 1986. Position-sensitive photomultiplier tubes for scintillation imaging. *IEEE Trans. Nucl. Sci.* 33:359–363.

Kwon, S.I., J.S. Lee, H.S. Yoon et al. 2011. Development of small-animal PET prototype using silicon photomultiplier (SiPM): Initial results of phantom and animal imaging studies. *J. Nucl. Med.* 52:572–579.

Larsson, S.A. 1980. Gamma camera emission tomography. Development and properties of a multi-sectional emission computed tomography aystem. *Acta Radiol. Suppl.* 363:1–75.

Levin, C.S. 2012. Promising new photon detection concepts for high-resolution clinical and preclinical PET. *J. Nucl. Med.* 53:167–170.

Levine, E.A., R.I. Freimanis, N.D. Perrier et al. 2003. Positron emission mammography: Initial clinical results. *Ann. Surg. Oncol.* 10:86–91.

Li, Z., M. Wedrowski, P. Bruyndonckx et al. 2010. Nonlinear least-squares modeling of 3D interaction position in a monolithic scintillator block. *Phys. Med. Biol.* 55:6515–6532.

Lim, C.B., D. Chu, L. Kaufman et al. 1975. Initial characterization of a multi-wire proportional chamber positron camera. *IEEE Trans. Nucl. Sci.* 22:388–394.

Limousin, O. 2003. New trends in CdTe and CdZnTe detectors for X- and gamma-ray applications. *Nucl. Instrum. Methods A* 504:24–37.

Ling, T., T.K. Lewellen, and R.S. Miyaoka. 2007. Depth of interaction decoding of a continuous crystal cetector module. *Phys. Med. Biol.* 52:2213–2228.

Liu, C.Y. and A.L. Goertzen. 2014. Multiplexing approaches for a 12 × 4 array of silicon photomultipliers. *IEEE Trans. Nucl. Sci.* 61(1):35–43.

Llosa, G., J. Barrio, C. Lacasta et al. 2010. Characterization of a PET detector head based on continuous LYSO crystals and monolithic, 64-pixel silicon photomultiplier matrices. *Phys. Med. Biol.* 55:7299–7315.

Madsen, M.T. 2007. Recent advances in SPECT imaging. *J. Nucl. Med.* 48:661–673.

Majewski, S., J. Proffitt, J. Brefczynski-Lewis et al. 2011. HelmetPET: A silicon photomultiplier based wearable brain imager. In *The 2011 IEEE Nuclear Science Symposium and Medical Imaging Conference Record (NSS/MIC)*, Valencia, Spain, pp. 4030–4034.

Majewski, S., A. Stolin, P. Martone et al. 2011. Dedicated mobile PET prostate imager. *J. Nucl. Med.* 52S:1945.

Martinez, J.D., J.M. Benlloch, J. Cerda et al. 2004. High-speed data acquisition and digital signal processing system for PET imaging techniques applied to mammography. *IEEE Trans. Nucl. Sci.* 51:407–412.

Mathews, A.J., S. Komarov, H.Y. Wu et al. 2013. Improving PET imaging for breast cancer using virtual pinhole PET half-ring insert. *Phys. Med. Biol.* 58:6407–6427.

Meier, D., S. Mikkelsen, J. Talebi et al. 2010. An ASIC for SiPM/MPPC readout. In *The 2010 IEEE Nuclear Science Symposium and Medical Imaging Conference Record (NSS/MIC)*, pp. 1653–1657.

Meng, L.J. 2006. An intensified EMCCD camera for low energy gamma ray imaging applications. *IEEE Trans. Nucl. Sci.* 53:2376–2384.

Miller, B.W., H.H. Barrett, L.R. Furenlid et al. 2008. Recent advances in BazookaSPECT: Real-time data processing and the development of a gamma-ray microscope. *Nucl. Instrum. Methods A* 591:272–275.

Mitchell, D., C.B. Hruska, J.C. Boughey et al. 2013. Tc-99m-sestamibi using a direct conversion molecular breast imaging system to assess tumor response to neoadjuvant chemotherapy in women with locally advanced breast cancer. *Clin. Nucl. Med.* 38:949–956.

Miyake, K.K., K. Matsumoto, M. Inoue et al. 2014. Performance evaluation of a new dedicated breast PET scanner using NEMA NU4-2008 standards. *J. Nucl. Med.* 55:1198–1203.

Moehrs, S., A. Del Guerra, D.J. Herbert et al. 2006. A detector head design for small-animal PET with silicon photomultipliers (SiPM). *Phys. Med. Biol.* 51:1113–1127.

Moliner, L., A.J. Gonzalez, A. Soriano et al. 2012. Design and evaluation of the MAMMI dedicated breast PET. *Med. Phys.* 39:5393–5404.

Moses, W.W. 2003. Time of flight in PET revisited. *IEEE Trans. Nucl. Sci.* 50:1325–1330.

Moses, W.W. 2004. Positron emission mammography imaging. *Nucl. Instrum. Methods A* 525:249–252.

Moses, W.W. 2007. Recent advances and future advances in time-of-flight PET. *Nucl. Instrum. Methods A* 580:919–924.

Moses, W.W. and S.E. Derenzo. 1994. Design studies for a PET detector module using a pin photodiode to measure depth of interaction. *IEEE Trans. Nucl. Sci.* 41:1441–1445.

Moses, W.W. and M. Ullisch. 2006. Factors influencing timing resolution in a commercial LSO PET camera. *IEEE Trans. Nucl. Sci.* 53(1):78–85.

Muehllehner, G. 1970. Rotating collimator tomography. *J. Nucl. Med.* 11:347.

Muehllehner, G. 1971. A tomographic scintillation camera. *Phys. Med. Biol.* 16:87.

Muehllehner, G., F. Atkins, and P.V. Harper. 1977. Positron camera with longitudinal and transverse tomographic ability. In *Medical Radionuclide Imaging*. Vienna, Austria: IAEA.

Muehllehner, G. and J.S. Karp. 2006. Positron emission tomography. *Phys. Med. Biol.* 51:R117–R137.

Mullani, N.A., D.C. Ficke, and M.M. Terpogossian. 1980. Cesium fluoride—New detector for positron emission tomography. *IEEE Trans. Nucl. Sci.* 27:572–575.

Mullani, N.A., C.S. Higgins, J.T. Hood et al. 1978. ETTt-IV—Design analysis and performance-characteristics. *IEEE Trans. Nucl. Sci.* 25:180–183.

Mullani, N.A., J. Markham, and M.M. Terpogossian. 1980. Feasibility of time-of-flight reconstruction in positron emission tomography. *J. Nucl. Med.* 21:1095–1097.

Mullani, N.A., W.H. Wong, R.K. Hartz et al. 1982. Design of TOFPET: A high resolution time-of-flight positron camera. In *Proceedings of Workshop on Time-of-Flight Tomography*, St. Louis, MO. Silver Spring, MD: IEEE Computer Society Press.

Murphy, P., J. Burdine, M. Moore et al. 1978. Single photon emission computed tomography (ECT) of body. *J. Nucl. Med.* 19:683–683.

Nagarkar, V.V., I. Shestakova, V. Gaysinskiy et al. 2006. A CCD-based detector for SPECT. *IEEE Trans. Nucl. Sci.* 53:54–58.

Nishikido, F., N. Inadama, E. Yoshida et al. 2013. Four-layer DOI PET detectors using a multi-pixel photon counter array and the light sharing method. *Nucl. Instrum. Methods A* 729:755–761.

Nutt, R. 2002. 1999 ICP Distinguished Scientist Award. The history of positron emission tomography. *Mol. Imaging Biol.* 4(1):11–26.

O'Connor, M.K., S.W. Phillips, C.B. Hruska et al. 2007. Molecular breast imaging: Advantages and limitations of a scintimammographic technique in patients with small breast tumors. *Breast J.* 13:3–11.

Olcott, P.D., A. Fallu-Labruyere, F. Habte et al. 2006. A high speed fully digital data acquisition system for positron emission tomography. In *The 2006 IEEE Nuclear Science Symposium and Medical Imaging Conference Record (NSS/MIC)*, pp. 1909–1911.

Pani, R., P. Bennati, M. Betti et al. 2006. Lanthanum scintillation crystals for gamma ray imaging. *Nucl. Instrum. Methods A* 567:294–297.

Pani, R., A. Soluri, R. Scafe et al. 1999. Multi-PSPMT scintillation camera. *IEEE Trans. Nucl. Sci.* 46:702–708.

Parodi, K. 2012. PET monitoring of hadrontherapy. *Nucl. Med. Rev.* 15:C37–C42.

Parodi, K., T. Bortfeld, and T. Haberer. 2008. Comparison between in-beam and offline positron emission tomography imaging of proton and carbon ion therapeutic irradiation at synchrotron- and cyclotron-based facilities. *Int. J. Radiat. Oncol. Biol. Phys.* 71:945–956.

Patton, J., A.B. Brill, J. Erickson et al. 1969. A new approach to mapping 3-dimensional radionuclide distributions. *J. Nucl. Med.* 10:363.

Peterson, T.E. and L.R. Furenlid. 2011. SPECT detectors: The anger camera and beyond. *Phys. Med. Biol.* 56:R145–R182.

Peterson, T.E. and S. Shokouhi. 2012. Advances in preclinical SPECT instrumentation. *J. Nucl. Med.* 53:841–844.

Peterson, T.E., S. Shokouhi, L.R. Furenlid et al. 2009. Multi-pinhole SPECT imaging with silicon strip detectors. *IEEE Trans. Nucl. Sci.* 56:646–652.

Peterson, T.E., D.W. Wilson, and H.H. Barrett. 2003. Application of silicon strip detectors to small-animal imaging. *Nucl. Instrum. Methods A* 505:608–611.

Phelps, M.E., E.J. Hoffman, S.C. Huang et al. 1978. ECAT—New computerized tomographic imaging-system for positron-emitting radiopharmaceuticals. *J. Nucl. Med.* 19:635–647.

Phelps, M.E., E.J. Hoffman, N.A. Mullani et al. 1975. Application of annihilation coincidence detection to transaxial reconstruction tomography. *J. Nucl. Med.* 16:210–224.

Popovic, K., J.E. McKisson, B. Kross et al. 2014. Development and characterization of a round hand-held silicon photomultiplier based gamma camera for intraoperative imaging. *IEEE Trans. Nucl. Sci.* 61:1084–1091.

Powolny, F., E. Auffray, S.E. Brunner et al. 2011. Time-based readout of a silicon photomultiplier (SiPM) for time of flight positron emission tomography (TOF-PET). *IEEE Trans. Nucl. Sci.* 58:597–604.

Pshenichnov, I., I. Mishustin, and W. Greiner. 2006. Distributions of positron-emitting nuclei in proton and carbon-ion therapy studied with GEANT4. *Phys. Med. Biol.* 51:6099–6112.

Rankowitz S., J.S. Robertson, W.A. Higinbotham et al. 1962. Positron scanner for locating brain tumors. *BNL 6049 and 1962 IRE Int. Conv. Rec.* 10:49.

Raylman, R.R., S. Majewski, M.F. Smith et al. 2008. The positron emission mammography/tomography breast imaging and biopsy system (PEM/PET): Design, construction and phantom-based measurements. *Phys. Med. Biol.* 53:637–653.

Robertson, J.S. and A.M. Niell 1962. Use of a digital computer in the development of a positron scanning procedure. In *The Fourth IBM Medical Symposium*, Endicott, NY.

Rogers, W.L., N.H. Clinthorne, L. Shao et al. 1988. Sprint-Ii—A 2nd generation single photon ring tomograph. *IEEE Trans. Med. Imaging* 7:291–297.

Roncali, E. and S.R. Cherry. 2011. Application of silicon photomultipliers to positron emission tomography. *Ann. Biomed. Eng.* 39:1358–1377.

Roncali, E., V. Viswanath, and S.R. Cherry. 2014. Design considerations for DOI-encoding PET detectors using phosphor-coated crystals. *IEEE Trans. Nucl. Sci.* 61:67–73.

Ronzhin, A., M. Albrow, S. Los et al. 2013. A SiPM-based TOF-PET detector with high speed digital DRS4 readout. *Nucl. Instrum. Methods A* 703:109–113.

Roy, T., J. Ratheesh, and A. Sinha. 2014. Three-dimensional SPECT imaging with $LaBr_3$:Ce scintillator for characterization of nuclear waste. *Nucl. Instrum. Methods A* 735:1–6.

Russo, P., G. Mettivier, R. Pani et al. 2009. Imaging performance comparison between A $LaBr_3$:Ce scintillator based and a CdTe semiconductor based photon counting compact gamma camera. *Med. Phys.* 36:1298–1317.

Rutherford, E. and H. Geiger. 1908. An electrical method of counting the number of alpha-particles from radioactive substances. *Proc. R. Soc. Lond. Ser. a—Containing Papers Math. Phys. Character* 81(546):141–161.

Sacco, I., P. Fischer, M. Ritzert et al. 2013. A low power front-end architecture for SiPM readout with integrated ADC and multiplexed readout. JINST 8:C01023. doi:10.1088/1748-0221/8/01/C01023.

Schaart, D.R., H.T. van Dam, S. Seifert et al. 2009. A novel, SiPM-array-based, monolithic scintillator detector for PET. *Phys. Med. Biol.* 54:3501–3512.

Scheiber, C. 1996. New developments in clinical applications of CdTe and CdZnTe detectors. *Nucl. Instrum. Methods A* 380:385–391.

Scheiber, C. 2000. CdTe and CdZnTe detectors in nuclear medicine. *Nucl. Instrum. Methods A* 448:513–524.

Scheiber, C. and J. Chambron. 1992. CdTe detectors in medicine—A review of current applications and future perspectives. *Nucl. Instrum. Methods A* 322:604–614.

Scheiber, C. and G.C. Giakos. 2001. Medical applications of CdTe and CdZnTe detectors. *Nucl. Instrum. Methods A* 458:12–25.

Schlyer, D., P. Vaska, D. Tomasi et al. 2007. A simultaneous PET/MRI scanner based on RatCAP in small animals. In *The 2007 IEEE Nuclear Science Symposium and Medical Imaging Conference Record (NSS/MIC)*, pp. 3256–3259.

Schulz, D. and P. Vaska. 2011. Integrating PET with behavioral neuroscience using RatCAP tomography. *Rev. Neurosci.* 22:647–655.

Schulz, V., B. Weissler, P. Gebhardt et al. 2011. SiPM based preclinical PET/MR insert for a human 3T MR: First imaging experiments. In *The 2011 IEEE Nuclear Science Symposium and Medical Imaging Conference Record (NSS/MIC)*, Valencia, Spain, pp. 4467–4469.

Seifert, S., G. van der Lei, H.T. van Dam et al. 2013. First characterization of a digital SiPM based time-of-flight PET detector with 1 mm spatial resolution. *Phys. Med. Biol.* 58:3061–3074.

Shah, K.S., R. Farrell, R. Grazioso et al. 2001. Large-area APDs and monolithic APD arrays. *IEEE Trans. Nucl. Sci.* 48:2352–2356.

Shah, K.S., J. Glodo, M. Klugerman et al. $LaBr_3$:Ce scintillators for gamma-ray spectroscopy. *IEEE Trans. Nucl. Sci.* 50:2410–2413.

Shakirin, G., H. Braess, F. Fiedler et al. 2011. Implementation and workflow for PET monitoring of therapeutic ion irradiation: A comparison of in-beam, in-room, and off-line techniques. *Phys. Med. Biol.* 56:1281–1298.

Shakirin, G., P. Crespo, H. Braess et al. 2007. Influence of the time of flight information on the reconstruction of in-beam PET data. In *The 2007 IEEE Nuclear Science Symposium and Medical Imaging Conference Record (NSS/MIC)*, pp. 4395–4396.

Shao, Y.P., X.S. Sun, K. Lou et al. 2014. In-beam PET imaging for on-line adaptive proton therapy: An initial phantom study. *Phys. Med. Biol.* 59:3373–3388.

Sharir, T., P.J. Slomka, and D.S. Berman. 2010. Solid-state SPECT technology: Fast and furious. *J. Nucl. Cardiol.* 17:890–896.

Shibuya, K., F. Nishikido, T. Tsuda et al. 2008. Timing resolution improvement using DOI information in a four-layer scintillation detector for TOF-PET. *Nucl. Instrum. Methods A* 593:572–577.

Shokouhi, S., D.W. Wilson, S.D. Metzler et al. 2010. Evaluation of image reconstruction for mouse brain imaging with synthetic collimation from highly multiplexed SiliSPECT projections. *Phys. Med. Biol.* 55:5151–5168.

Siegel, S., R.W. Silverman, Y.P. Shao et al. 1996. Simple charge division readouts for imaging scintillator arrays using a multi-channel PMT. *IEEE Trans. Nucl. Sci.* 43:1634–1641.

Song, T.Y., H.Y. Wu, S. Komarov et al. 2010. A sub-millimeter resolution PET detector module using a multi-pixel photon counter array. *Phys. Med. Biol.* 55:2573–2587.

Spanoudaki, V.C. and C.S. Levin. 2014. Investigating the temporal resolution limits of scintillation detection from pixelated elements: Comparison between experiment and simulation. *Phys. Med. Biol.* 56:735–756.

Sportelli, G., N. Belcari, N. Camarlinghi et al. 2014. First full-beam PET acquisitions in proton therapy with a modular dual-head dedicated system. *Phys. Med. Biol.* 59:43–60.

Stankova, V., J. Barrio, J.E. Gillam et al. 2012. Multichannel DAQ system for SiPM matrices. In *The 2012 IEEE Nuclear Science Symposium and Medical Imaging Conference Record (NSS/MIC)*, Anaheim, CA, pp. 1069–1071.

Stolin, A., S. Majewski, G. Jaliparthi et al. 2014. Evaluation of imaging modules based on SensL array SB-8 for nuclear medicine applications. *IEEE Trans. Nucl. Sci.* 61:2433–2438.

Stratos, D., G. Maria, F. Eleftherios et al. 2013. Comparison of three resistor network division circuits for the readout of 4 × 4 pixel SiPM arrays. *Nucl. Instrum. Methods A* 702:121–125.

Streun, M., G. Brandenburg, H. Larue et al. 2002. Coincidence detection by digital processing of free-running sampled pulses. *Nucl. Instrum. Methods A* 487(3):530–534.

Surti, S. and J.S. Karp. 2008. Design considerations for a limited angle, dedicated breast, TOF PET scanner. *Phys. Med. Biol.* 53:2911–2921.

Surti, S., A. Kuhn, M.E. Werner et al. 2007. Performance of Philips gemini TF PET/CT scanner with special consideration for its time-of-flight imaging capabilities. *J. Nucl. Med.* 48:471–480.

Tanaka, E., H. Toyama, and H. Murayama. 1984. Convolutional image-reconstruction for quantitative single photon-emission computed-tomography. *Phys. Med. Biol.* 29:1489–1500.

Tashima, H., T. Yamaya, E. Yoshida et al. 2012. A single-ring OpenPET enabling PET imaging during radiotherapy. *Phys. Med. Biol.* 57:4705–4718.

Ter-Pogossian, M.M. 1977. Basic principles of computed axial tomography. *Semin. Nucl. Med.* 7:109–127.

Ter-Pogossian, M.M., D.C. Ficke, J.T. Hood, Sr. et al. 1982. PETT VI: A positron emission tomograph utilizing cesium fluoride scintillation detectors. *J. Comput. Assist. Tomogr.* 6:125–133.

Ter-Pogossian, M.M., N.A. Mullani, D.C. Ficke et al. 1981. Photon time-of-flight-assisted positron emission tomography. *J. Comput. Assist. Tomogr.* 5:227–239.

Ter-Pogossian, M.M., M.E. Phelps, E.J. Hoffman et al. 1975. A positron-emission transaxial tomograph for nuclear imaging (PETT). *Radiology* 114:89–98.

Thompson, C.J., A.L. Goertzen, E.J. Berg et al. 2012. Evaluation of high density pixellated crystal blocks with SiPM readout as candidates for PET/MR detectors in a small animal PET insert. *IEEE Trans. Nucl. Sci.* 59:1791–1797.

Thompson, C.J., K. Murthy, I.N. Weinberg et al. 1994. Feasibility study for positron emission mammography. *Med. Phys.* 21:529–538.

Thompson, C.J., Y.L. Yamamoto, and E. Meyer. 1979. Positome. 2. High-efficiency positron imaging device for dynamic brain studies. *IEEE Trans. Nucl. Sci.* 26:583–589.

Tornai, M.P., C.N. Archer, A.G. Weisenberger et al. 2001. Investigation of microcolumnar scintillators on an optical fiber coupled compact imaging system. *IEEE Trans. Nucl. Sci.* 48:637–644.

Truman, A., A.J. Bird, D. Ramsden et al. 1994. Pixellated Csi(Tl) arrays with position-sensitive PMT readout. *Nucl. Instrum. Methods A* 353:375–378.

van Dam, H.T., G. Borghi, S. Seifert et al. 2013. Sub-200 ps CRT in monolithic scintillator PET detectors using digital SiPM arrays and maximum likelihood interaction time estimation. *Phys. Med. Biol.* 58:3243–3257.

van Eijk, C.W.E. 2002. Inorganic scintillators in medical imaging. *Phys. Med. Biol.* 47:R85–R106.

van Loef, E.V.D., P. Dorenbos, K. Kramer et al. 2001. Scintillation properties of $LaCl_3:Ce^{3+}$ crystals: Fast, efficient, and high-energy resolution scintillators. *IEEE Trans. Nucl. Sci.* 48:341–345.

van Velden, F.H.P., R.W. Kloet, B.N.M. van Berckel et al. 2009. HRRT versus HR plus human brain PET studies: An interscanner test-retest study. *J. Nucl. Med.* 50:693–702.

Vaska, P., C. Woody, D. Schlyer et al. 2007. The design and performance of the 2nd-generation RatCAP awake rat brain PET system. In *The 2007 IEEE Nuclear Science Symposium and Medical Imaging Conference Record (NSS/MIC)*, pp. 4181–4184.

Vecchio, S., F. Attanasi, N. Belcari et al. 2009. A PET prototype for "In-Beam" monitoring of proton therapy. *IEEE Trans. Nucl. Sci.* 56:51–56.

Wang, Y.J., Z.M. Zhang, D.W. Li et al. 2012. Design and performance evaluation of a compact, large-area PET detector module based on silicon photomultipliers. *Nucl. Instrum. Methods A* 670:49–54.

Watanabe, M., K. Shimizu, T. Omura et al. 2002. A new high-resolution PET scanner dedicated to brain research. *IEEE Trans. Nucl. Sci.* 49:634–639.

Wernick, M. and J. Aarsvold. 2004. *Emission Tomography: The Fundamentals of PET and SPECT*. Amsterdam, the Netherlands: Elsevier Academic Press.

Wienhard, K., M. Schmand, M.E. Casey et al. 2002. The ECAT HRRT: Performance and first clinical application of the new high resolution research tomograph. *IEEE Trans. Nucl. Sci.* 49:104–110.

Williams, M.B., A.R. Goode, V. Galbis-Reig et al. 2000. Performance of a PSPMT based detector for scintimammography. *Phys. Med. Biol.* 45:781–800.

Wilson, C.T.R. 1911. On a method of making visible the paths of ionizing particles through a gas. *Proc. R. Soc. Lond. A* 85:285–288.

Wrenn, F.R., M.L. Good, and P. Handler. 1951. The use of positron-emitting radioisotopes for the localization of brain tumors. *Science* 113:525–527.

Wu, H.Y., D. Pal, T.Y. Song et al. 2008. Micro insert: A prototype full-ring PET device for improving the image resolution of a small-animal PET scanner. *J. Nucl. Med.* 49:1668–1676.

Xi, D.M., C.-M. Kao, W. Liu et al. 2013. FPGA-only MVT digitizer for TOF PET. *IEEE Trans. Nucl. Sci.* 60:3253–3261.

Xi, W.Z., J. Seidel, J.W. Kakareka et al. 2010. MONICA: A compact, portable dual gamma camera system for mouse whole-body imaging. *Nucl. Med. Biol.* 37:245–253.

Xie, Q., Y.B. Chen, J. Zhu et al. 2013. Implementation of LYSO/PSPMT block detector with all digital DAQ system. *IEEE Trans. Nucl. Sci.* 60:1487–1494.

Xie, Q., C.-M. Kao, Z. Hsiau et al. 2005. A new approach for pulse processing in positron emission tomography. *IEEE Trans. Nucl. Sci.* 52:988–995.

Yamamoto, S., M. Honda, T. Oohashi et al. 2011. Development of a brain PET system, PET-hat: A wearable PET system for brain research. *IEEE Trans. Nucl. Sci.* 58:668–673.

Yamamoto, S., M. Imaizumi, E. Shimosegawa et al. 2010. Development of a compact and high spatial resolution gamma camera system using $LaBr_3(Ce)$. *Nucl. Instrum. Methods A* 622:261–269.

Yamaya, T., T. Inaniwa, S. Minohara et al. 2008. A proposal of an open PET geometry. *Phys. Med. Biol.* 53:757–773.

Yamaya, T., T. Mitsuhashi, T. Matsumoto et al. 2011. A SiPM-based isotropic-3D PET detector X'tal cube with a three-dimensional array of 1 mm^3 crystals. *Phys. Med. Biol.* 56:6793–6807.

Yasillo, N.J., R.N. Beck, and M. Cooper. 1990. Design considerations for a single tube gamma-camera. *IEEE Trans. Nucl. Sci.* 37:609–615.

Yoon, H.S., G.B. Ko, S.I. Kwon et al. 2012. Initial results of simultaneous PET/MRI experiments with an MRI-compatible silicon photomultiplier PET scanner. *J. Nucl. Med.* 53:608–614.

Yoshida, E., S. Kinouchi, H. Tashima et al. 2012. System design of a small OpenPET prototype with 4-layer DOI detectors. *Radiol. Phys. Technol.* 5:92–97.

Yoshida, E., H. Tashima, H. Wakizaka et al. 2013. Development of a single-ring OpenPET prototype. *Nucl. Instrum. Methods A* 729:800–808.

Zeniya, T., H. Watabe, T. Aoi et al. 2006. Use of a compact pixellated gamma camera for small animal pinhole SPECT imaging. *Ann. Nucl. Med.* 20:409–416.

Zhou, J. and J.Y. Qi. 2009. Theoretical analysis and simulation study of a high-resolution zoom-in PET system. *Phys. Med. Biol.* 54:5193–5208.

Zhou, J.A. and J.Y. Qi. 2011. Adaptive imaging for lesion detection using a zoom-in PET system. *TEEE Trans. Med. Imaging* 30:119–130.

Zhu, X. and G. El Fakhri. 2013. Proton therapy verification with PET imaging. *Theranostics* 3:731–740.

Zhu, X.P., S. Espana, J. Daartz et al. 2011. Monitoring proton radiation therapy with in-room PET imaging. *Phys. Med. Biol.* 56:4041–4057.

Ziemons, K., E. Auffray, R. Barbier et al. 2005. The ClearPET (TM) project: Development of a 2nd generation high-performance small animal PET scanner. *Nucl. Instrum. Methods A* 537:307–311.

9 Review of Detectors Available for Full-Field Digital Mammography

Nico Lanconelli and Stefano Rivetti

CONTENTS

9.1 From Analog to Digital Mammography 233
9.2 Computed Radiography 238
    9.2.1 Dual-Side Reading 239
    9.2.2 Line-Scan Reading 240
    9.2.3 Columnar Photostimulable Phosphor 242
9.3 Flat-Panel Detectors 242
    9.3.1 Indirect Conversion Detector 242
        9.3.1.1 TFT Readout 243
        9.3.1.2 CMOS Readout 244
    9.3.2 Direct Conversion Detectors 245
        9.3.2.1 TFT Readout 246
        9.3.2.2 Optical Readout 246
        9.3.2.3 TFT with Hexagonal Array 246
9.4 Slot-Scanning Systems 249
    9.4.1 Indirect Conversion Approach with Tiled CCD 249
    9.4.2 Direct Conversion with Photon-Counting Detectors 249
References 251

9.1 FROM ANALOG TO DIGITAL MAMMOGRAPHY

Mammography is considered the best tool for the early detection of breast cancer and puts the highest demands on the imaging systems, among all the medical applications. In fact, both high spatial and contrast resolutions are needed for visualizing small structures as microcalcifications and for differentiating breast tissues with very similar X-ray absorption properties, such as tumor opacities. With these requirements, mammography necessitates very small detecting elements and high signal-to-noise ratio (SNR): it is one of the most challenging imaging technologies and is the most recent X-ray-based imaging technology converted to the digital world. Conventional screen-film (analog) imaging is still far from perfect; approximately 10% of breast cancers that are detected by breast self examination or physical examination are not visible by screen film [1]. This is particularly a problem with radiographically dense breasts, affecting between 15% and 50% of women, depending on the definition of density used [2,3]. In such cases, a false negative can result when the cancer is obscured by overlying and surrounding fibroglandular tissue.

In conventional analog mammography, the film acts as an image acquisition detector as well as a storage and display device. The film can still be considered the state of the art in providing excellent spatial resolution of high-contrast structures. It can provide spatial resolution outcomes that can be as high as 15–20 lp/mm but with a very low associated contrast. However, one must also take into account that noise can limit the reliability of detection, especially for small or subtle structures. All radiological images contain random fluctuations or noise owing to the statistical nature of the X-ray quantum absorption. The ideal situation for any imaging system, with respect to the noise, is when the X-ray quantum noise is the dominant source of random fluctuations. Additional sources of noise come from the structure of the fluorescent screen and the granularity of the film emulsion used to record the image. Generally, mammographic screen-film systems are not quantum limited, and particularly for fine breast structures, the noise is dominated by fluctuations associated with the imaging system [4]. In screen-film systems, a trade-off is made between the spatial resolution and the detection efficiency of the X-ray image. To achieve high-resolution characteristics, mammographic screens must be kept relatively thin in order to limit the blur resulting from the lateral diffusion of light in the screen. Such a design would result in low detection efficiency and requires increased radiation dose to achieve the desired image quality. There is a further trade-off to manage between the dynamic range (film latitude) and the contrast resolution (film gradient): to have high contrast resolution, the dynamic range has to be reduced. The exposure dynamic range is the ratio between the X-ray fluence providing the maximum signal that the detector can accommodate and the fluence that provides a signal equivalent to the noise of the detector. A wide dynamic range is needed to include information from both dense regions, the fatty tissue and the skin line. The range of screen films in mammography is limited to a factor of about 25–50. This may be a problem because, depending on the composition of the breast, the maximum range of transmitted exposure can be 100:1 or more. Then, even though adequate attenuation contrast is provided by the X-ray beam, the final contrast displayed to the radiologist may be reduced severely because of the limited exposure dynamic range of the film. This is a concern, particularly in patients with dense breasts.

Digital mammography has emerged during the last decade as the new technology to be used for early detection of breast cancer. It has the potential to overcome many of the limitations of screen-film mammography, as limited latitude, reduced contrast, and low efficiency. The transition from screen-film to digital mammography has been very slow, since it started with the computed radiography (CR) systems about 25 years ago, but the Food and Drug Administration approved digital mammography systems only at the beginning of this century. With digital mammography, acquisition, display, and storage are performed independently, allowing for optimization of each. The expected improvement in image quality offers potential advantages in terms of more accurate detection and diagnosis.

Digital mammography can provide better dynamic range and linearity compared to film, leading to better contrast resolution. This should give a better perception of the diagnostic findings and may help to discover more subtle features indicative of cancer. Besides, the improved features increase the ability to distinguish between potential cancers and harmless tissue abnormalities. With digital imaging, it is also much easier to extract quantitative information from the images, allowing an exhaustive manipulation and analysis to be achieved with software tools. Digital detectors are able to give a dynamic range wider than the limited one obtained with the screen-film combination. In practice, high dynamic ranges produce a large ratio between the X-ray attenuation of the most radiopaque and the most radiolucent regions present in the same image, allowing an increased contrast on the final image. Better contrast properties give the capability to differentiate structures of interest with lower relative contrast to the background compared to film. Besides, postprocessing techniques can be used for further increasing the contrast of the lesions and providing better visibility of the breast structures. Digital detectors are able to provide a dynamic range up to 5000 (up to 50 times the dynamic range of typical screen film).

To be suitable for mammography, a detector must be able to depict as much of the breast tissue as possible. In other words, its field of view should be sufficiently large to include the entire breast. In this case, the term full-field digital mammography (FFDM) is considered, and this requirement is usually satisfied with detectors that can acquire images of 18 cm × 24 cm (only for small breasts) or 24 cm × 30 cm (this format can cover all the breast sizes). It is also essential that the detector can be positioned as close to the chest wall as possible, in order to avoid losing any part of the inner breast tissue. Digital detectors should guarantee the absence of any kind of artifacts, such as the presence of "ghost" images or dead pixels (or dead area) within the detector field of view. Some of the principal features of analog and digital mammography are summarized in Table 9.1.

Digital detectors usually provide detective quantum efficiency (DQE) outcomes better than those coming from screen film. This improved performance can be exploited for reducing the dose to the patient, keeping the same SNR, or

TABLE 9.1
Main Characteristics of Analog and Digital Mammography Systems

Characteristics	Analog	Digital
Exposure time	2–3 s	1–6 s
Average dose per view	Less than 3.0 mGy	Comparable to analog or less
Detector type	Screen-film cassette	Solid-state device (linear or 2D)
Detector size	18 × 24 cm or 24 × 30 cm	18 × 24 cm or 24 × 30 cm
Spatial resolution	15–20 lp/mm (33–25 μm)	5–13 lp/mm (100–40 μm)
Image contrast resolution	Approximately 0.04	Comparable to analog or better
Dynamic range (latitude)	25–100	1000–5000
Noise	Quantum and film granularity	Quantum and electronic
Image development	Wet processing	Digital or (eventually) laser printers
Image display	Light box	LCD or CRT (dated technology)

for increasing the SNR, keeping the dose constant. The inherent superior contrast and low noise properties should compensate for lower limiting spatial resolution in digital mammography. In addition to the potential for equivalent or superior image quality, an FFDM system opens to mammography the world of processing and management of digital data. A major breakthrough in the organization of mammography centers will be the possibility to review, to handle, and to store the digital images. Digital image management allows sites to reduce the volume required to archive data and to facilitate the access to the patient files. With a soft copy reading, a substantial cost saving can be made with the elimination or the reduction of film and chemistry. Film storage and handling are no longer required nor are the associated resources to conduct these tasks. Finally, a soft copy workstation makes possible the transfer of images to off-site experts and between remote locations and health centers.

During the last couple of decades, many approaches based on different technologies were considered for the development of detectors for FFDM: some of them reached the market and are currently used in practical clinical systems [5,6]. An outline of the current technologies available for FFDM is shown in Figure 9.1.

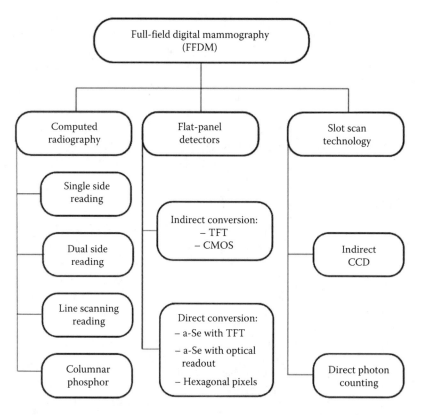

FIGURE 9.1 Current approaches to clinical systems for full-field digital mammography.

Three major types of detectors for FFDM can be identified, according to the technologies used [7]:

1. CR detectors, the first technology available for obtaining a digital version of the conventional screen-film mammograms. CR is based on photostimulable phosphor (PSP) plates and was intended to replace the screen-film cassette in a conventional system for analog mammography.
2. Flat-panel (FP) detectors, grouped into direct and indirect conversion, in accordance to the mechanism involved in the X-ray detection. Detectors based on the indirect conversion employ a scintillator as the primary detector of X-rays and a photodetector, as amorphous silicon (a-Si), optically coupled to a scintillation screen. Structured thallium-activated cesium iodide (CsI:Tl) scintillators are commonly used to this purpose. The readout can be achieved through arrays of thin-film transistors (TFTs) or complementary metal-oxide-semiconductor (CMOS) elements. Detectors based on the direct conversion technology employ a photoconducting layer (PCL) (usually an amorphous selenium [a-Se] layer) that converts X-ray energy to electric charge without the intermediate production of scintillation light. The readout is usually achieved by means of arrays of TFTs or can also be realized through an optical switch.
3. Linear detectors used as slot-scan systems with fan beam collimated beams. Also these systems can be grouped into direct and indirect conversion (as seen for FP detectors), even if the readout technology here is typically based on charge-coupled device (CCD) or photon-counting electronics.

There are a certain number of units for FFDM available on the market from different manufacturers. We can group these systems in accordance to their detector (some systems employed the same detector):

- CR systems based on BaFBr:Eu of BaF(BrI):Eu phosphor and 50 μm pixel size: Fujifilm FCR Profect, Carestream DirectView, Agfa CR, and Philips PCR Eleva
- CR systems based on BaFI:Eu phosphor and 43.8 μm pixel size: Konica Pureview and Konica Regius
- FP systems based on indirect conversion, CsI scintillator on a-Si TFT array with 100 μm pixel size: General Electric Senographe
- FP systems based on direct conversion, a-Se on TFT array with 85 μm pixel size: Siemens Inspiration, Philips MammoDiagnost DR, IMS Giotto, and Planmed Nuance
- FP systems based on direct conversion, a-Se on TFT array with 70 μm pixel size: Hologic Selenia and Siemens Mammomat
- FP systems based on direct conversion, a-Se with optical readout and 50 μm pixel size: Fujifilm Amulet
- FP systems based on direct conversion, a-Se with TFT array with hexagonal layout and 50 μm pixel size: Fujifilm Amulet Innovality
- Slot-scanning systems based on photon-counting linear detectors with 50 μm pixel size: Philips MicroDose

9.2 COMPUTED RADIOGRAPHY

CR was the first digital technology made available for FFDM and is characterized by detectors based on photostimulable luminescence [8]. On one hand, CR has many of the advantages of digital detectors, such as high dynamic range, contrast enhancement, archival, and the ability to provide FFDM with a mammographic system designed for screen-film use. On the other hand, it is a low-productivity technology, since the delay between the detector's exposure and readout is much similar to screen film. In this technique, an image phosphor plate, typically barium fluorobromide or iodide compound doped with trace amounts of europium (BaFBr:Eu^{2+}, BaFI:Eu^{2+}), is laid on a suitable substrate in the form of a portable cassette.

The photostimulated luminescence (PSL) is based on bromine vacancies, which act as traps for the electrons that migrate to the conduction band of the crystal. This migration happens when the image plate (IP) is exposed to X-rays, every time an X-ray interacts with the phosphor. The latent image (i.e., the excited electron distribution due to the trapping effect) remains stable for several hours and the number of trapped electrons is proportional to the intensity of the X-ray beam at each point of the IP. This stored signal can be subsequently readout with a proper stimulation of the IP with a laser beam (usually red light), as shown in Figure 9.2. The stimulating beam is focused and moved in a way that scans, point by point, the entire IP. In the readout step, the electrons stimulated by laser can return to the valence band by emitting light, giving rise to the well-known PSL. This emitted light is then detected by an optical collecting system and a photomultiplier tube, thereby generating a digital signal and finally the image, as illustrated in Figure 9.3.

An optical filter is used to avoid that the stimulating light interferes with the measurement. The coordinates of each pixel of the acquired image are determined by the

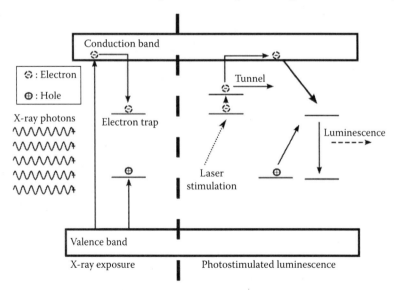

FIGURE 9.2 Simplified scheme of the energy levels involved in the photostimulated luminescence mechanism.

Review of Detectors Available for Full-Field Digital Mammography

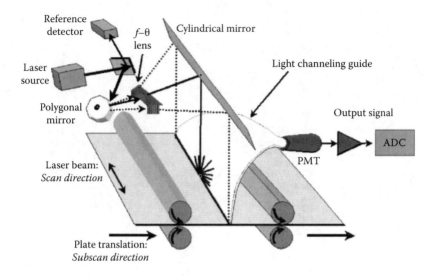

FIGURE 9.3 The main components involved in the readout process for computed radiography plates: the stimulating laser source, a beam splitter, oscillating beam deflector lens, stationary reflecting mirror, light collection guide, photomultiplier tube, and analog-to-digital converter system.

time at which the laser beam strikes a given location. Hence, the size and the shape of the laser spot and the distance between the sample pitch define the spatial sampling of the digital image. The readout of CR plates presents considerable challenges when high spatial resolution and DQE are necessary, as in mammography. In the standard CR system implementation, phosphor crystal is unstructured and deposited as "powder." In this case, the scattering of light within the phosphor layer produces a degradation of the spatial resolution. The readout of emitted light is achieved from only the upper surface of the IP, allowing acquisition of images with a pixel size of around 100 μm and typical limiting spatial resolution of about 3.5 lp/mm.

9.2.1 Dual-Side Reading

Due to the strict requirements of mammography, CR systems have demanded the enhancement of the efficiency of phosphor plates' readout. This has been achieved with an application of a dual-side reading approach [9]. This technology requires the phosphor deposition on a clear support medium, thus allowing the light emitted during the scanning process to be detected also on the "back side" of the storage phosphor plate (Figure 9.4). A second light guide, coupled to a second photomultiplier tube, is employed in this case, in order to collect this "back" light. This approach allows an improvement of both the overall sensitivity and the DQE of the system [10], if compared to the traditional single side reading, without increasing too much the cost of the unit. The two images (from the upper and back sides) are simultaneously acquired by the CR reader and added together to produce the final image. Dual-side reading permits to achieve images with very small pixel size, nominally at 50 μm [11].

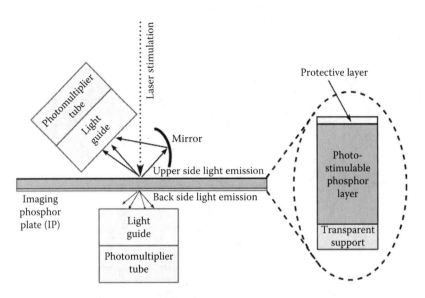

FIGURE 9.4 Simplified scheme of the photostimulable image plate used for the dual-side reading technology.

In the dual-side reading technique, the final image is achieved through an additive process, where the *front* and *back* images are combined in Fourier space (Figure 9.5). This process is very critical for the quality of the final image and must be optimized at all the spatial frequencies. The filter shape used in the frequency domain is designed in order to maximize the efficiency throughout all frequencies, thus allowing a significantly improved image quality, if compared with that reached with the conventional single side reading system.

9.2.2 Line-Scan Reading

An advanced readout method for PSP realized in the last few years is the line-scan reader, originally developed to obtain a faster readout of the IPs. Its working principle is based on the stimulation of the phosphor plate one line at a time, instead of the conventional point-by-point stimulation. The so-generated photoluminescence signal is then acquired by a linear array of photodetectors based on CCD elements. In fact, the scanning module consists of three main components:

1. Several linear laser units for the excitation of the plates.
2. An array of optical light collection lenses deployed along the length of the scan unit.
3. A high-sensitivity linear array of photosensitive elements. These CCDs are positioned in a way to be able to capture the resultant PSL signal simultaneously, one row at a time. Since an entire line is stimulated and read, in this case a lens array is required to focus the light along each point of the stimulated IP to a corresponding point on the CCD array (Figure 9.6).

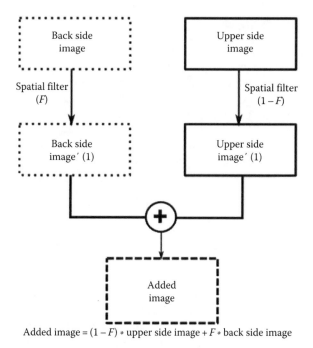

FIGURE 9.5 The final image with the dual-side reading technique is achieved through a combination of the two images (back side and upper side).

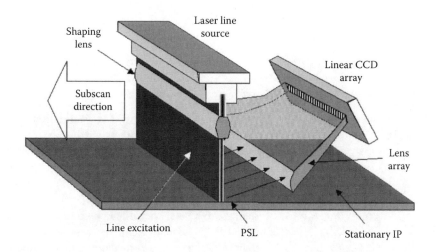

FIGURE 9.6 The main components of a computed radiography system based on a line-scan detector: the laser source, shaping lens, photostimulated luminescence lens array, and charge-coupled device (CCD) array move as a unit over the stationary imaging plate. The lens array aims to focus the light emerging from the image plate onto the corresponding detector elements of the CCD array.

9.2.3 COLUMNAR PHOTOSTIMULABLE PHOSPHOR

Another remarkable advancement that recently appeared in the CR systems for mammography is connected to the use of columnar structured storage phosphors based on cesium bromide [12,13]. In fact, this crystal can be grown to form needle-like or columnar structures. The main goal of using columnar phosphors is to reduce the light spread of the PSL emission, since in this case the light cannot move laterally within the phosphor, as it happens with the conventional *powder* scintillator. In practice, the phosphor needles act as fiber optics, making it possible to have a thicker detector with minimal loss of spatial resolution. Indeed, the columnar phosphor can provide a good combination of both detection efficiency and spatial resolution, characteristics that are usually achieved through a trade-off when unstructured phosphor materials are considered.

9.3 FLAT-PANEL DETECTORS

FP detectors, sometimes called direct radiography systems, is a generic way to indicate a kind of large-area stationary detectors that make use of different materials for converting X-rays first in a digital signal and then in an image. The requirement for this sort of systems is to produce an image in a few seconds without the necessity of manually handling the detector. An FP is composed of a matrix of individual detector elements arranged in rows and columns, with a spacing dimension from 50 to 100 µm. FP detectors are usually classified in direct or indirect conversion, in accordance to the X-ray conversion type exploited [14]. Another classification criterion for FPs is related to the readout method applied for sampling the signal and then producing the final image. The most common method is based on TFTs, but other promising technologies are now accessible. In the next few sections, a brief description of the FPs developed for FFDMs is given, for all the systems currently available in the market.

9.3.1 INDIRECT CONVERSION DETECTOR

Indirect conversion refers to the fact that the incoming X-rays are first converted into visible light through a scintillator phosphor: the light is successively detected by a sensor, typically a crystalline or a-Si photodiode, and converted in electric charges. The charge signal is finally sampled through a TFT or CMOS readout. The most common scintillator used in clinical FPs is CsI:Tl. Scintillator-based FPs employed in FFDM are characterized by a pixel size of 100 µm or less [15–17]. A compromise must be realized between the efficiency and the spatial resolution: Indeed, on the one hand, thick phosphors can increase the X-rays' absorption efficiency, but, on the other hand, they also contribute to a degradation of the spatial resolution. To optimize the trade-off between efficiency and spatial resolution, the phosphor can be deposited in a columnar structure that acts as a fiber-optic light guide for the visible light produced by X-ray interactions. This effect is achieved through the difference in refractive index (around 15%) between the phosphor and the air that fills the space among columns, thus allowing a reduction of

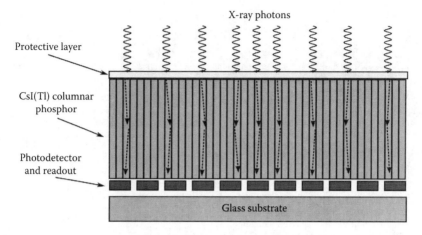

FIGURE 9.7 Sketch of a typical indirect conversion flat-panel detector, based on thallium-activated cesium iodide columnar phosphor. The lateral light diffusion is reduced and usually confined within one single (or a few) *column* of the scintillator.

the lateral light diffusion and an increase of the spatial resolution. Figure 9.7 shows a sketch of a typical indirect conversion detector for FFDM.

9.3.1.1 TFT Readout

Active matrix flat-panel imagers is the name typically used for an FP detector based on a TFT active matrix array (AMA) readout: this kind of detector appeared in the market at the very beginning of this century and is, up to date, one of the most widespread. The key element of this detector is the TFT that acts as an electronic switch. An array of a-Si TFTs is deposited upon the detector substrate with lithographic etching and material evaporation techniques. Each detector element (corresponding to the image pixel) includes many components: at least a TFT, a charge collection electrode, and a capacitor for storing charges must be present for each cell element. The electronic interconnections, including gate and drain lines, are connected to each of the TFTs to control the on/off status of each pixel and to provide the connection to the charge amplifiers, respectively. All the TFTs are off during the X-ray exposure, in order to collect the induced charges, proportional to the incident X-ray fluence (Figure 9.8). Once the exposure has ended, the readout of the array occurs one row at a time, by activating the respective gate line, which turns on the TFTs and allows the stored charge to flow along the columns from each del capacitor via drain lines to the corresponding charge amplifier [18]. Banks of amplifiers simultaneously amplify the charge, convert them to a proportional voltage, digitize signals in parallel from each row of the detector matrix, and produce a corresponding row of integer values in the digital image matrix (Figure 9.8). This process is repeated for each row of the matrix, giving rise to the final digital image. The detector readout is very fast and governed by the intrinsic lag characteristics of the X-ray photodiode and the switching speed of the TFT electronics. This readout type (for both indirect and direct conversion) makes FPs ready for advanced mammography applications such as tomosynthesis.

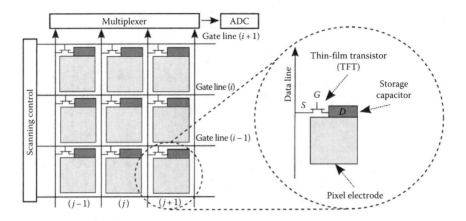

FIGURE 9.8 Diagram of a generic thin-film transistor readout employed in full-field digital mammography detectors. The charge distribution residing on the flat-panel storage capacitor is read out by scanning the arrays row by row. The final image is achieved through the peripheral electronics, which multiplexes the parallel columns to a serial digital signal.

9.3.1.2 CMOS Readout

CMOS image sensors first appeared in the late 1960s with the so-called passive pixel architecture [19]. During the end of the last century, CMOS *active* pixel sensor (APS) technology has been introduced in medical imaging applications and digital X-ray detectors based on CMOS reemerged as an alternative to CCD [20]. APS elements integrate the signal and operate by resetting the photoelement (usually a photodiode) in each pixel, allowing charge to accumulate and finally sensing the charge value. In contrast to the original *passive* architecture, the term *active* refers to the presence of at least one source follower transistor in each pixel, which buffers and/or amplifies the accumulated signal [21]. This allows the transmission of the signal onto a common readout bus as a voltage rather than as a charge. A CMOS imager is an indirect conversion system, consisting of a CMOS sensor sensitive to visible light photons optically coupled to a scintillator. CMOS detectors have the potential for low-cost mass production and low power consumption that could offer an alternative. Currently, there are FP detectors based on CMOS readout available on the market that can be used for FFDM [22,23].

The CMOS sensor is optically coupled to a structured (i.e., columnar) CsI:Tl scintillator, which acts as converter for the incoming X-ray photons. A fiber-optic plate is attached to the sensor's surface to reduce or remove the possibility of direct absorption of X-rays in the CMOS sensor. The image is captured after the pixel integration start level is set. During this phase, the pixel voltage level rises because of the light signal affecting the junction of the pixel photodiode. The acquired charge signal is proportional to the incoming light intensity in the pixel. After the integration is finished, the sensor is read pixel by pixel by using the row and column transistors for multiplexing only one pixel from the array at time. The pixel current is converted to voltage signal and then digital signal by analog-to-digital conversion. Then several digital processing steps are used to produce the final output image. The APS architecture also leads to an improved readout speed and SNR, due to the decreased read noise.

CMOS detectors are based on single crystalline technology, whereas other kinds of detectors are based on a polycrystalline or amorphous form of silicon. The mobility of charge carriers is higher when the semiconductor is in its crystalline structure and lower in the polycrystalline and amorphous form. This is one of the main reasons why CMOS detectors are advantageous over a-Si ones: the higher readout speed. In fact, CMOS imagers can reach a readout speed about 10 times faster than conventional a-Si TFT arrays [24]. The higher mobility also allows building electronic components with a dense structure, so readout and drive electronics can be integrated into the CMOS, without the need of implementing them as external circuits. This can decrease the cost of the overall architecture. A further advantage of CMOS detectors is their low noise, thanks to the presence of an active circuit. In fact, for conventional a-Si imagers, the noise generated in each pixel is amplified by the dataline capacitance and resistivity that can be very high for large arrays. The CMOS circuitry has the potential of eliminating this effect, thus reducing the overall noise of the detector.

The recent advances in the performance of CMOS with APS elements led the potential for diagnostic mammography, the major limitation being the size of the single chip. In fact, to cover the entire field of view required for mammography, tiling with various small subimagers is required. At present, the maximum size of a single CMOS crystalline wafer is around 20 cm × 20 cm, less than the size of a mammogram, and therefore, tiling of some wafers is needed for achieving the desired size. The pixel pitch of the available CMOS imagers is 75 µm, well in line to the other existing detectors for FFDM.

9.3.2 Direct Conversion Detectors

Direct detection does not make use of the scintillator, and the conversion is made directly from X-rays to electrons after interaction in an a-Se photoconductive layer (Figure 9.9). When an interaction of the incident X-rays occurs with the thin a-Se layer (100–200 µm thick), an electric charge is created in the material in the form of

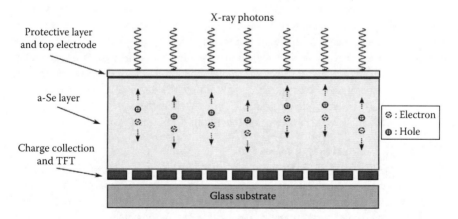

FIGURE 9.9 Sketch of a direct conversion full-field digital mammography detector based on an a-Se layer. Very high spatial resolution can be achieved, thanks to the inherent nature of the direct conversion process.

electron–hole pairs [25]. The charge signal can then be collected onto the readout surface, by applying an electric field between the electrodes placed on the upper and lower surfaces of the a-Se layer. The lateral spreading of the signal is dramatically reduced in direct conversion detectors, since there is no production of visible light, allowing an increase of the overall spatial resolution of the detector.

9.3.2.1 TFT Readout

In the last decade, considerable improvement has been realized in the manufacturing of a-Se layers and also in advanced charge-readout methods. The most common readout for direct conversion FPs can be generated from a layer of a-Si in a very similar way to that described for the phosphor FP systems. The main difference is that the photodiodes are here replaced by a set of simple electrode pads that collect the charge within the a-Se layer. The created electron–hole pairs are drifted by means of a bias voltage applied to the photoconductor. In this way, the charges can be collected and stored in a micro capacitor present within each detecting element and subsequently readout by the AMA, one of the most common readout methods used in clinical systems. Currently available FFDM detectors have pixel size in the range of 70–85 µm with a thickness of the a-Se layer of about 250 µm, which resulted to provide very good spatial resolution and DQE.

9.3.2.2 Optical Readout

During the last few years, a different way for the readout of the signal generated in the a-Se layer appeared on the market. This approach is based on an optical switch, instead of the conventional TFT readout employed in the majority of the direct conversion FPs. In this case, the detector includes two a-Se layers: the first one where the X-rays are directly converted into electron–hole pairs and the second one acting as an optically controlled switch. The entire detector consists of six main components: a negative top electrode, a thick X-ray PCL (with a thickness of less than 200 µm), an electron trapping layer (ETL), a thin readout PCL, stripe electrodes, and an optical source [26–28]. A strong electric field is applied to the electrodes, for guiding the generated electrons toward the ETL, where they are temporarily stored and a latent electron image is formed. Once the X-ray exposure has ended, the negative voltage is turned off and the top electrode becomes grounded. In this way, trapped electrons induce positive charges on the stripe electrodes that generate electric field in the readout PCL. In the subsequent readout phase, a linear optical source is used to irradiate the lower side of the detector, making possible the generation of electron–hole pairs in the readout PCL. These charges are drifted and finally collected on the stripe electrodes, where they can be detected as signal charges (Figure 9.10). This detector is on the market with a pixel size of 50 µm.

9.3.2.3 TFT with Hexagonal Array

A detector dedicated for FFDM that uses a TFT layout based on hexagonal geometry instead of the conventional square one recently appeared on the market. The possibility of having hexagonal arrays is very interesting because of the higher sampling efficiency, if compared to a traditional square grid, and the isotropic resolution. In nature, for example, this arrangement is present in few animal vision systems, where the fovea can be described by a regular hexagonal tessellation [29]. Some hints might be useful for understanding the benefit of using hexagonal geometry in X-ray detectors.

Review of Detectors Available for Full-Field Digital Mammography

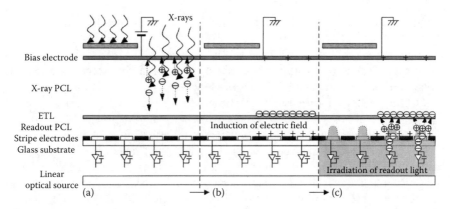

FIGURE 9.10 Sketch of the detector of a flat-panel system based on optical readout. The main components of the detector are a negative top electrode, a thick X-ray photoconducting layer (PCL), an electron trapping layer, a thin readout PCL, stripe electrodes, and an optical source. The optical switch scheme imaging processes are also shown from left to right: (a) conversion, (b) accumulation to (c) readout and behaviors of charges in the detector.

The advantages can be seen both on the acquisition process and on the sampling of the signal. In fact, for the hexagonal TFTs, the electric field distribution for each detector element is much more uniform, if compared to the square TFT. The main reason is because the square TFT presents a weaker electric field in the regions in the proximity of the corners [30]. This higher uniformity of the electric field corresponds to a greater collection of the electric charges and thus an increase on the system's efficiency.

In terms of signal sampling, Nyquist frequency is superior in the hexagonal grid, when the same element area is considered. Indeed, with the application of the hexagonal grid, a wider spectrum can be sampled without coming across aliasing, with the same numbers of elements [31]. In other words, using the hexagonal structure, the sampling of the same signal requires about 15% less number of elements for obtaining the same spatial resolution, as that achieved with square sampling [32]. This effect is closely related to the distance between neighbor's elements: in a square grid, each element has eight neighbors, and there are two different types of distances (corners and axes). The distance between adjacent elements in the diagonal direction is around 1.4 times of that in the axial directions (Figure 9.11). On the contrary, in

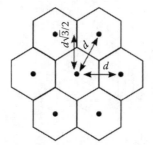 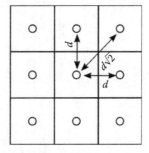

FIGURE 9.11 The distances for a hexagonal and a square grid are shown. The area of a single element is $d^2\sqrt{3}/2$ for the hexagonal geometry and d^2 for the square one.

the hexagonal case, each element has six adjacent pixels, and each element is equidistant from its six neighbors along the six sides of the element. The centroid of the central element is at the same distance from the centroids of the six adjacent pixels (Figure 9.11), and the equidistance between elements has an impact on the grid symmetry and consequently on the angular resolution.

The hexagonal matrix needs to be converted into a square grid, in order to generate a standard digital image reproducible on a dedicated monitor. This can be realized by means of simple averaging techniques or more sophisticated interpolation methods [33].

Nowadays, there is only one system for FFDM available on the market that employs this particular TFT's geometry. This system is based on direct conversion of the incident radiation in an a-Se layer, and an anticrystallization organic layer is used for improving the durability and reliability of the detector [30]. The TFT element used in the FFDM implementation has an area for each hexagonal element similar to that achieved with a 68 μm square pixel. The hexagonal TFT grid is converted into a 50 μm square pixel image. Moreover, the hexagonal TFT is slightly irregular: the distance between adjacent TFT centroid is 75 μm on the x-axis and 73 μm on the diagonal directions, so that the element pitch on the y-axis results to be 62.5 μm. In this geometry, the least common multiple used for converting the hexagonal TFT array into square pixels is 150 μm on the x-axis and 250 μm on the y-axis making the arrangement process illustrated in Figure 9.12 quite easy.

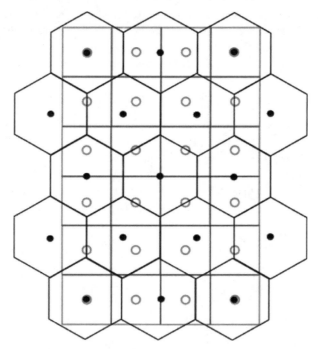

FIGURE 9.12 The square grid superimposed on the hexagonal one is shown here. Thanks to high regularity of the hexagonal lattice, the centroids of the square grid correspond exactly to the centroids of the hexagonal grid each four pitches on the horizontal direction and every six pitches along the vertical axis.

9.4 SLOT-SCANNING SYSTEMS

Some FFDM devices use a slot-scan projection geometry: the image acquisition is performed by a narrow beam that scans the breast. Despite a long acquisition time, the slot-scan geometry is able to provide excellent and efficient scatter removal. In fact, since the scattered radiation is greatly reduced by this geometry, these systems, differently from all the FFDM systems described in the previous sections, operate without a grid, allowing a significant dose advantage for the patient. The time required for a complete scan is few seconds, and so problems associated with mechanical movement of the breast, blurring artifacts, and X-ray tube heating arise. With this approach, only a fraction of the emitted X-ray beam is used at each step, so that the load for the X-ray tube is higher, with respect to the conventional geometry. In fact, X-ray tubes with a tungsten target with high heat storage capacity are preferred to the typical molybdenum or rhodium [34].

9.4.1 INDIRECT CONVERSION APPROACH WITH TILED CCD

In this approach, a slot-scan detector is employed, together with a CsI:Tl scintillator optically coupled to an array of four CCDs through fiber-optic plate. The detector operates in time delay integration mode: continuous measurement of the signal at each point of the breast is achieved through the scanning [35–37]. As in indirect conversion, the scintillator convert X-rays in visible light, transmitted to the CCD by the fibers with minimal loss of spatial resolution. The CCD elements convert light into charge signals, which are shifted along each row in alignment with the scanning direction to allow the exposure signals to be integrated on each detector element [38,39]. The total scanning time for the entire breast is around 5–6 s, but each point is irradiated for only 200 ms. The size of the detector is around 20 cm × 1 cm, and the X-ray beam is collimated into a narrow slot to match the detector size. A precise synchronization of the scanning system and readout electronics is required for avoiding blurring and artifacts. FFDM clinical units following this approach were developed during the first years of the century and commercially available for a few years. However, today, these systems are no longer available on the market.

9.4.2 DIRECT CONVERSION WITH PHOTON-COUNTING DETECTORS

Another slot-scanning approach available for FFDM makes use of X-ray photon-counting detectors [40]. An X-ray-sensitive solid-state device connected to photon-counting electronics is used [41]. The imager consists of a large number of crystalline linear silicon-strip detectors in an edge-on geometry (i.e., with their long axis parallel to the direction of the X-ray beam). The Si-strip detectors are similar to the ones used for tracking particles in high-energy physics experiments. The energy of the incident X-ray photons is converted into electron–hole pairs directly within the Si layer. The concept of single photon counting with energy discrimination allows rejecting the scattered photons and electronic noise. Another photon-counting detector employs a high-pressure gaseous ionization chamber as an X-ray absorber, and a signal is formed by pulses of ions created in the gas [42]. Prototypes based on GaAs as the sensitive layer are, up to date, still working in progress [43]. Historically, it has been very difficult to implement photon-counting detectors for FFDM, given the high photon fluences involved in mammography (more than 10^5 photons/mm^2). At this fluence rate, effective discrimination and counting of signal from each interaction is very challenging.

The main characteristic of this approach is the ability to identify individual X-ray photons. The counting of each single photon enables the measurement of the pulse height generated by each detected X-ray. In this way, it becomes practicable the rejection of pulses with low amplitude that are generated from noise in the detector or in the electronic components. With photon-counting capabilities, it is also possible to eliminate the noise due to fluctuations in the amount of charge produced per quanta at different energies. The signal intensity in each pixel is not proportional to the incident energy but to the number of photons counted. The possibility of realizing pulse height analysis offers some interesting options for improving the quality of the acquired images. For instance, one can generate images corresponding to a particular energy, or the images can be weighted according to energy content. To this end, since it is known that low-energy photons carry more contrast information, a higher weighting can be assigned to them, with respect to X-rays with high energy that may be associated to a lower weighting. Such a process can lead to enhancements, with respect to the conventional integration method. One of the major drawbacks of the photon-counting approach is the complex readout electronics required.

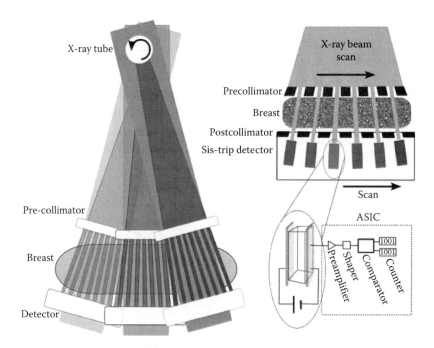

FIGURE 9.13 Sketch of a slot-scan projection geometry full-field digital mammography system based on photon-counting detector: the gantry is rotated to acquire an image of the entire breast. The X-ray tube and beam are shown on the left and on the right (top) the X-ray collimation system and a simplified detection head's scheme (bottom).

Figure 9.13 illustrates an example of a photon-counting FFDM system. The X-ray beam is collimated to a fan beam and directed toward the so-called precollimator. This first collimator device splits the incoming beam into few equidistant linear beams. Beneath the breast support, a second collimator is positioned, namely, the postcollimator, and few linear silicon-strip detectors are placed in a way to match the linear beams exiting the breast. The fan beam, pre- and postcollimators, and the silicon-strip detectors move together with a continuous motion. The movement takes place along an arc with the axis of rotation collinear with the X-ray tube focal spot. The presence of the pre- and postcollimations is a very important feature of this unit, since it can produce an environment with very low scattered radiation. The crystal silicon detector operating on edge is 500 μm thick and provides an absorption efficiency greater than 90% with high fill factor and a pixel size of 50 μm. Each strip is wire bonded to a separate pulse counter. Counting systems do not require traditional analog to digital converter (ADC), but the electronics must be designed to manage high rates of incident photons. The application-specific integrated circuit (ASIC) contains a complete chain, preamplifier, shaper, comparator, and digital counter and can handle count rates up to 2 MHz, allowing the acquisition of an entire image in a few seconds.

REFERENCES

1. H.C. Burell, S.E. Pinder, A.R. Wilson, A.J. Evans, L.J. Yeoman, C.W. Elston, and I.O. Ellis, The positive predictive value of mammographic signs: A review of 425 nonpalpable breast lesions, *Clin. Radiol.* 51, 277–281 (1996).
2. L. Tabar and P.B. Dean, Mammographic parenchymal pattern: Indicator for breast cancer? *J. Am. Med. Assoc.* 247, 185–189 (1982).
3. I.T. Gram, E. Funkhouser, and L. Tabar, The Tabar classification of mammographic parenchymal patterns, *Eur. J. Radiol.* 24, 131–136 (1997).
4. S.A. Feig and M. Yaffe, Digital mammography, computer-aided diagnosis, and telemammography, *Radiol. Clin. North Am.* 33, 1205–1230 (1995).
5. L. Lanca and A. Silva, Digital radiography detectors—A technical overview: Part 2, *Radiography* 15, 134–138 (2009).
6. E.D. Pisano and M.J. Yaffe, Digital mammography, *Radiology* 234(2), 353–362 (2005).
7. A. Karellas and M.L. Giger, Syllabus, Advances in Breast Imaging: Physics, Technology, and Clinical Applications, RSNA Categorical Course in Diagnostic Radiology Physics, RSNA (2004).
8. J.A. Rowlands, The physics of computed radiography, *Phys. Med. Biol.* 47(23), R123–R166 (2002).
9. S. Rivetti, N. Lanconelli, R. Campanini, M. Bertolini, G. Borasi, A. Nitrosi, C. Danielli, L. Angelini, and S. Maggi, Comparison of different commercial FFDM units by means of physical characterization and contrast-detail analysis, *Med. Phys.* 33(11), 4198–4209 (2006).
10. S. Rivetti, B. Canossi, R. Battista, N. Lanconelli, E. Vetruccio, C. Danielli, G. Borasi, and P. Torricelli, Physical and clinical comparison between a screen-film system and a dual side reading mammography dedicated computed radiography system, *Acta Radiol.* 10, 1109–1118 (2009).
11. K.A. Fatterly and B.A. Schueler, Performance evaluation of a dual side read dedicated mammography computed radiography system, *Med. Phys.* 30(7), 1843–1854 (2003).

12. A. Mackenzie and I.D. Honey, Characterization of noise sources for two generations of computed radiography systems using powder and crystalline photostimulable phosphors, *Med. Phys.* 34(8), 3345–3357 (2007).
13. S. Rivetti, N. Lanconelli, M. Bertolini, G. Borasi, D. Acchiappati, and A. Burani, Comparison of different computed radiography systems: Physical characterization and contrast detail analysis, *Med. Phys.* 37(2), 440–448 (2010).
14. G. Borasi, A. Nitrosi, P. Ferrari, and D. Tassoni, On site evaluation of three flat panel detectors for digital radiography, *Med. Phys.* 30(7), 1719–1731 (2003).
15. S. Vedantham, A. Karellas, S. Suryanarayanan, D. Albagli, S. Han, E.J. Tkaczyk, C.E. Landberg et al., Full breast digital mammography with an amorphous silicon-based flat panel detector: Physical characteristics of a clinical prototype, *Med. Phys.* 27(3), 558–567 (2000).
16. C. Ghetti, A. Borrini, O. Ortenzia, R. Rossi, and P.L. Ordonez, Physical characteristics of GE Senographe essential and DS digital mammography detectors, *Med. Phys.* 35(2), 456–463 (2008).
17. N. Lanconelli, S. Rivetti, P. Golinelli, R. Sansone, M. Bertolini, and G. Borasi, Physical and psychophysical characterization of a GE Senographe DS clinical system, *Proc. SPIE* 6510, 65104K (2007).
18. J.A. Rowlands and J. Yorkston, Flat panel detectors for digital radiography, in *Handbook of Medical Imaging*, Vol. 1, J. Beutel, H.L. Kundel, and R.L. Van Metter, eds., SPIE Press, Bellingham, Washington D.C., Chapter 4 (2000).
19. R.H. Dyck and G.P. Weckler, Integrated arrays of silicon photodetectors for image sensing, *IEEE Trans. Electron Devices* 15, 196–201 (1968).
20. H.K. Kim, G. Cho, S.W. Lee, Y.H. Shin, and H.S. Cho, Development and evaluation of a digital radiographic system based on CMOS image sensor, *IEEE Trans. Electron Devices* 48, 662–666 (2001).
21. S. Mendis, S.F. Kemeny, and F.R. Fossum, CMOS active pixel image sensor, *IEEE Trans. Electron Devices* 41, 452–453 (1994).
22. A.C. Konstantinidis, M.B. Szafraniec, R.D. Speller, and A. Olivo, The Dexela 2923 CMOS X-ray detector: A flat panel detector based on CMOS active pixel sensors for medical imaging applications, *Nucl. Instrum. Methods Phys. Res. A* 689, 12–21 (2012).
23. A.C. Konstantinidis, M.B. Szafraniec, L. Rigon, G. Tromba, D. Dreossi, N. Sodini, P.F. Liaparinos et al., X-ray performance evaluation of the Dexela CMOS APS X-ray detector using monochromatic synchrotron radiation in the mammographic energy range, *IEEE Trans. Nucl. Sci.* 60, 3969–3980 (2013).
24. G. Zentai, Comparison of CMOS and a-Si flat panel imagers for X-ray imaging, *2011 IEEE International Conference on Imaging Systems and Techniques (IST) Proceedings*, Penang, Malaysia, pp. 194–200 (2011).
25. R.S. Saunders Jr., E. Samei, J.L. Jesneck, and J.Y. Lo, Physical characterization of a prototype selenium-based full field digital mammography detector, *Med. Phys.* 32(2), 588–599 (2005).
26. N. Rezsnik, P.T. Komljenovic, S. Germann, and J.A. Rowlands, Digital radiography using amorphous selenium: Photoconductively activated switch (PAS) readout system, *Med. Phys.* 35(3), 1039–1050 (2008).
27. K. Irisawa, K. Yamane, S. Imai, M. Ogawa, T. Shouji, T. Agano, Y. Hosoi, and T. Hayakawa, Direct-conversion 50 μm pixel-pitch detector for digital mammography using amorphous selenium as a photoconductive switching layer for signal charge readout, *Proc. SPIE* 7258, 72581I (2009).
28. S. Rivetti, N. Lanconelli, M. Bertolini, G. Borasi, P Golinelli, and E. Gallo, Physical and psychophysical characterization of a novel clinical system for digital mammography, *Med. Phys.* 36(11), 5139–5148 (2009).
29. D.H. Hubel, *Eye, Brain and Vision*, W.H. Freeman & Co., London, U.K. (1988).

30. Y. Okada, K. Sato, T. Ito, Y. Hosoi, and T. Hayakawa, Newly developed a-Se mammography flat panel detector with high-sensitivity and low image artifact, *Proceedings of SPIE 8668, Medical Imaging 2013: Physics of Medical Imaging*, Orlando, FL, p. 86685V (2013).
31. X. He and W. Jia, Hexagonal structure for intelligent vision, *First International Conference on Information and Communication Technologies ICICT 2005*, pp. 52–64 (2005).
32. R. Vitulli, U. Del Bello, P. Armbruster, S. Baronti, and L. Santurti, Aliasing effects mitigation by optimized sampling grids and impact on image acquisition chains, *IEEE Geoscience and Remote Sensing Symposium IGARSS'02*, Vol. 2, Toronto, Canada, pp. 979–981 (2002).
33. X. He, J. Li, and T. Hintz, Comparison of image conversions between square structure and hexagonal structure, *Lecture Notes in Computer Science*, Vol. 4678, pp. 262–273 (2007).
34. A. Noel and F. Thibault, Digital detectors for mammography: The technical challenges, *Eur. Radiol.* 14, 1990–1998 (2004).
35. G.M. Besson et al., Design and evaluation of a slot-scanning full-field digital mammography system, *Proc. SPIE* 4682, 457–468 (2002).
36. Z. Jing, W. Huda, J.K. Walker, and W.Y. Choi, Detective quantum efficiency of a CsI:Tl scintillator based scanning slot X-ray detector for digital mammography, *Proc. SPIE* 3336, 583–591 (1998).
37. J.G. Mainprize, N.L. Ford, S. Yin, T. Tumer, and M.J. Yaffe, A slot scanned photodiode-array/CCD hybrid detector for digital mammography, *Med. Phys.* 29(2), 214–225 (2002).
38. S.Z. Shen, A.K. Bloomquist, G.E. Mawdsley, and M.J. Yaffe, Effect of scatter and an antiscatter grid on the performance of a slot-scanning digital mammography system, *Med. Phys.* 33(4), 1108–1115 (2006).
39. C.J. Lai, C.C. Shaw, W. Geiser, and L. Chen, Comparison of slot scanning digital mammography system with full-field digital mammography system, *Med. Phys.* 35(6), 2339–2346 (2008).
40. M. Aslund and B. Cederstrom, Physical characterization of a scanning photon counting digital mammography system based on Si-strip detectors, *Med. Phys.* 34(6), 1918–1925 (2007).
41. E. Fredenberg, M. Hemmendorff, B. Cederstrom, M. Aslund, and M. Danielsson, Contrast-enhanced spectral mammography with a photon-counting detector, *Med. Phys.* 37(5), 2017–2029 (2010).
42. S. Thunberg, T. Francke, J. Egerstrom, M. Eklund, L. Ericsson, T. Kristoffersson, V.N. Peskov et al., Evaluation of a photon counting mammography system, *Proc. SPIE: Phys. Med. Imag.* 4682, 202–208 (2002).
43. S.R. Amendolia, M.G. Bisogni, P. Delogu, M.E. Fantacci, G. Paternoster, V. Rosso, and A. Stefanini, Characterization of a mammographic system based on single photon counting pixel arrays coupled to GaAs X-ray detectors, *Med. Phys.* 36(4), 1330–1339 (2009).

10 Grating-Based Phase-Contrast X-Ray Imaging Technique

Salim Reza

CONTENTS

10.1	Introduction	256
10.2	X-ray and Its Interaction with Matter	256
10.3	Imaging with X-ray	257
10.4	Phase-Contrast X-ray Imaging	258
10.5	Grating-Based PCXI	260
10.6	Gratings	261
10.7	Phase Stepping	262
10.8	Image Reconstruction	263
10.9	Optimization	264
10.10	Limitations	265
References		266

> It seemed at first a new kind of invisible light.
> It was clearly something new, something unrecorded.
> There is much to do, and I am busy, very busy.
>
> —W.C. Röntgen, November 8, 1895.

Phase-contrast X-ray imaging (PCXI) has recently received much attention from researchers within different disciplines because of its ability to image low-absorbent soft materials and also materials with homogeneous density distribution. In this method, an image of an object is constructed based on the phase shift of the wave passing through it. The enhanced edge effect is also a significant advantage in cases where the observation of the inner structures of any object is of interest. Typical grating-based PCXI systems include a source grating, a phase grating, and an absorption grating. These three gratings are used to transform the phase shift into periodic intensity patterns, which are recorded by detectors.

In this chapter, the basic principles of the PCXI technique, PCXI using gratings, and image reconstruction procedure are discussed.

10.1 INTRODUCTION

Since its discovery in 1895 by Röntgen [1], X-rays have been used in medical imaging, material inspections, security applications, research in numerous academic disciplines, and many more, to look inside objects. X-ray imaging has become a standard tool for in vivo observation because of its nondestructive properties.

In a conventional X-ray imaging technique, an image of an object is achieved based on the absorption of the fraction of the X-ray beam, which passes through it. This limits the efficiency of absorption-based X-ray imaging when the objects of interest are very soft or thin and show little absorption. This is because the objects that are too soft or thin, such as biological tissues and cellulose fiber layer, are nearly transparent to X-rays. They produce low contrast or no contrast in their X-ray images. PCXI can be very efficient in imaging these types of objects.

PCXI is also suitable for imaging objects with a homogeneous density distribution throughout its volume. In the cases of such objects, the region of interest appears contrastless in their absorption-based X-ray images.

10.2 X-RAY AND ITS INTERACTION WITH MATTER

X-rays are electromagnetic waves, and the X-ray photons have a wavelength range of 10^{-2} to 100 Å [2]. X-rays are divided into two groups: soft X-ray and hard X-ray. The soft X-rays have an energy smaller than ~12.4 keV and a wavelength longer than 1 Å. The hard X-rays have wavelengths that are shorter than 1 Å and a photon energy larger than ~12.4 keV [3].

X-ray photons interact with matter in three main ways: photoelectric absorption, Compton scattering, and pair production [4].

In the process of photoelectric absorption, a photon interacts with an atom within the interacting material. The photon is completely absorbed by the atom, and an electron is knocked out, generally from the K-shell of the atom. The removed electron has an energy that can be calculated by subtracting the binding energy of the electron in its original shell from the energy of the interacting photon. The vacancy left behind by the electron may be filled up by capturing a free electron from the interacting material or by obtaining an electron from the other shells of the atom, by generating characteristic X-ray photons.

The Compton scattering or the Compton effect was first observed and explained by Compton [5]. When a photon with a particular wavelength collides with an electron in an atom, it scatters away from its original direction with a longer wavelength. This is an inelastic process because the scattered photons experience a decrease in their energy. The change in the wavelength can be obtained by

$$\lambda_s - \lambda_p = \frac{2h}{mc} \sin^2 \frac{1}{2}\theta$$

where
  λ_s and λ_p are the wavelengths of the scattered and the primary photons, respectively
  h is Planck's constant
  m is the mass of the electron
  c is the velocity of light
  θ is the angle between the directions of the scattered and the primary photons

The pair production process occurs when the energy of an incident X-ray photon is more than twice the rest-mass energy of an electron, which is 1.02 MeV. In this process, the incident photon is annihilated and an electron–positron pair is produced. All the extra energy of the photon above 1.02 MeV is shared by the electron and the positron as their kinetic energy.

In addition to the interaction processes discussed earlier, X-ray photons may interact with all the electrons in an atom through coherent or Rayleigh scattering, leaving the atom unaffected.

10.3 IMAGING WITH X-RAY

The conventional X-ray imaging technique is based on the concept of X-ray transmission [6]. In this technique, X-rays are generated and emerged from a source, pass through an object to be imaged, and are then detected. The differences in the X-ray attenuation in different areas within the object form the image. The difference in thickness and density within the volume of the object causes a variation in the intensity of the passing X-ray beam. X-ray attenuation is explained by the Beer–Lambert law:

$$I = I_0 e^{-\mu t}$$

where
  I and I_0 are the intensities of the X-ray beam, after and before passing the object, respectively
  t is the thickness of the material
  μ is the linear attenuation coefficient

μ, however, depends on the type of material and the energy of the passing X-ray beam.

In the traditional method of X-ray imaging for medical and industrial applications, X-rays are detected on a film after passing the object to be imaged. The film is coated with light-sensitive silver halide, typically silver bromide. The film can be exposed directly to the beam, but in that case, a long exposure time is required due to the low sensitive nature of this type of film. In order to reduce the exposure time and, thus, to reduce the radiation dose on the objects, an intensifying scintillator screen is often placed before the film to convert the X-ray to visible light. Figure 10.1 shows an image of human fingers, acquired using X-ray. The bones inside the fingers are visible.

FIGURE 10.1 X-ray image of human fingers presented in W.C. Röntgen's article in 1896. One finger with a ring.

X-ray images can also be acquired, stored, and displayed digitally using X-ray detectors and computers. X-rays produce electron–hole pairs inside the detector material by interacting with the material in the ways discussed in the previous section. These can be detected and then processed as an electrical signal [7]. When coupled with scintillating plates, charge-coupled devices (CCD) can also be used for digital X-ray imaging. One example of an X-ray detector is the single-photon processing Medipix detector [8], developed by an international collaboration of researchers, hosted by European Organization for Nuclear Research (CERN).

10.4 PHASE-CONTRAST X-RAY IMAGING

It is important to discuss the concept of the complex refractive index while studying PCXI. X-rays or any other electromagnetic waves are affected by the complex refractive index of a material while passing through it [9]. The complex refractive index is a mathematical expression regarding the propagation of electromagnetic waves in a material. It can be written as

$$n = 1 - \delta + i\beta$$

where
 n is the complex refractive index
 δ represents a phase shift
 β represents absorption

The real part of this index δ can be expressed as

$$\delta = \frac{2\pi \rho_a Z r_0}{k^2}$$

where
ρ_a is the atomic number density
Z is the atomic number
r_0 is the classical electron radius
k is the magnitude of the wave vector

The imaginary part of the index β can be calculated as

$$\beta = \frac{\rho_a \sigma_a}{2k}$$

where σ_a is the absorption cross section.

When an X-ray photon passes through a material, it becomes refracted due to a decrease in the real part of the refractive index δ. β attenuates the incident beam [10]. The distribution of the decrement in δ within an object produces contrast in the phase-contrast images of that object [11].

The real part of the refractive index δ and the imaginary part β depend on the wavelength of an incoming photon differently. For high-energy hard X-rays, the cross section for elastic scattering is much larger than that for absorption [12]. Elastic scattering causes the phase shift. PCXI at a certain wavelength can have some orders of magnitude of higher sensitivity than attenuation-based imaging at that wavelength [13].

X-ray imaging based on the phase shift rather than on only the absorption can offer an increased contrast in the image, when objects of interest show very low absorption in comparison to phase shift. One example of such objects can be biological tissue. Soft tissues with a 50 μm thickness can hardly attenuate a 17.5 keV X-ray, but the phase shift can be near π [14].

The advantages of PCXI over attenuation-based imaging led researchers to conduct a continuous investigation with regard to this method. As high intensity and high beam coherency are required for PCXI, this method was limited to only synchrotron radiation sources in the beginning [15]. Recent developments on optical instruments and methods for X-ray imaging have enabled phase-contrast radiography with a conventional tube source in a laboratory.

Several techniques have been developed over the years to produce an X-ray image of an object based on phase shift, namely, crystal interferometer [16], propagation-based imaging [17], analyzer-based imaging [18], and grating interferometry. The details of the grating interferometry-based PCXI are discussed in this chapter.

10.5 GRATING-BASED PCXI

The grating interferometer consists of a source grating g_0, a phase grating g_1, and an analyzer/absorption grating g_2. However, with a microfocus X-ray source, for which the spatial coherency in the beam is sufficient for PCXI, the interferometer can work without g_0 [19]. Figure 10.2 shows a grating interferometer using a microfocus X-ray source.

While working with conventional X-ray sources with a large focal spot, the source grating g_0 is placed immediately after the source. It is an arrangement of transmission slits, which creates an array of periodically repeating line sources [20]. The phase grating g_1 splits the incoming beam and divides it into positive first and negative first diffraction orders [21]. This causes a phase shift of π to the passing beam and attenuates it by a negligible amount. The periodic phase modulation in the incident beam caused by g_1 is transformed into an intensity modulation on the plane of g_2, through the Talbot effect [22]. The g_2 has the same periodicity as the fringes produced by g_1 and is placed just before the detector. The g_2 transforms the position of the fringe into an intensity variation. The detector requires the assistance of the transmission properties of g_2, because generally, the detector resolution is not sufficient to resolve the few microns spacing of the interference fringes. In the future, a high-resolution detector with a smaller pixel size will enable the construction of a grating interferometry without g_2. g_2 is scanned along the direction x_t, so that the intensity in each pixel in the detector oscillates in relation to x_t.

The distance between g_1 and g_2 for a plane wave d_m can be calculated by

$$d_m = m \frac{p_1^2}{8\lambda}$$

where
p_1 is the period of g_1
λ is the wavelength
m is an odd integer representing the order of the fractional Talbot distance

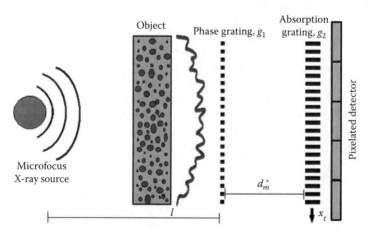

FIGURE 10.2 Grating interferometer.

For a setup using a spherical wave, this distance will be

$$d_m^* = \frac{l}{l-d_m} d_m$$

where l is the distance between the source and g_1. The period of g_2, p_2 should be the same as the periodicity of the fringes produced by g_1 on the plane of g_2. The periodicity of the fringes, for the plane wave [23], is

$$p_2 = \frac{p_1}{2}$$

For a spherical wave, the period of g_2, p_2^* can be obtained by

$$p_2^* = \frac{l+d}{l} p_2$$

where d is a general term for the distance between g_1 and g_2 for any beam geometry. For spherical wave, $d = d_m^*$. So, p_2^* can be rewritten as

$$p_2^* = \frac{l}{l-d_m} p_2$$

10.6 GRATINGS

The performance of the grating interferometry depends on the quality of the gratings. g_1 works as a π-phase shifter with a duty cycle of 0.5. This grating ensures that no undiffracted part of the beam remains. The phase-contrast signal is generated by the diffracted parts of the beam. The beam absorption at g_2 should be as high as possible.

The gratings are fabricated using photolithography, silicon deep etching, and electroplating techniques [24]. The processing of g_0 is simpler as compared to the processing of g_1 and g_2, since it has larger pitch than others. The area of g_0 has to be sufficiently large, so that it covers the whole focal spot of the X-ray source. g_1 should be processed in such a way that it shows a low absorption with few micron-sized π-phase shifting structures. These structures can be deep etched into a silicon wafer [25]. The period of g_2 is even smaller than that of g_1, thus making it particularly difficult to process g_2 [26]. The beam transmission through g_2 should not exceed 25%. As gold is a good X-ray absorber, it is often electroplated to fill the grooves in a silicon grating in order to fabricate g_2.

10.7 PHASE STEPPING

The grating interferometry is designed to detect the deflection in the beam after passing through an object. When an object is placed in front of the phase grating g_1, it attenuates the intensity of the beam due to its absorption properties and also deflects the beam by its refractivity [27]. The deflection angle is given by [28]

$$\alpha(x_t) = \frac{\lambda}{2\pi} \frac{\partial \Phi(x_t)}{\partial x_t}$$

where $\Phi(x_t)$ is the phase profile in the wavefront in relation to the transverse direction x_t.

The deflection angle α causes a local displacement of $\alpha \cdot d_m^*$ in the interference pattern at a distance of d_m^* from g_1. In this manner, the phase shift $\Phi(x_t)$ caused by the object's refractive index transforms into an oscillated intensity pattern on the plane of g_2. The absorption grating g_2 assists in resolving the pattern, since the limited spatial resolution of the detector makes it difficult to detect the pattern directly.

In order to record the intensity pattern, g_2 is scanned over its minimum one period along the direction x_t. The g_2 is stepped toward x_t, and at every position, an image is taken. The intensity in every pixel in the detector oscillates as a function of the position of g_2.

Figure 10.3 shows the oscillated intensity data at a single pixel in the detector for different g_2 positions. The minimum scanning distance from the first data point to the last one is deemed to be one period of g_2. The number of scan points may vary, but since the phase retrieval process requires a fully established mathematical system, the minimum number should be three [29]. It is common practice among researchers to use eight scan points per period. In order to reduce the statistical errors, it is better to increase the number of scan points. However, statistical errors also depend on the photon counts in each pixel of the detector in all the scan images.

For the image reconstruction method described in the next section, the entire scanning process must be performed twice in order to obtain a phase-contrast image of an object. One scanning should be conducted with the object in front of the phase grating g_1 and one without the object in order to record reference images. All the scan positions for the images with the object should be exactly the same as those for the reference images. Thus, in this manner, two intensity datasets are obtained for

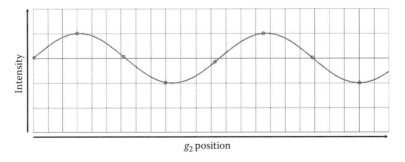

FIGURE 10.3 Intensities at a single pixel at different g_2 positions.

every pixel. Two sine curves can be fitted to those data, and using the properties of the curves in all pixels, the phase-contrast image can be constructed.

10.8 IMAGE RECONSTRUCTION

There are several methods for constructing a phase-contrast X-ray image of an object using the scan images obtained by grating interferometry. In this chapter, the sine-fitting method is discussed.

Figure 10.4 shows two sine signals, which are fitted to the two sets of intensity data at a single pixel. The curve at the top shows the intensity data for the reference images without an object, and the second curve shows the intensities from the images with an object.

Both curves have an amplitude A, phase Φ, and offset O. The absorption image ABS is the ratio of the offsets of the two curves [30]:

$$ABS = -\log\left(\frac{O_{OBJ}}{O_{REF}}\right)$$

The differential phase image PHS can be obtained by

$$PHS = \Phi_{REF} - \Phi_{OBJ}$$

One advantage of the grating interferometry is that the dark-field image [31] of the object can also be obtained using the same setup and the same image reconstruction method, with no additional equipment being required. The dark-field image of an object is constructed based on only those X-rays that are scattered while passing through that object. It is a very useful and efficient method for contrast enhancing. A dark-field signal is very sensitive to the microstructures and the granularity within objects. The dark-field images of two objects with the same absorption but with different inner structures are significantly different, while their absorption images are almost identical. The dark-field image DKF can be calculated as

$$DKF = \frac{V_{OBJ}}{V_{REF}}$$

where $V_{OBJ} = A_{OBJ}/O_{OBJ}$ and $V_{REF} = A_{REF}/O_{REF}$ are the visibilities of the two signals.

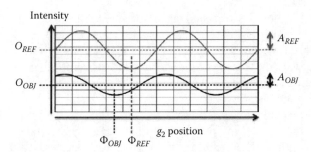

FIGURE 10.4 Two sets of intensities at a single pixel.

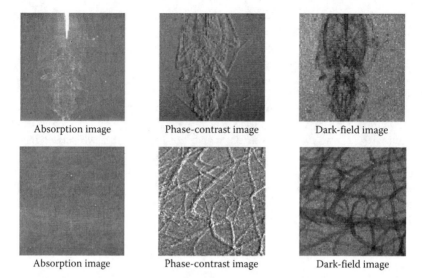

FIGURE 10.5 (**See color insert.**) Reconstructed absorption, differential phase-contrast, and dark-field images of a bug (top) and lichen (bottom).

The whole process of calculating the absorption image, the differential phase image, and the dark-field image has to be repeated for every pixel in the detector in order to obtain the complete matrices of the three images.

Figure 10.5 shows the reconstructed absorption, differential phase-contrast, and dark-field images of a bug and lichen, obtained using grating interferometry.

10.9 OPTIMIZATION

One way to optimize the performance of grating interferometry is to enhance the visibility by fine-tuning the setup. Visibility can be defined as

$$V = \frac{I_{max} - I_{min}}{I_{max} + I_{min}}$$

where I_{max} and I_{min} are the maximum and minimum values in the intensity modulation at a single pixel. Visibility for the complete detector matrix can be achieved by calculating the visibility for every pixel in the detector. The visibilities in the interferometry with the same instruments but with different Talbot orders are different. The imaging should be conducted at the Talbot order where the interferometry provides maximum visibility.

The detector should be carefully selected depending on the photon energy used in the interferometry. As the photons with the design energy take part to create interference and thus act as information careers in the interferometry, the detector should have high quantum efficiency for that photon energy. Figure 10.6 shows the

Grating-Based Phase-Contrast X-Ray Imaging Technique

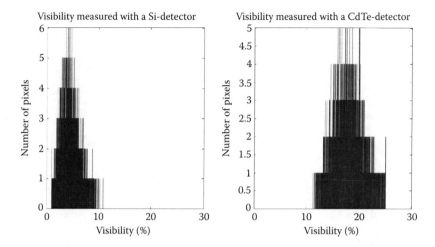

FIGURE 10.6 Visibility measured with two different detectors: (a) Si detector and (b) CdTe detector.

difference in the visibilities in an interferometry, operating at 38 keV, measured with 300 μm thick Si- and CdTe-based single-photon processing pixel detectors. As Si has much lower detection efficiency than CdTe for the photons with an energy of 38 keV, the visibility in the interferometry is recorded to be much higher with the CdTe detector than that with the Si detector.

The relative angle between the grating lines of g_1 and those of g_2, about the optical axis, should be kept as small as possible. When they are not aligned perfectly, moiré fringes [32] appear in the acquired images. The moiré fringes become more dense as the relative angle increases.

All the equipment in interferometry should be mechanically stable. The setup environment should experience as low a vibration as possible.

10.10 LIMITATIONS

As in other X-ray imaging systems, the spatial resolution in the acquired images by means of grating interferometry is limited by the pixel size in the detector [33]. Detectors with a smaller pixel size can offer a better spatial resolution in the image. The resolution is also limited by the period of g_2. Unresolved microstructures within objects reduce the visibility of the interferometry.

PCXI with grating interferometry requires multiple exposures, so the objects under investigation are exposed to radiation for a longer time as compared to other imaging systems. This thus limits the widespread applications of grating interferometry in the medical sector, even though the image quality is better than that obtained when using conventional attenuation-based imaging. X-ray sources with high brilliancy, such as with a liquid-metal-jet anode X-ray tube [34], may assist in reducing the exposure times involved.

REFERENCES

1. W. C. Röntgen (1896), On a new kind of rays, *Nature*, 53, 274–277.
2. G. Margaritondo (1988), *Introduction to Synchrotron Radiation*, Oxford University Press, Oxford, U.K.
3. B. M. Weon et al. (2006), Phase contrast X-ray imaging, *International Journal of Nanotechnology*, 3, 280–297.
4. G. F. Knoll (2010), *Radiation Detection and Measurement*, 4th edn, John Wiley & Sons, Inc.
5. A. H. Compton (1923), A quantum theory of the scattering of X-rays by light elements, *The Physical Review*, 21(5), 483.
6. D. Pfeiffer (2010), Requirements for medical imaging and X-ray inspection, *Handbook of Optics*, Vol. V—Atmospheric Optics, Modulators, Fiber Optics, X-Ray and Neutron Optics by Michael Bass, 3rd edn., McGraw-Hill Professional.
7. L. Rossi et al. (2006), *Pixel Detectors, From Fundamentals to Applications*, Springer.
8. X. Llopart et al. (2002), Medipix2: A 64-k pixel readout chip with 55-µm square elements working in single photon counting mode, *IEEE Transactions on Nuclear Science*, 49(5), 2279–2283.
9. P. C. Diemoz et al. (2012), Theoretical comparison of three X-ray phase-contrast imaging techniques: Propagation-based imaging, analyzer-based imaging and grating interferometry, *Optics Express*, 20(3), 2789–2805.
10. D. Hahn et al. (2012), Numerical comparison of X-ray differential phase contrast and attenuation contrast, *Biomedical Optics Express*, 3(6), 1141–1148.
11. M. Willner et al. (2013), Quantitative X-ray phase-contrast computed tomography at 82 keV, *Optics Express*, 21(4), 4155–4166.
12. T. Weitkamp et al. (2005), X-ray phase imaging with a grating interferometer, *Optics Express*, 13(16), 6296–6304.
13. W. Yashiro et al. (2010), On the origin of visibility contrast in X-ray Talbot interferometry, *Optics Express*, 18(16), 16890–16901.
14. C. David et al. (2006), Quantitative phase imaging and tomography with polychromatic x rays, *Proceedings of the Eighth International Conference on X-ray Microscopy*, IPAP Conference Series 7, pp. 346–348.
15. T. Donath et al. (2009), Phase-contrast imaging and tomography at 60 keV using a conventional X-ray tube source, *Review of Scientific Instruments*, 80, 053701.
16. U. Bonse and M. Hart (1965), An X-ray interferometer, *Applied Physics Letters*, 6(8), 155–156.
17. A. Snigirev et al. (1995), On the possibilities of X-ray phase contrast microimaging by coherent high-energy synchrotron radiation, *Review of Scientific Instruments*, 66, 5486.
18. V. N. Ingal and E. A. Beliaevskaya (1995), X-ray plane-wave topography observation of the phase contrast from a non-crystalline object, *Journal of Physics D: Applied Physics*, 28, 2314–2317.
19. T. Donath et al. (2009), Inverse geometry for grating-based X-ray phase-contrast imaging, *Journal of Applied Physics*, 106, 054703.
20. F. Pfeiffer et al. (2008), Hard-X-ray dark-field imaging using a grating interferometer, *Letters to Nature Materials*, 7, 134–137.
21. M. Engelhardt et al. (2008), The fractional Talbot effect in differential X-ray phase-contrast imaging for extended and polychromatic X-ray sources, *Journal of Microscopy*, 232, 145–157.
22. A. Momose et al. (2003), Demonstration of X-ray talbot interferometry, *Japanese Journal of Applied Physics*, 42, L866–L868.

23. M. Engelhardt et al. (2007), High-resolution differential phase contrast imaging using a magnifying projection geometry with a microfocus X-ray source, *Applied Physics Letters*, 90, 224101.
24. C. David et al. (2007), Fabrication of diffraction gratings for hard X-ray phase contrast imaging, *Microelectronic Engineering*, 84, 1172–1177.
25. C. David et al. (2001), Wet etched diffractive lenses for hard X-rays, *Journal of Synchrotron Radiation*, 8, 1054–1055.
26. W. Yashiro et al. (2006), *Optimal Design of Transmission Grating for X-Ray Talbot Interferometer.* International Centre for Diffraction Data, Newtown Square, PA.
27. C. Kottler et al. (2007), A two-directional approach for grating based differential phase contrast imaging using hard X-rays, *Optics Express*, 15(3), 1175–1181.
28. M. Born and E. Wolf (1993), *Principles of Optics*, 6th edn., Pergamon Press, Oxford, England.
29. P. Bartl et al. (2009), Simulation of X-ray phase-contrast imaging using grating-interferometry, *IEEE Nuclear Science Symposium Conference Record,* Orlando, FL.
30. T. Weber et al. (2011), Noise in X-ray grating-based phase-contrast imaging, *Medical Physics*, 38(7), 4133–4140.
31. B. Schwarzschild (2008), Dark-field imaging is demonstrated with a conventional hard-X-ray source, *Physics Today*, 61, 12.
32. H. Itoh (2011), Two-dimensional grating-based X-ray phase-contrast imaging using Fourier transform phase retrieval, *Optics Express*, 19(4), 3339.
33. J. Thim et al. (2011), Suitable post processing algorithms for X-ray imaging using oversampled displaced multiple images, *Journal of Instrumentation*, 6, 2.
34. O. Hemberg et al. (2004), Liquid-metal-jet anode X-ray tube, *Optical Engineering*, 43(7), 1682–1688.

11 Emerging Concept in Nuclear Medicine
The Voxel Imaging PET (VIP) Project

Mokhtar Chmeissani, Gianluca De Lorenzo, and Machiel Kolstein

CONTENTS

11.1 Introduction 270
    11.1.1 Voxel Imaging PET (VIP) 274
    11.1.2 Pixel CdTe...276
    11.1.3 VIP-PIX Chip 278
11.2 Image Reconstruction...280
    11.2.1 Introduction 280
    11.2.2 Simple Back-Projection 281
    11.2.3 Filtered Back-Projection 282
    11.2.4 Ordered Subset Expectation Maximization (OSEM)......................283
    11.2.5 Origin Ensemble (OE) Algorithm 284
    11.2.6 Image Quality Criteria..285
    11.2.7 Available Software..285
11.3 Voxel Imaging PET...286
    11.3.1 Current Limitations in PET Design..286
    11.3.2 Evaluation of the VIP Design 288
    11.3.3 The VIP Scatter Fraction 288
    11.3.4 VIP Counting Performance 289
    11.3.5 VIP Resolution..291
    11.3.6 VIP Image Quality 292
    11.3.7 VIP Minimum Detectable Lesion Size...295
    11.3.8 Simulation and Image Reconstruction of Real 3-D Human Head Phantom 298
    11.3.9 VIP Image Reconstruction Algorithms..298
11.4 Positron Emission Mammography with VIP..299
    11.4.1 VIP PEM Design 299
    11.4.2 Evaluation of Imaging Performance...301

11.5 VIP Compton Camera 303
    11.5.1 Introduction to SPECT and Compton Camera 303
        11.5.1.1 SPECT 303
        11.5.1.2 Compton Camera 305
    11.5.2 VIP Compton Camera Design 307
    11.5.3 VIP Compton Camera Performance 309
        11.5.3.1 Optimization of Design 309
        11.5.3.2 Detector Sensitivity 311
        11.5.3.3 Spatial Resolution 314
    11.5.4 VIP Compton Camera Image Reconstruction Results 314
11.6 Conclusion 321
Acknowledgments 322
References 322

11.1 INTRODUCTION

Positron emission tomography (PET) is a fundamental nuclear medicine diagnostic tool for molecular imaging. Although the first time a PET scanner was used for oncology was in 1982 [1], it took a long time before PET could be used as a full-fledged operational diagnostic device. In the 1970s and early 1980s, the lack of interest of the radiology community made it difficult to establish a successful market for PET and companies that fabricated PET devices had difficulties in surviving. Two of the main obstacles were the need to have a cyclotron machine and a team to make the radiotraces in situ, and to have a image quality comparable to the emerging CT and later MRI techniques. A breakthrough along this path was the development of the new radiotracer 2-deoxy-2-(^{18}F)fluoro-D-glucose (^{18}F-FDG) containing the positron-emitting radioactive isotope ^{18}F with a long lifetime [2,3] and the introduction of the dual imaging PET + CT technique by T. Beyer et al. [4]. With the introduction of ^{18}F-FDG, it became possible to acquire radiotracers from dedicated centers. Since ^{18}F-FDG has a relatively long lifetime of 90 min, it is possible to transport it within a range of a city, and this has helped in boosting the PET market.

Although the spatial resolution of the current state-of-the-art PET systems is a few millimeter, the device is unique in its sensitivity to detect the molecular functionality. Scientist's interest in PET and single positron emission computed tomography (SPECT) has increased due to the applicability of these devices in molecular imaging and neuroscience, particularly with dual imaging modality such as PET/CT and PET/MRI. A large number of scientists in the field of detector instrumentation, data acquisition, and image reconstruction have embarked on R & D projects to improve the performance of these diagnostic devices. As a figure of merit, more than half of the abstracts accepted for the IEEE-NSS-MIC-2013 were related to PET and SPECT, though the world market of nuclear medicine represent only 7% of the total world market for diagnostic imaging compared to 34% for X-ray and 21% for CT [5].

One way to improve PET or SPECT images is to combine them with CT scan images, as shown in Figure 11.1. The CT scan provides information about the body

Emerging Concept in Nuclear Medicine

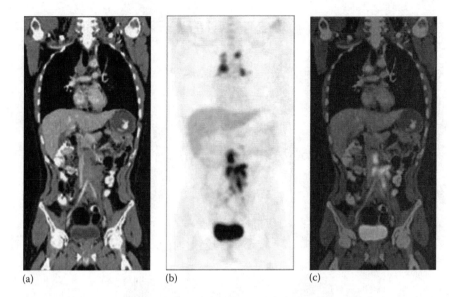

(a) (b) (c)

FIGURE 11.1 (See color insert.) (a) CT scan of full body, (b) PET scan of full body, and (c) PET + CT scan fused images. There is an obvious improvement of the image quality when PET and CT scan are fused together. (Image taken from: www.dh.org/about-pet-ct.)

density and helps to correct for the attenuation of the 511 keV photons. At the same time, it helps to provide a good body map for the hot spots' activities detected by the PET scanner. Today 97% of the PET scanners are equipped with a CT scanner. Similarly, one can do the same by combining PET with MRI as shown in Figure 11.2.

PET images can be improved by improving the intrinsic resolution of the PET at the level of the detector, data acquisition, and image reconstruction. The image reconstruction is still a wide open field that has a lot of room for improvement both at the conceptual level and at the level of processing time, as discussed in Section 11.2. For the data acquisition, nowadays data are corrected for body movements and in particular for the breathing cycle and heart beats. Also, the old technique of using measurements of the time of flight (TOF) of the two photons that make up the PET event [6] has been reintroduced in PET data acquisition. Results with the current state-of-the-art electronics [9] show that the signal/noise ratio can be improved by a factor of 2.

In PET images, the line of response (LOR) is defined as the straight line connecting the two cells of the scanner ring that have detected the 511 keV photons. The quality of PET images suffers mainly due to wrongly aligned LORs from scattered photons, as depicted in Figure 11.3. Reducing the number of events from scattered photons is a difficult but essential task in order to achieve better image quality.

Scintillator crystals currently employed in PET scanners have limited energy resolution due to nonuniform light output along the crystal length and limited photo-electron yield [8]. Properties of commonly used scintillator crystals for PET are shown in Table 11.1. For this reason, most of the PET scanners use 10- to 20-mm-long crystals to keep the variance of the light output within the target limit. The large energy resolution makes it very difficult to eliminate scattered events. Because of the wide range

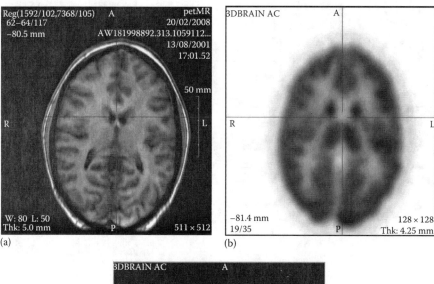

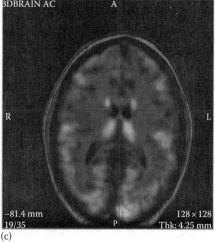

FIGURE 11.2 (See color insert.) (a) MRI scan of a brain, (b) PET scan of the same brain. (c) PET + MRI fused images. There is an obvious improvement of the image quality when PET and MRI scans are fused together. (Image taken from: en.wikipedia.org/wiki/PET-MRI.)

in energy of both true events and scattered events, placing a hard cut on the detected energy will remove the scattered events as well as a significant part of true events.

On the other hand, with room temperature semiconductor detectors, such as CdTe, it is much easier to apply an energy cut to reduce the number of scattered events in the PET image event sample, as will be shown in the following sections. Table 11.2 lists potential semiconductors that can be used in radiation detection. The small fluctuation and the large electric signal created by the detector per 1 keV energy deposition make it possible to do energy spectroscopy.

Another limitation that current PET scanners suffer from is the low detection efficiency. Improving the detection efficiency provides us with a dilemma because

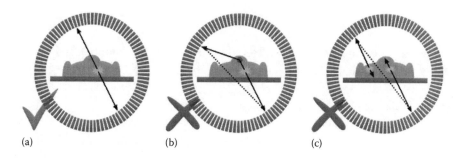

FIGURE 11.3 (a) LOR of a true PET event, (b) LOR of a scattered event, and (c) LOR caused by a random coincidence. With good energy resolution, LORs from scattered events can be eliminated.

TABLE 11.1
List of Scintillator Crystals That Are Commonly Used in PET Scanners

	Effective Atomic Number (Z)	Density (g/cm³)	Attenuation Length at 511 keV (cm)	#γ/ keV	Decay Constant (ns)	Wave length (nm)	Energy Resolution % at 511 keV
NaI(Tl)	51	3.7	2.9	38	230	410	6.6
BaF₂ (fast)	54	4.9	2.2	2	0.8	220	11.4
GSO	59	6.7	1.4	10	60	430	8.5
BGO	75	7.1	1.1	8	300	480	20
LSO	66	7.4	1.2	30	40	415	10

Sources: The data is compiled from Dorenbos, P., *Nucl. Instrum. Methods Phys. Res. A*, 486, 208, 2002; L'Annunziata, M.F., *Handbook of Radioactivity Analysis*, Academic Press, 2012; Bailey, D.L. et al., *Positron Emission Tomography: Basic Sciences*, Springer Science + Business Media; Saha, G.B., *Basics of PET Imaging, Physics, Chemistry, and Regulations*, Springer ed., 2005.
Overall, the photon yields, and hence the generated electronic signals, are relatively low when compared to signals from high Z semiconductor radiation detectors.

although, on the one hand, a longer crystal gives better detection efficiency for 511 keV photons, on the other hand, the variance of the light output increases with crystal length. Therefore, with long crystals one can achieve good detection efficiency but at the price of a large variance of light output and hence a large variance in energy resolution. For this reason, a compromise must be adopted to keep both the detection efficiency and the variance of light output within the acceptable limit to obtain the best images at the lowest radiation dose. Additionally, since the scintillator crystal is a rectangular parallelepiped, it is not possible to make seamless PET scanners from such elements and, consequently, a good fraction of events go undetected. Although it is possible to fabricate crystals in a trapezoidal parallelepiped shape, the light yield would not be uniform that would make the energy resolution worse.

The large volume of the crystal is one of the main limitations of current PET scanners. For instance, in the case of [7], it is 4 × 4 × 22 mm³, and in case of [15], it

TABLE 11.2
List of Semiconductor Radiation Detectors That Are Commonly Used in X-ray and Gamma Ray Detection and Imaging

	Effective Atomic Number (Z)	Density (g/cm³)	Attenuation Length at 511 keV (cm)	Band Gap (eV)	e–h Pairs/keV	$(\mu\tau)_{e,h}$ (cm²/V)
Si	14	2.3	5.0	1.1	261	>1, >1
Ge	32	5.3	2.4	0.7	410	>1, >1
GaAs	31, 33	5.3	2.2	1.4	205	8.0×10^{-5}, 4.0×10^{-6}
CdTe	52, 48	5.9	1.8	1.5	179	3.3×10^{-3}, 2×10^{-4}
HgI$_2$	80, 53	6.4	1.3	2.1	137	10^{-4}, 4×10^{-5}
TiBr	83, 35	7.5	1.0	2.7	106	1.6×10^{-5}, 1.5×10^{-6}

Sources: The data are compiled from Semiconductor Detector Material Properties, eV Products Inc., webpage: http://ww.evproducts.com/pdf/material prop.pdf; Bencivelli, W. et al., *Nucl. Instrum. Methods*, A310, 210, 1991.

The number of e–h pairs generated per 1 keV is computed from the work ionizing energy $W_i = 2.8E_g + 0.75$ eV, which is the energy needed to create an e–h pair.

is 2.1 × 2.1 × 10 mm³. This adds uncertainty to the estimation of the points of interaction within the crystal and this is translated in its turn to uncertainty in the LOR.

Ronald Nutt formulated Nutt's law that states that the number of crystals in PET devices will double every 2 years [16] (in analogy with Moore's law that predicts the number of transistors on integrated circuits doubles every 2 years). This prediction was consistent with reality from the mid-1970s until the early 2000s. In 2002, the ECAT HRRT PET scanner was introduced, which consisted of 119808 crystals. However, since then there has not been much progress in the growth of number of channels in PET devices. For instance, in 2011, Hitachy introduced a brain PET scanner [17] with 152024 CdTe coplanar detectors, where each detector has a size of 4 mm × 7,5 mm × 1 mm and every 2 detectors are connected to one readout channel. This number of detectors used in this PET represent an increase of about factor 1.5 with respect to ECAT-HRRT, over a period of one decade. The Voxel Imaging PET (VIP) project [18] presents a new conceptual design of a high granular brain PET detector with 6.3 million channels using pixel CdTe detectors coupled to dedicated readout ASICs. Each pixel has a size of 1 mm × 1 mm × 2 mm and is connected to a dedicated reachout channel. The development of the VIP scanner with such a high number of channels injects new life into Nutt's law.

11.1.1 Voxel Imaging PET (VIP)

VIP is a novel design for a future PET scanner [18], that employs pixelated CdTe detectors coupled to a dedicated readout electronics via a bump-bonding process. The aim of this novel design is to overcome the current intrinsic limitations of the PET scanners

Emerging Concept in Nuclear Medicine

mainly caused by the use of scintillator crystals. The excellent energy resolution of CdTe helps in rejecting scattered events. Additionally, it will be possible to provide true high granular three-dimensional (3D) information about the impact point of the 511 keV photon, making it possible to achieve minimum smearing of the LOR. Moreover, CdTe semiconductors provide direct conversion of the 511 keV into an electronic signal, whereas in the case of the scintillator crystal the conversion is indirect, causing major losses in the electric signal and hence limiting the energy resolution. Finally, one can cut the CdTe detector into a trapezoidal parallelepiped shape without an impact on the detector response, thus making it possible to construct a seamless PET scanner as shown in Figure 11.4 and consequently increase the detection efficiency of the scanner,

The VIP detector design concept can be extended to a Compton camera imaging device. A more detailed description of the general principle of Compton cameras and

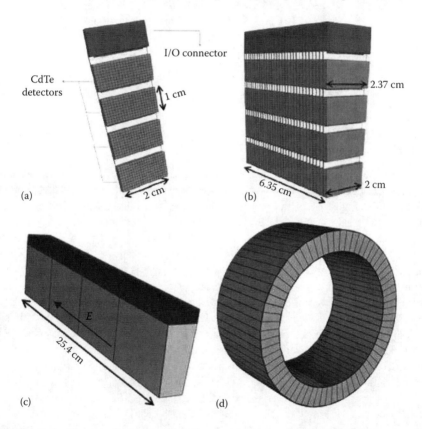

FIGURE 11.4 (See color insert.) Detailed depiction of the design of the VIP scanner. (a) One single VIP unit module, consisting of a single layer with CdTe pixel detectors mounted on a thinned readout ASIC (not visible) and both together mounted on thin kapton PCB. (b) A stack of such layers to form a module block. The number of layers per stack is flexible. (c) How a set of module blocks form a sector of the PET scanner. (d) How the full scanner can be made seamless by using the trapezoidal shape of the VIP unit module. (Reprinted from Mikhaylova, E. et al., Simulation of pseudo-clinical conditions and image quality evaluation of PET scanner based on pixelated CdTe detector, *IEEE NSS MIC Conf. Rec.*, 2716–2722, 2011. Copyright 2011 IEEE with permission.)

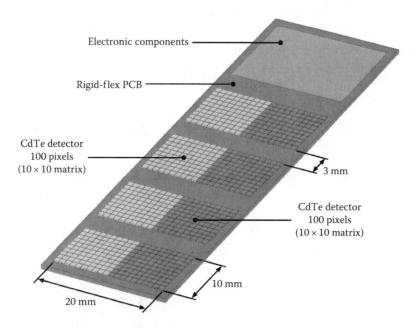

FIGURE 11.5 One single unit module layer with rectangular pixel CdTe detectors mounted on thin kapton PCB.

the VIP design of such a device can be found in Section 11.5. Compton camera needs high energy and spatial resolution for both detector planes, the scatterer and the absorber, as depicted in Figure 11.29. As shown in this figure, instead of using LORs as in PET imaging to locate the origin of the event, with Compton camera image reconstruction one searches for the origin of each event on the surface of the cone that can be reconstructed from the energy and position measurements of the photon hits in the detector planes. The precision of the opening angle of the Compton cone is essential for a good image reconstruction and directly depends on the precision of the hit measurements. Up till now, Compton cameras have not been used for medical imaging and the VIP project will be the first to make such device a reality. For the Compton camera, we use a similar VIP unit module design as described before, but in a rectangular shape as shown in Figure 11.5. From such kernel sensor layers, one can stack and build a large and high granular detector for the Compton camera as described in more detail in Section 11.5. The same rectangular design for the VIP unit module can be used also for designing a positron emission mammography (PEM) scanner as described in more detail in Section 11.4.

11.1.2 Pixel CdTe

CdTe and CdZnTe are good candidates, as semiconductor detectors, for the VIP PET design. Other semiconductor detectors, listed in Table 11.2 are less favorable either because they have low Z, and/or a low value for the mobility-lifetime product $\mu\tau$ for electrons and holes. The material dependent mobility μ and lifetime τ of electrons and holes are important factors as a measure of the quality of the semiconductor material. Since the drift length of a charge carrier l_e is given by $l_e = E\mu\tau$, the product μτ should be maximized [12]. Both mobilities and lifetimes are sufficiently high for CdTe to guarantee a sufficiently long drift length of the charge carriers [12,13].

Regardless of the type of the semiconductor detector that we use, it is important to have enough detector thickness to achieve a good stopping power for the 511 keV photons. For example, with 4 cm of CdTe thickness, one can insure that 70% of the 511 keV photons are fully absorbed. However, using a single crystal CdTe detector with the same shape and geometry as commonly used for PET scintillator crystals is not possible. First, the drift time of the e–h pair in the semiconductor would be too long to have a fast trigger to match the needs of PET data acquisition (in the nano seconds scale). Secondly, to acquire a 4 cm long CdTe detector of spectroscopy grade is very expensive because of the low production yield of homogeneous compound semiconductor CdTe material.

In the VIP design, as shown in Figure 11.4, the semiconductor detector is placed in such a way that its edge is pointing to the axis of the PET scanner. The detector thickness is only 2 mm and hence the trigger time is no longer an issue. It is not the thickness of each detector but the total number of CdTe detectors that establishes the length of CdTe that the 511 keV photon will traverse. A 2 mm thick CdTe biased at 1000 V/mm can achieve a trigger time, fit for a PET application [19]. Using Schottky contact electrodes (Pt/CdTe/Al/Au/Ni/Au/AlN) allows the user to polarize the detector at high voltage (>500 V/mm) and with small leakage current (around 60 pA/mm^2 with a bias of 500 V/mm at room temperature). The VIP Schottky CdTe detector has a pixel pitch of 1 mm and thickness of 2 mm. The choice of 1 × 1 mm^2 pixel size can be justified by the mean free path of the emitted positron in human tissue of 0.6 mm (using the most commonly used radiotracer ^{18}F-FDG). The mean free path of the positron makes that the LORs of the back-to back 511 keV photons, caused by the annihilation of the positron with an electron in the source environment, are not perfectly collinear, as shown in Figure 11.6. Hence, given that the LORs will not be perfectly collinear, a smaller pixel pitch would not improve the PET image spatial resolution. A 2 mm thickness of

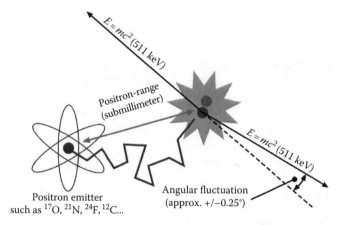

FIGURE 11.6 The mean free path of the positron emitted by the radionuclide in the human tissue depends on its initial energy. It is about 1.1, 1.5, 2.5, and 0.6 mm for radionuclides ^{11}C, ^{13}N, ^{15}O, ^{18}F, respectively [11]. The positron undergoes multiple scattering before it gets annihilated with a free electron and, hence, the mean free path should be as small as possible. Therefore, using ^{18}F as a radiotracer is to be preferred.

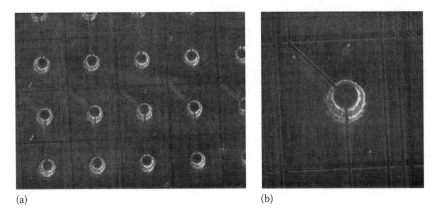

(a) (b)

FIGURE 11.7 (See color insert.) (a) shows the pixel CdTe detector bonded to glass substrate. The pixel pitch is 1 mm and the gap between neighboring electrodes is 50 μm. In the pixel center one can see the 250 μm BiSn solder ball between the detector and the glass substrate. (b) shows a more detailed picture of the pixel CdTe detector and the bump bonding connection.

the CdTe detector is a good tradeoff between having a fast trigger for a PET application and reducing the number of readout channels.

Using the fact that the mobility of the holes is 10 times less than electrons inside CdTe and by processing the signal of the pixel CdTe by a fast and slow shaper [20], one can determine the depth of interaction along the 2 mm CdTe thickness with a precision better than 0.2 mm.

Figure 11.7 shows how each pixel on the CdTe detector is connected to a dedicated readout channel via a solder bump ball of BiSn (48%–52%) with melting point at 138°C. A good review on low-temperature solder alloys can be found in [21]. Bonding the pixel CdTe detector to the VIP-PIX readout chip at low temperature is essential to avoid any degradation in the detector. BiSn, eutectic solder, allows the user to do the solder reflow bonding process at temperatures less than 160°C. To ensure good solder contacts, the reflow temperature is recommended to be about 20° above the melting point and in a formic acid oven.

Another possible low-temperature bonding solder is indium where, to connect the detector and the chip, a cold compression technique is used, which implies using a force on both detector and chip that could be as high as 100 kg. It is also possible to do bonding at the melting point of indium but this could cause the oxidization of the indium if it is not done in an oxygen-free ambient. To ensure good contact, it is recommended to deposit the indium bumps on both chip and substrate.

11.1.3 VIP-PIX Chip

The VIP-PIX chip is designed with 0.25 μm TSMC technology. It has 100 channels and each has a preamplifier, discriminator, shaper, peak-hold, and a 10 bit SAR ADC with energy resolution <0.7 keV [22]. The schematic and the layout of the pixel cell are shown in Figure 11.8 and the pixel-channel properties are shown in Table 11.3. The chip also has a 10 bits TDC with a time resolution of less than 1.0 to provide the PET event a time stamp and a temperature sensor to adjust the

Emerging Concept in Nuclear Medicine

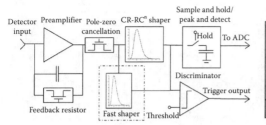

FIGURE 11.8 (See color insert.) Schematic depiction and layout of the VIP-PIX pixel cell. (Reprinted from Macias-Montero, J.-G., Sarraj, M., Chmeissani, M. et al., A 2D 4 × 4 channel readout ASIC for pixelated CdTe detectors for medical imaging applications, *IEEE NSS and MIC Conf. Rec.*, 2013. Copyright 2013 IEEE with permission.)

TABLE 11.3
VIP Pixel Performance Properties

Input charge maximum range	± 17 to ± 70 fC
Gain for both polarities	10, 16, 20, and 40 mV/fC
Detector leakage compensation	up to 500 pA/pixel
Minimum threshold (from 80 pixels)	8 keV
ENC 40 mV/fC	152 e-RMS
Supply voltage	2.5 V
Power consumption	190 μW/pixel

Source: Reprinted from Macias-Montero, J.-G., Sarraj, M., Chmeissani, M. et al., A 2D 4 × 4 channel readout ASIC for pixelated CdTe detectors for medical imaging applications, *IEEE NSS MIC Conf. Rec.*, 2013. Copyright 2013 IEEE with permission.

cooling of the CdTe to control the leakage current. The dimensions of the chip are 10 mm × 12.5 mm, and it is schematically shown in Figure 11.9.

The Under-Bump-Metal (UBM) layer is deposited on the input of each pixel on the VIP-PIX ASIC, and followed later by the process of solder bump deposition as shown in Figure 11.10. As a general rule, the diameter of the UBM pad should be 80% of the solder ball diameter. The UBM consists of based metal of TiW/Cu done by sputtering and then Cu-Ni-Au by electroplating. The active part of the VIP-PIX is less than 20 um thick out of 725 μm of the total thickness of the ASIC. Therefore, to reduce unwanted passive material in the VIP detector, the passive material is removed by thinning the ASIC down to 50 μm. However, the process of thinning the ASIC requires that the ASIC has no solder bumps and hence the solder bump deposition will take place in the final stage. After the ASIC is populated with the solder bumps, it undergoes a reflow process in a formic acid oven to remove any possible oxidization of the solder balls. This is the last processing stage before doing the flip-chip and bonding process to the pixel CdTe detector.

For every triggered pixel, the surrounding 8 pixels are read as well and the data of 132 bits that includes the time stamp of the event are stored in the shift register. The dataflow from the pixel to the processing unit (PC or Server) is based on a bus

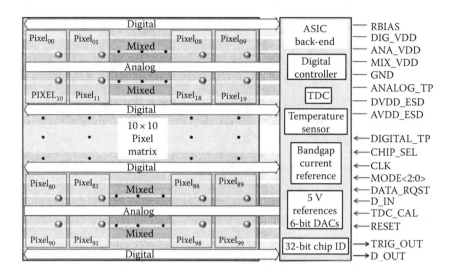

FIGURE 11.9 (See color insert.) Depiction of the entire VIP-PIX chip pixel array. (Reprinted from Macias-Montero, J.-G., Sarraj, M., Chmeissani, M. et al., A 2D 4 × 4 channel readout ASIC for pixelated CdTe detectors for medical imaging applications, *IEEE NSS and MIC Conf. Rec.*, 2013. Copyright 2013 IEEE with permission.)

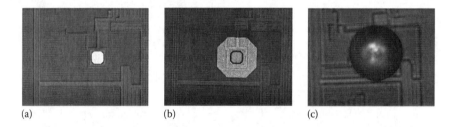

FIGURE 11.10 Detail of a ASIC pixel pad. (a) shows the bare aluminum input pad of a pixel. (b) shows the UBM deposited on the input pad. (c) shows the solder ball covering the UBM pad.

protocol that manages the hand shaking between the VIP-PIX, the FPGA, and the PC. The data are stored in list mode for further quasi offline analysis, grouping the potential PET events based on the coincidence of their time stamps within a time window of 20 ns. At nominal dose, the expected trigger rate per pixel is about 10 Hz.

11.2 IMAGE RECONSTRUCTION

11.2.1 INTRODUCTION

The different characteristics of the VIP detector proposals (PET, PEM, and Compton gamma camera) have consequences for the choice of the optimal image reconstruction algorithm. With PET and PEM detectors, the source of radiation for an event lies on the LOR, that is, the line connecting the two back-to-back photons causing a hit in

Emerging Concept in Nuclear Medicine

the detector. With the Compton camera, the radiation source is expected to lie on the Compton cone. An additional challenge of image reconstruction with the proposed VIP detector designs is their large number of signal channels, which makes some of the standard reconstruction algorithms unpractical, unless additional techniques are applied.

11.2.2 Simple Back-Projection

With image reconstruction, we have to distinguish between the image space and the projection space. The image space is defined by the field of view (FOV), that is, the part of space where the original source of activity is located. The measured data can be considered as the projection of the original activity unto the detectors, that is, the projection space.

A parallel projection of an image object is the collection of all activity along LORs with the same orientation along angle φ (Figure 11.11). This is also known as a Radon transform, defined as a series of line integrals through an image presented by $f(x, y)$, at a certain angle φ and at different offsets r from the origin. A sinogram is the collection of all projections $p(r, \varphi)$ over the full angular sphere, binned in angle φ and with offset r, as depicted in Figure 11.12.

In simple back-projection, all data projections $p(\varphi, r)$ are projected back along their corresponding LORs onto the image space. The final resulting image corresponds to the sum of all backprojected views from the sinogram. Since the value of each projection is smeared equally over all bins in the FOV that lie on the LOR, this leads to a blurred final image, as depicted in Figure 11.13. The advantage of using sinograms is that a 1D Fourier transform can be applied to them, which is used in projection (FBP), as explained later.

In the case of detectors without full angular coverage, the back-projection approach will not work very well. This is especially true for Compton cameras, where the data cannot be easily represented in sinograms and entire cones would have to be projected into the image space causing extremely blurred images.

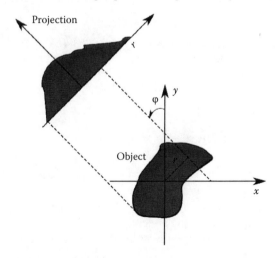

FIGURE 11.11 Depiction of a Radon transform.

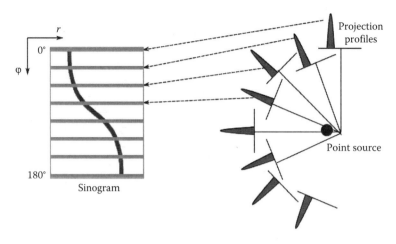

FIGURE 11.12 Depiction of a sinogram.

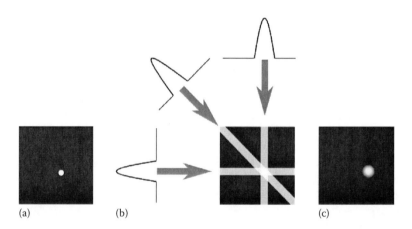

FIGURE 11.13 Schematic depiction of how simple back projection creates a blurred image. (a) is the original image. (b) shown how the views from three different angles are back projected onto the FOV, with their contents smeared along the LOR. (c) the final blurred image when all projections from all angles are back projected onto the FOV.

11.2.3 Filtered Back-Projection

As explained before, simple back-projection will result in blurred images. To avoid this, a Fourier transform can be applied to the projection data. After the Fourier transform, a high-frequency filter can be applied before applying an inverse 1D Fourier transform and projecting the resulting filtered sinogram back onto the image space. This technique is called FBP, as depicted in Figure 11.14.

The main drawback of FBP is the requirement of an angular coverage of at least 180° and the use of sinograms, which makes its usage for PEM and Compton gamma cameras complicated if not impossible.

Emerging Concept in Nuclear Medicine

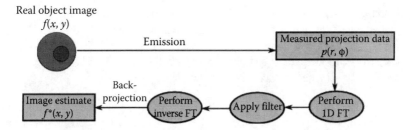

FIGURE 11.14 Schematic overview of FBP. A Fourier transform is applied to the measured data, and subsequently a low-frequency filter is applied. After applying an inverse Fourier transform, the data is back-projected onto an estimate of the original image.

11.2.4 Ordered Subset Expectation Maximization (OSEM)

With maximum likelihood expectation maximization (MLEM), an image estimate is forward projected onto the detector and the projected data is compared with the real measured data. Next, a cost function is used to update the image estimate. These steps are repeated iteratively until a minimum error has been reached, as depicted in Figure 11.15. Ordered subset expectation maximization (OSEM) [24] works the same as MLEM, but it is optimized by doing an update after each subset of the total amount of data has been processed.

OSEM, however, still requires an impractical large memory and central processor unit (CPU)-time consumption for detectors with a large number of channels. This is because OSEM uses a system matrix that maps probabilities for signals coming from voxels in the FOV to be detected in a particular measurement bin. For the VIP PET or PEM design, the number of measurement bins corresponds to the number of signal channels, and—especially for PET—is already unpractically large for OSEM. For the Compton camera, each measurement bin corresponds to a particular combination of a signal channel in the scatter detector, a signal channel in the absorber and a certain bin in the scatter energy range, and clearly the total number of measurement bins would be impossibly large.

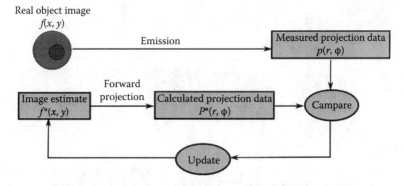

FIGURE 11.15 Schematic overview of MLEM from [85] (licensed under CC BY). Iteratively, image estimates are forward projected unto the detector and compared with the actual measured data. A cost function is used to update the image estimate.

List-mode (LM) OSEM [25] would significantly reduce this because it sums over all detected events instead of looping over measurement bins. To find the FOV bins that lie on the Compton cone (Compton camera) or LOR (PET or PEM) of each event, a fast list-mode back-projection algorithm can be used [26].

Each LM-OSEM iteration consists of two steps:

- Forward projection of the image estimate λ onto the detector: $\sum_{k=1}^{M_{FOV_i}} t_{ik}\lambda_k$, where t_{ik} is the transition probability for event i to have originated from FOV bin k and M_{FOV_i} is the number of bins in the FOV that are intersected by the cone of event i.
- Back-projection of the measured data, weighted with the forward projected data, providing an update correction for the image estimate.

These steps combine into the following equation to update the image:

$$\lambda_j^{l+1} = \frac{\lambda_j^l}{s_j} \sum_{i=1}^{N_{events}} \frac{t_{ij}}{\sum_{k=1}^{M_{FOV_i}} t_{ik}\lambda_k} \tag{11.1}$$

The transition probability t_{ik} and the sensitivity s_j for FOV bin j (i.e., the probability for this bin to produce a detected event) depend on physics and geometry considerations. The calculation of these variables requires the determination of the probability of each event to occur, starting from a particular FOV bin, by determining its particular cross section. To reduce the cost in CPU time and memory, these variables are set to one for the results presented here.

11.2.5 Origin Ensemble (OE) Algorithm

With OE [27,28], at initialization, for each event a random position is assigned on the Compton cone or PET or PEM LOR and the event density matrix D stores the number of events for each FOV voxel location L. All random positions on the cone or LOR, also have to be inside the FOV. For each iteration $i + 1$, the following steps are done for each event (Figure 11.16):

- A new random location L_{i+1} is selected on the cone or LOR of this event and inside the FOV.
- The location L_{i+1} is accepted with probability P, comparing the event density $D_{L_{i+i}}$ at the new location with the density D_{L_i} at the old location for this event:

$$P(Y_i \to Y_{i+1}) = \min\left\{1, \frac{(D_{L_{i+1}} + 1)}{D_{L_i}}\right\}$$

where the current set of origins of events is described by the vector Y_i.
- When the new location for the event is accepted, the density matrix D is immediately updated accordingly before moving onwards to the next event.

Due to the stochastic nature of the algorithm, various trial runs have to be executed and the final result is an average of these trial runs.

Emerging Concept in Nuclear Medicine

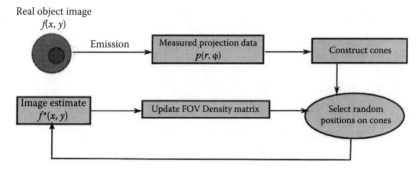

FIGURE 11.16 Schematic overview of OE. In each iteration, the image estimates are updated by random estimates of possible origins of the event.

Resolution recovery mechanism: Because of the finite spatial resolution of the detectors, the precise location of the radiation source will never be reached without the addition of an additional resolution recovery mechanism [28]. To counteract the geometry resolution due to the finite size of the detector voxels, the hit positions are varied uniformly in the range of the voxel size (± 0.5 mm in x, z or ± 1 mm in y). In the case of the Compton camera, an additional smearing of the energy deposition in the scatter detector to recover from the energy resolution, and to the Compton angle, to counteract the Doppler broadening effect, is applied.

11.2.6 IMAGE QUALITY CRITERIA

Various criteria can be defined to quantify the image quality and to evaluate and optimize the image reconstruction algorithms and their parameters.

The modulation transfer function (MTF) is a measure of the contrast in an image and can be used to optimize the FOV pixel size (i.e., the pixel size of the final image).

Image quality metrics such as the bias, the variance and the average mean square error (MSE) can be used to optimize parameters of image reconstruction algorithms. Bias and variance are measures of the accuracy and the consistency of the image reconstruction algorithm and the average MSE is a combination of both [29].

Another measure of the image quality is the recovery coefficient (RC), which is defined as the ratio between the observed concentration within a well-defined region in the final image and the real radioactivity concentration.

11.2.7 AVAILABLE SOFTWARE

All simulation data was obtained with the Geant4-based Architecture for Medicine-Oriented Simulations (GAMOS) package [30]. Some implementations of image reconstruction algorithms are already available. The SSRB-FBP2D utility from GAMOS provides a FBP implementation, which can be used for PET. The open source Software for Tomographic Image Reconstruction (STIR) toolkit [31] provides an implementation of OSEM (from on delabeled as STIR-OSEM). However, STIR-OSEM needs a representation of the data in sinograms and, hence, cannot be used for Compton cameras. For a faster version of OSEM that can also be used for Compton cameras and PEM without the need to present the data in sinograms, the VIP project developed its own implementation of the LM-OSEM algorithm and also its own implementation of OE.

11.3 VOXEL IMAGING PET

11.3.1 Current Limitations in PET Design

The vast majority of the state-of-the-art scanners used for clinical applications employ scintillating crystals coupled with either photomultiplier tubes (PMT), avalanche photodiodes (APD), or silicon photomultipliers (SiPM). A summary of the relevant properties of the most commonly used scintillating crystals for PET applications is listed in Table 11.1. Since their first appearance in the 1950s, scintillator-based designs have undergone a tremendous development and have become in the last 25 years a standard diagnosis tool for the study of the cancer metabolism. Nevertheless, scintillating crystal detectors present a number of disadvantages that hamper the progress of the PET performance toward the physics fundamental limits.

The crystals usually have a parallelepiped shape few tens of millimeter thick and organized in planar matrices with few millimeter pixel pitch. Though such a design offers an excellent stopping power for 511 keV photons, consecutive matrices, commonly arranged in a ring fashion around the FOV, are separated by gaps through which part of the radiation can escape undetected. The interaction position is known with a precision driven by the size of a single detector element and this is particularly relevant along the radial direction since typical crystals have a tangential section of 4×4 mm^2 and a radial thickness of 20 mm. Even if the position resolution can be improved by using smaller crystals and depth of interaction (DOI) measurement techniques [32–34], in practice it is very difficult to reach a volumetric resolution smaller than $3 \times 3 \times 3$ mm^3 in the large size PETs used for whole body (WB) or head clinical screening [15]. Furthermore, the energy resolution of the most commonly used scintillating crystals is intrinsically limited to ~10% at 511 keV (Table 11.1). This reduces the capacity of the scanner to discriminate between true and scatter events thus limiting the purity of the collected data and the resulting signal-to-noise ratio (SNR) in the reconstructed image. Up to 50% of the total events collected by a clinical scanner in a normal operation mode is due to scatter events [9].

The combination of all these contributions increases the image blurring and degrades the PET performance away from the physical limits. State-of-the-art head-PETs currently offer a PSF around 2.5 mm FWHM and above [35]. Researchers are developing TOF PET where the excellent time resolution of the scintillating crystals is exploited to improve the information of each collected event, but such a solution can only partly compensate for the limitations listed earlier.

Another option is to substitute scintillating crystals with semiconductor materials with relatively high atomic number, for example, CdTe. Together with an adequate stopping power for 511 keV radiation with just few centimeters of material, these detectors offer two main advantages with respect to scintillators: an excellent energy resolution of ~1% for 511 keV at room temperature [36], and an excellent spatial resolution on the photon interaction point with the 3D stack of finely pixelated detectors. This is the reason behind the rising interest towards the many designs based on CdTe detectors currently being proposed and evaluated. Available results in literature include the characterization of single detector components [19], the evaluation of small-scale prototypes [37–39], and the simulation of full detectors [40], as well as studies on the image reconstruction algorithms to use with finely granulated detectors [41]. A full brain PET using CdTe semiconductor detectors has also been reported on [17].

In this context, the VIP design constitutes a competitive solution as indicated by the results obtained with MC techniques [18].

The proposed PET scanner has a modular design based on the VIP CdTe module described in detail in Section 11.1.1. It resembles the cylindrical geometry of a typical head-PET with a gantry aperture of 420 mm diameter and 254 mm axial length. A distinctive characteristic of the VIP is that the module can be given a trapezoidal shape to form a scanner ring without cracks to boost the system sensitivity. As shown in Figure 11.4, the ring is obtained by grouping detector units of increasing complexity.

First, 30 VIP unit modules are stacked to constitute a module block. Second, a stack of 4 consecutive blocks connected to the same bus form a detector section. Finally, 66 sections are grouped in a circle to obtain a seamless ring shape with a total of 6,336,000 individual channels for a density of 450 channels/cm^3. The key feature of the VIP design is that each channel acts as a self-contained detector with an independent trigger and signal readout. For each trigger, the readout provides a digitized value of the energy, the time stamp, and the position of the channel where the detection happened. The electronics specifically developed for this purpose is described in detail in Section 11.1.3.

In summary, the VIP design presents the following properties:

- The stopping power of 4 cm CdTe that corresponds to a ~70% probability for the complete absorption of a 511 keV photon, and the resulting ~50% probability for the absorption of an annihilation photon pair. These values are compatible to those obtained by the commonly used clinical PETs.
- An energy resolution of ~1% FWHM for 511 keV photons at room temperature. Scintillating crystals are intrinsically limited to values at least six times larger. The excellent energy resolution makes the VIP design virtually immune from scatter event contamination.
- A volumetric resolution of $1 \times 1 \times 2$ mm^3 on the position of the photon impact point obtained with the electronic pixelization of the CdTe detectors. Such resolution can be further reduced by implementing DOI reconstruction algorithms in the front end electronics as discussed in Section 11.1.2.

A full PET scanner based on the combination of these key properties will provide images of unprecedented quality and resolution and considerably reduce the screening time, or alternatively the administered radiotracer dose per patient.

11.3.2 Evaluation of the VIP Design

The VIP design has been thoroughly evaluated with MC techniques with the aim to compare the expected performance of the proposed device with state-of-the-art scanners currently used for clinical applications as well as for research purposes. The simulation is done with GAMOS [30]. Several protocols have been agreed on to provide a common framework for the evaluation and the comparison of different PET systems. The most widely accepted of these standards are those recommended by the National Electric Manufacturers Association (NEMA) such as the NEMA NU 2-2001 [42] for WB and head PET scanners, and the NEMA NU 4-2008 [43] for small animal PETs. These protocols propose a number of tests to assess the different aspects of a detector performance, both in terms of counting performance and image quality. In particular, standard recipes are provided for the measurement of the scatter fraction, the counting rate, the spatial resolution, the image noise and contrast, and the minimum detectable lesion size.

In the following, a summary of the main results obtained for the VIP is provided together with a direct comparison to the analogous results available in literature for the best PET scanners currently available on the market.

11.3.3 The VIP Scatter Fraction

The scatter fraction (SF) is defined as the ratio between the amount of scatter events and the total collected events when contamination from random events can be neglected. The SF is the measurement of the maximum signal purity reachable with a PET scanner in a given imaging condition. LORs reconstructed from scatter events lose the correlation with the activity distribution in the FOV and contribute to the final image noise. Though some correction strategies can be applied to reduce the effect [44], a low SF is a desirable quality of a PET scanner and an index of the goodness of the device.

The SF depends on three main factors: the amount and density of the passive material in the FOV, the geometrical coverage of the solid angle, and finally the energy resolution of the detector. While the material in the FOV and the geometry of the scanner are usually fixed for a given imaging application, the energy resolution is a property of the scanner and depends on the used technology. As a general rule, the better the energy resolution, the narrower the energy acceptance window and the lower the SF.

The tests proposed by the NEMA NU 2-2001 and NEMA NU 4-2008 protocols differ in the size and the material of the phantoms to be used. The former requires a 190-mm-long cylinder filled with nonradioactive water and with a diameter of 200 mm. The active part of the phantom is a 185-mm-long line source with a diameter of 2 mm, placed along the cylinder axis with a variable radial offset and filled with an FDG solution. In the case of NEMA NU 4-2008, a smaller cylindrical phantom is described, 70 mm long and with 25 mm diameter. The phantom is made of high-density polyethylene and the active line source is placed at a radial distance of 10 mm from the axis. In both the cases, the activity of the source is required to be as low as possible, to assure negligible random event rate.

According to the NEMA NU 2-2001, the SF of the VIP is smaller than 4% while common values for the state-of-the-art WB and head PETs lay between 40% and 50% [45]. The NEMA NU 4-2008 test yields a considerably smaller SF for the VIP system, a negligible 0.73%, while the results for the state-of-the-art small animal PETs are only a little smaller than their WB equivalent, with the best values between 20% and 30% [47–49]. The superior signal purity of the VIP design is mostly due to the excellent energy resolution of the CdTe detectors and the resulting very narrow energy acceptance window of just 10 keV compared to the several hundreds of keV typically employed in scintillator-based PETs.

11.3.4 VIP Counting Performance

The prototype version of the VIP CdTe detectors and electronics, whose specifics are used in the simulation studies, can yield the event timing information with a time resolution of less than 10 ns across the whole energy spectrum. The full signal processing takes a total of 150 us per channel. This means that the time coincidence window can be set as low as 20 ns and each single channel can tolerate a maximum trigger rate of around 6 kHz.

We studied the counting performance of the VIP system in a realistic scenario where saturation effects due to the high activity in the FOV can jeopardize the system performance, as reported in Ref. [18].

The quantity generally used to measure the counting performance of a PET is the noise equivalent count (NEC) rate, (R_{NEC}), defined as the count rate that would have resulted in the same SNR in the absence of scatter and random events:

$$R_{NEC} = \frac{R_t^2}{R_t + R_s + R_r} \tag{11.2}$$

where R_t, R_s, and R_r are the true, scatter, and random count rate, respectively.

In other words, the NEC rate is an approximate measure of the useful coincidence event rate. The NEC rate is always lower than the total rate registered by a scanner during a data acquisition. It increases linearly with activity as long as the contribution from random events is negligible and no pile-up effect is observed. At higher activities, the losses due to random and pile-up contamination slows down the NEC accretion. The R_{NEC} reaches its maximum R_{NEC}^{peak} at a characteristic activity value A_{NEC}^{peak} and then gradually decreases down to zero when saturation effects become dominant. The A_{NEC}^{peak} indicates the optimal working point for a given PET system while the R_{NEC}^{peak} indicates the maximum achievable rate. The performance of different PETs can be compared by contrasting NEC curves and the R_{NEC}^{peak} and A_{NEC}^{peak} values.

The NEMA NU 2-2001 prescribes a standard scenario for this comparison in the case of WB and head PETs. The test is based on the same cylindrical phantom as for the SF test, but this time the phantom is entirely filled with an FDG solution. The phantom active volume is 5966 mL. The total, true, random, scatter, and NEC rates are measured at different levels of activity. The rate curves obtained with the simulation of the VIP are reported in Ref. [18]. Table 11.4 reports the A_{NEC}^{peak} and the R_{NEC}^{peak} values for the VIP system compared to those obtained with the G-PET [45,46]. The NEC for the VIP system peaks at a specific activity of 5.3 kBq/mL corresponding to a total activity of 31.6 MBq in the FOV.

Additionally, the counting performance test is repeated following the NEMA NU 4-2008 guidelines. Also, in this case the same phantom as for the SF is employed. The rate curves of the VIP for the NEMA NU 4-2008 test are reported in Ref. [18]. The comparison of the NEC peak values for the NEMA NU 4-2008 test is summarized in Table 11.5.

Both tests show the superior counting performance of the VIP system compared to state-of-the-art devices based on scintillating crystals. The simulation confirms that the VIP design is not affected by the large time resolution and the large signal processing time of the CdTe detectors. This result is due to the combination of two factors. On the one hand, the excellent energy resolution suppresses the SF and boosts the signal purity. On the other hand, the high number of channels (more than 6 M in the full ring) makes the system immune from event losses due to photon pile-up within the same channel. In fact, as shown in Figure 11.17, for a total activity of 31.6 MBq in the FOV, the maximum rate experienced by the innermost channels is 35 Hz, well below the 6 kHz limit.

TABLE 11.4
Counting Rate Measurements according to NEMA NU 2-2001

Parameter	VIP	G-PET [45,46]
$R_{NEC,peak}$ (kcps)	122	60
$A_{NEC,peak}$ (kBq/mL)	5.3	7.40
$R_{t,peak}$ (kcps)	152	132
$A_{t,peak}$ (kBq/mL)	9.43	13.69

Sources: Reprinted from Mikhaylova, E., De Lorenzo, G., Chmeissani, M. et al., Simulation of the expected performance of a seamless scanner for brain PET based on highly pixelated CdTe detectors, *IEEE Trans. Med. Imaging,* 2014. Copyright 2014 IEEE with permission.

TABLE 11.5
Counting Rates Measurements according to NEMA NU 4-2008

Parameter	VIP	ClearPET [47]	rPET-1 [47]
$R_{NEC,peak}$ (kcps)	908	73.4	29.2
$A_{NEC,peak}$ (MBq/mL)	1.6	0.51	1.35
$R_{t,peak}$ (kcps)	989.338	n.a.	n.a.
$A_{t,peak}$ (MBq/mL)	1.6	n.a.	n.a.

Sources: Reprinted from Mikhaylova, E., De Lorenzo, G., Chmeissani, M. et al., Simulation of the expected performance of a seamless scanner for brain PET based on highly pixelated CdTe detectors, *IEEE Trans. Med. Imaging,* 2014. Copyright 2014 IEEE with permission.

Emerging Concept in Nuclear Medicine

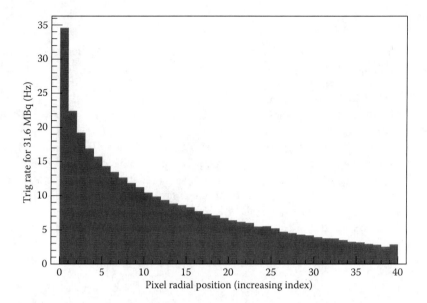

FIGURE 11.17 (See color insert.) The trigger rate per single channel as a function of the channel radial position from, position 0 (the closest to the FOV center) to position 39 (the farthest from the FOV center).

11.3.5 VIP Resolution

The image resolution of a PET can be expressed in terms of the value of the point spread function (PSF) defined as the full width at half maximum (FWHM) of the gaussian reconstructed activity distribution of an emitting point-like source. As a general rule, the smaller the PSF the better the image resolution.

As already discussed earlier regardless of the detector, the PSF is intrinsically limited by two factors: the positron range in the imaged volume, and the acollinearity of the photons emitted in a positron electron annihilation. There are several other detector-related factors contributing further to the PSF of a PET [11], namely the size of the detector elements, the optical decoding used in scintillator-based designs, the parallax error due to the penetration of the photon in the detector, the FOV sampling error due to the finite number of possible LORs, and finally the statistical noise depending on the signal purity and the counting performance of the scanner. All these detector-related factors are expected to be mostly negligible in the VIP system due to the fine granularity of the CdTe detectors and the high signal purity. In fact, the simulated imaging scan of an ideal point-like source yields a PSF close to the 1 mm physical limit [18].

A further improvement can be obtained by implementing the DOI measurement algorithm in the ASIC front-end electronic to virtually reduce the size of the single detector unit as discussed in Section 11.1.2. Also, the definition of a smaller pixel pitch is very much possible, which is a feature that is not applicable to crystal scintillators.

11.3.6 VIP IMAGE QUALITY

Alongside with the excellent counting performance and a resolution close to the intrinsic limit set by physics, the VIP design is expected to provide images with high contrast and low noise, the two main parameters in the evaluation of the image quality. This is a natural consequence of the small SF and the excellent PSF.

The NEMA NU 4-2008 defines a standard test for the evaluation of the image quality based on a dedicated cylindrical phantom reproducing three different imaging conditions: a region of uniform emission, a region with cold insets in a uniform warm background and a third region with hot rods of variable radius in a cold background. The geometry of the phantom is reproduced in Figure 11.18. The image quality test is simulated in two different conditions: the standard one prescribed by the NEMA protocol that requires the phantom to be placed in the center of the FOV with no extra volumes in the surrounding, and a more challenging one with the phantom placed in the center of a 150 mm radius water sphere occupying most of the FOV. A total of 1.5 million coincidences are collected for each test, corresponding to a few

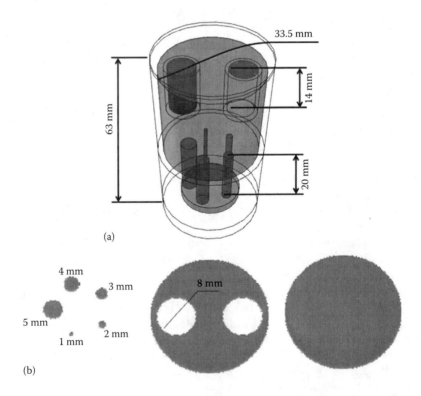

FIGURE 11.18 (a) Geometry of the NEMA NU 4-2008 image quality phantom. (b) Emission point map of three different sections with five hot rods on cold background (left), two cold insets in warm background (center), and uniform emission (right). (Reprinted from Mikhaylova, E. et al., Simulation of pseudo-clinical conditions and image quality evaluation of PET scanner based on pixelated CdTe detector, *IEEE NSS MIC Conf. Rec.*, 2716–2722, 2011. Copyright 2011 IEEE with permission.)

Emerging Concept in Nuclear Medicine

second screening. We have shown in [18] that the reconstructed images for the two scenarios look identical. The reconstructed images for the two scenarios are shown in Figure 11.19. The protocol requires to calculate several parameters to assess the image quality, namely the recovery coefficients (RC) (see Section 11.2.6) for each rod, the minimum, the maximum, and the mean uniformity values of the uniform part of the phantom, and the spill-over ratio (SOR) of the two cold regions. The values of all the calculated coefficients for the VIP scanner in the two tests are summarized in Table 11.6. The results obtained with the phantom in water show no significant deterioration of the image noise and contrast with respect to the phantom-in-air case despite the fact that the same number of coincidences are collected in the two scenarios. This is a unique feature of the VIP scanner, not achievable by standard devices based on scintillating crystals and it is again due to the excellent energy resolution of the CdTe detectors.

Figure 11.20 compares the RC values obtained in the standard scenario with the VIP scanner to those obtained with four representative small animal PETs: the rPET-1 [47], the Clear-PET [47], the Inveon DPET [48], and the LabPET-8™ [49]. It must be noticed that the VIP results are based on the collection of only 1.5 million coincidences corresponding to a scan time of just 19 s, whereas the images of the other scanners are obtained with a 20 min scan, as indicated by the NEMA standard.

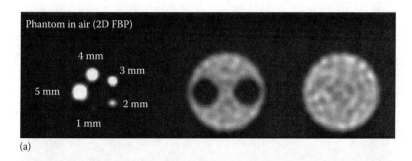
(a)

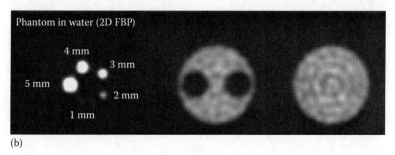
(b)

FIGURE 11.19 (See color insert.) Reconstructed, images of the three sections of the NEMA NU 4-2008 image quality phantom for the standard procedure with the phantom in air (a) and for the modified scenario with the phantom in water (b). (Reprinted from Mikhaylova, E., De Lorenzo, G., Chmeissani, M. et al., Simulation of the expected performance of a seamless scanner for brain PET based on highly pixelated CdTe detectors, *IEEE Trans. Med. Imaging*, 2014. Copyright 2014 IEEE with permission.

TABLE 11.6
Image Quality Parameters Comparison for NEMA NU 4-2008 Phantom Placed in Air and in Water

Parameter	In Air	In Water
RC(%STD) 1 mm	0.106(36.2%)	0.0929(26.5%)
RC(%STD) 2 mm	0.352(16.7%)	0.323(20%)
RC(%STD) 3 mm	0.621(16.4%)	0.578(18.4%)
RC(%STD) 4 mm	0.821(18.7%)	0.797(15.9%)
RC(%STD) 5 mm	0.939(17.1%)	0.924(16.9%)
Uniformity max.	1.13	1.2
Uniformity min.	0	0
Uniformity mean	0.764	0.798
Uniformity %STD	15.9%	15.6%
SOR(%STD) water	0.184(24.3%)	0.168(22.9%)
SOR(%STD) air	0.211(26.4%)	0.207(25.9%)

Sources: Reprinted from Mikhaylova, E., De Lorenzo, G., Chmeissani, M. et al., Simulation of the expected performance of a seamless scanner for brain PET based on highly pixelated CdTe detectors, *IEEE Trans. Med. Imaging,* 2014. Copyright 2014 IEEE with permission.

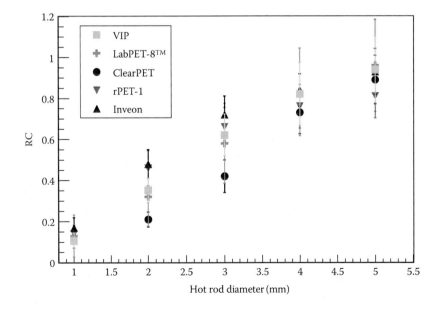

FIGURE 11.20 Comparison of RC of five rods of different size ranging between 1 and 5 mm for 5 different scanners.

Emerging Concept in Nuclear Medicine

It is clear that, despite its big size, the VIP is performing as good or even better than the very small PETs used only for research purposes and currently delivering the world's best images, though with a limited FOV.

11.3.7 VIP Minimum Detectable Lesion Size

The performance of a PET scanner is ultimately measured by the size of the minimum detectable lesion as a function of the tumor to noise ratio (TNR), defined as the ratio between the radiopharmaceutical uptake of tumoral and normal tissues. Results from the previous test shows how the VIP can resolve objects down to 1 mm diameter in a cold background.

The task becomes more challenging when the hot sources are submerged by a warm background. To assess the performance of WB and head scanners in such a scenario, the NEMA NU 2-2001 defines the so-called torso phantom, a plastic container resembling the shape of a human torso. To simplify the simulation, the phantom is given a cylindrical symmetry with a diameter of 240 mm and a total axial length of 180 mm as shown in Figure 11.21.

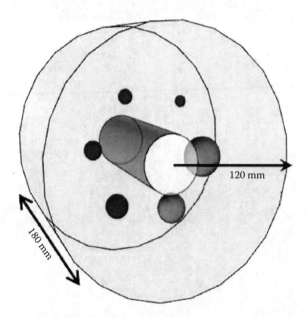

FIGURE 11.21 A plastic rod runs along the axial center to mimic a lung insert. Six spheres of decreasing radius are placed in a circle around the lung insert, the two biggest spheres are empty while the remaining four are filled with an FDG solution. The phantom is filled with an FDG solution of a lower activity to provide a warm background for the four hot spheres. The total activity is 370 MBq to resemble a typical WB screening. (Reprinted from Mikhaylova, E. et al., Simulation of pseudo-clinical conditions and image quality evaluation of PET scanner based on pixelated CdTe detector, *IEEE NSS and MIC Conf. Rec.*, 2716–2722, 2011. Copyright 2011 IEEE with permission.)

Two tests are performed with the simulated VIP with a TNR equal to 8:1 and 4:1 respectively. In both the cases, the images are reconstructed with 10 million coincidences corresponding to a data taking of just 20 s with a total luminosity of 37 MBq in the FOV. Results in Figure 11.22 show that the smallest sphere of 10 mm is already visible with such a small number of coincidences even in the case of a 4:1 TNR. A few image quality parameters are calculated for the different spheres as required by the NEMA prescriptions. The values of the contrast for the cold spheres (CC) and the hot spheres (HC), and the value of the background variation (BV) for the uniform region are summarized in Table 11.7. For comparison, the real measurements obtained with the HR + PET scanner [50] are shown in Table 11.8. The

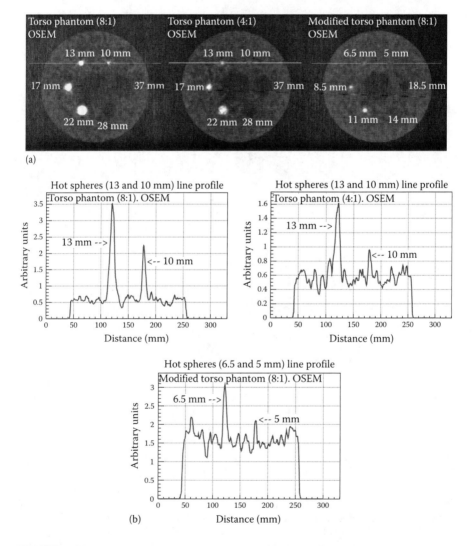

FIGURE 11.22 (a) Reconstructed images for the torso phantom NEMA NU 2-2001 image quality test. (b) emission distribution corresponding to the two smallest hot spheres.

TABLE 11.7
Image Quality Parameters of Reconstructed Torso Phantom

mm		8:1(%)	4:1(%)		8:1(%)	4:1(%)
37	Cold cont.	73.1	73.9	Backgr. var.	3.0	2.8
28	Cold cont.	64.2	60.6	Backgr. var.	4.5	3.7
22	Hot cont.	66.6	24.9	Backgr. var.	5.8	4.3
17	Hot cont.	55.3	21.5	Backgr. var.	6.9	4.9
13	Hot cont.	38.8	13.6	Backgr. var.	8.0	5.8
10	Hot cont.	20.7	5.6	Backgr. var.	9.4	7.4
Avg. residual error		6.6	6.1			
mm			Modified			
18.5	Cold cont.	52.4%	Backgr. var.	3.9%		
14	Cold cont.	34.8%	Backgr. var.	5.5%		
11	Hot cont.	27.2%	Backgr. var.	6.8%		
8.5	Hot cont.	16.6%	Backgr. var.	8.0%		
6.5	Hot cont.	8.8%	Backgr. var.	9.2%		
5	Hot cont.	2.6%	Backgr. var.	10.0%		
Avg. residual error			2.1%			

TABLE 11.8
Quality Parameters of Reconstructed Torso Phantom of HR+ Siemens ECAT PET Scanner

mm		8:1(%)	4:1(%)		8:1	4:1
37	Cold cont.	56.4%	54.1%	Backgr. var.	5.3%	6.2%
22	Hot cont.	43.1%	46.2%	Backgr. var.	—	—
10	Hot cont.	12.6%	1.4%	Backgr. var.	10.6%	13.4%

Source: Herzog, H. et al., *IEEE Trans. Nucl. Sci.*, 51, 2662, 2004.

VIP scanner is clearly outperforming the reported results of the commercial device. Higher CC and HC and comparable BV are observed, which is especially significant if one considers that the total screening time for the HR+ test is 6 min, almost 20 times longer than in the VIP case (with a total luminosity of 370 MBq in the FOV, i.e., 10 times larger).

A more stringent test is performed by reducing the diameter of the sphere to half the original size. The smallest sphere with 5 mm diameter can be distinguished from the background in the 8:1 TNR configuration (Figure 11.22) with a total of 30 million coincidences corresponding to a 60 s scan. The image of the 5 mm diameter sphere is significantly affected by the lack of statistic and its reconstructed activity is at the level of the background fluctuations. Nevertheless, the ability of the VIP scanner to locate such a small hot object in the warm background using a relatively small amount of coincidences confirms its excellent potential in terms of spatial resolution and image contrast.

11.3.8 Simulation and Image Reconstruction of Real 3-D Human Head Phantom

In order to assess the expected imaging VIP performance in pseudo-clinical conditions, a realistic 3-D human brain phantom is simulated. The brain phantom is created using GEANT4 from a set of 20 DICOM (Digital Imaging and Communications in Medicine) files [51] representing the voxelized digital images of 20 axial slices of a real brain obtained with a CT scan, divided into about 1 million voxels. Each voxel represents different human head material such as gray matter, white matter, water, skull, skin, etc., with densities corresponding to typical values for adults. The phantom is filled with a ^{18}F radioactive source of 111 MBq total activity that is 30% of a typical injected dose for body studies (370 MBq). The gray matter to white matter specific activity ratio is 3:2, corresponding to a realistic distribution in case of studies with a ^{18}F positron source. One hundred million total coincidences are collected. The image was reconstructed using the 2-D FBP method without applying any data correction. The pixel size is 0.89 mm and the slice thickness is 6.83 mm, which correspond to the dimensions of the DICOM image voxels.

An example of a DICOM file is shown in Figure 11.23a. Figure 11.23b demonstrates a slice of the simulated 3-D phantom that corresponds to this DICOM image. The brighter gray regions correspond to higher ^{18}F metabolism. The resulting image is presented in Figure 11.23c, showing the same chosen brain slice as in Figures 11.23a and b after scanning the whole reconstructed phantom. The brain structure and its ^{18}F metabolism are clearly visible from the simulation results.

11.3.9 VIP Image Reconstruction Algorithms

The optimization, evaluation, and comparison of image reconstruction algorithms for the VIP was reported in Ref. [52]. For a comparison of FBP, OE, STIR-OSEM, and LM-OSEM with PET, the NEMA NU 4-2008 image quality phantom as described in Section 11.3.6 was used (Figure 11.18) and the image quality metrics explained in Section 11.2.6 were calculated. By comparing the values for the bias,

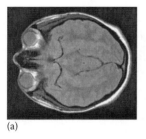
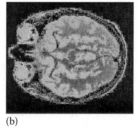
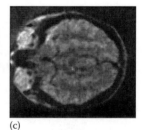

(a)   (b)   (c)

FIGURE 11.23 (See color insert.) (a) Example of a DICOM file. (b) A slice of the simulated 3-D brain phantom corresponding to the DICOM file on the left. (c) The same slice after the whole brain reconstruction. (Reprinted from Mikhaylova, E. et al., Simulation of pseudo-clinical conditions and image quality evaluation of PET scanner based on pixelated CdTe detector, *IEEE NSS MIC Conf. Rec.*, 2716-2722, 2011. Copyright 2011 IEEE with permission.)

TABLE 11.9
Comparison of Image Quality Metrics for FBP (Hamming cut-off 0.15), OE (20 iterations), STIR-OSEM (2 subsets, 2 iterations) and LM-OSEM (2 subsets, 4 iterations)

	Frequency at MTF = 0.5 (lp/cm)	<MSE>	RC (at 5 mm rod)
FBP	3.5	33.5	1.12
OE	3.2	109.2	0.32
STIR-OSEM	2.4	76.84	0.2
LM-OSEM	4.0	29.8	0.875

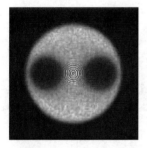  

FIGURE 11.24 Images of a NEMA-NU4 phantom, from PET data, obtained, with LM-OSEM, as published in Ref. [52] (licensed under CC BY).

variance and average MSE for different parameters for the four different algorithms, the optimal configuration for each of the algorithms was found.

Next, the images obtained with optimal parameters for each of the image reconstruction algorithms were compared. The image quality results, summarized in Table 11.9, show that FBP and LM-OSEM give the best performance for PET with this phantom. The images obtained with LM-OSEM are shown in Figure 11.24.

11.4 POSITRON EMISSION MAMMOGRAPHY WITH VIP

11.4.1 VIP PEM Design

A positron emission mammograph (PEM) is an organ dedicated PET scanner for breast screening [53,54]. PEMs have a restricted FOV to achieve higher cancer detection performance in terms of both sensitivity and specificity with respect to the conventional whole-body PET scanners [55]. Extra benefits include lower needed dose, lower cost, and enhanced device portability. The increasing interest in PEMs is also due to their recent employment in PET driven breast biopsies to exploit the advantage of functional over anatomic imaging in discriminating a benign process such as a scar from malignancy [56].

Several designs have been proposed and developed [57–59] with different geometrical solutions to limit the FOV around the imaged breast and axilla. The most commonly available commercial devices employ arrays of scintillating crystal detectors mounted on two parallel paddles whose distance can be regulated for breast immobilization [58] in a mammographylike fashion, thus offering an ideal instrument for PET-driven breast biopsies and PET-CT image co-registration [60]. Such devices can provide excellent in-plane* spatial resolution down to 2 mm FWHM [61] and high contrast images with 4 mm and 6 mm minimum detectable lesion size for 10:1 and 4:1 tumor to normal tissue ratio (TNR) respectively [58]. A number of clinical trials have been conducted over the past few years to show the huge potential of PEM devices in improving breast cancer treatment [62–64]. In particular, PEM showed higher sensitivity with respect to conventional PET/CT in diagnosis and characterization of small breast tumors (<2 cm diameter), though such sensitivity is significantly reduced for very small lesions (<1 cm diameter) [63]. Improving the tumor detectability limit for mm size lesions would dramatically increase the medical impact of PEMs and provide high specificity metabolic images in a region where only anatomic images are currently available.

In the case of a coplanar PEM, the main limitation is due to limited precision of the DOI inside the shintillator crystal [65]. With the two paddles touching the compressed breast during the screening, the FOV is entirely filled with background activity and the distance between top and bottom paddles is of the order of few centimeters. This configuration maximizes the effects of the parallax error in deteriorating the resolution along the vertical axis connecting the two paddles. Moreover, the poor energy resolution of the commonly used scintillators can also be a limiting factor in the screening of dense breast, when the contamination from scatter events becomes important.

The two effects contribute to the deterioration of the detector sensitivity and image quality. In particular, the large uncertainty in DOI has the biggest impact on the capability of detecting small size tumors, regardless of the chosen geometry [66]. This is due to the partial volume effect [10] that produces a loss of intensity and the smearing of the activity distribution around those high uptake regions whose volume is smaller than twice the detector resolution. In the particular case of coplanar scanners, a large parallax error results in a dramatic deterioration of the resolution up to 8 mm FWHM along the vertical axis [58], a factor of 4 worse than the spatial resolution in the in-plane. This poses a serious constraint on the minimum detectable tumor size and on the correct assessment of the malignancy of small lesions that are probably the two most important factors to determine the effectiveness of any breast cancer treatment [67]. Drawbacks to the limited cross-plane† resolution are also evident when using PEMs for PET-driven breast biopsies for which the correct vertical position must be guessed on statistical extrapolation and repeated sampling (up to 12 trials) and post-biopsy confirmation scans are necessary to correctly locate the lesion [56].

The use of finely pixelated CdTe detectors instead of scintillating crystals in a coplanar PEM is particularly appealing because of the small size of such a detector resulting in a much lower cost and complexity with respect to a full PET. The VIP module described in Section 11.1.1 fits perfectly to a coplanar PEM geometry where a stack of the needed number of modular units can easily fill the active region of the two paddles. With the 1% energy resolution at 511 keV and the $1 \times 1 \times 2$ mm^3

---

* Parallel to the paddle surfaces.
† Perpendicular to the paddle surfaces.

Emerging Concept in Nuclear Medicine

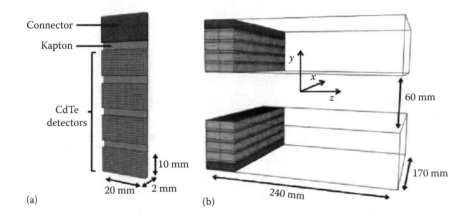

FIGURE 11.25 Basic unit detector (a) and full detector (b) geometrical specifications. (Depictions from Uzun, D. et al., *J. Instrum.*, 9, 2014, under CC BY.)

volumetric resolution in the detection of the annihilation photons, the VIP approach would be an ideal solution for the future PEM generation.

The VIP-PEM is based on the VIP unit detector module (Figure 11.5 and 11.25a) described in Section 11.1.1. Following the typical coplanar design [61], the VIP mammograph consists of two parallel paddles, each one hosting one sliding detector head. The two heads are made of 160 modules each, arranged along two parallel lines of 80 modules for a total of 128,000 channels per head. The coordinate system is chosen as in Figure 11.25b and centered in the center of the FOV. For the tests described in the following, the distance between the two paddles along the y-axis is fixed at 60 mm, but can in principle be varied arbitrarily. The head section is 170 mm wide along the x-axis and 40 mm wide along the z-axis, and the two detector heads must slide axially for a complete scan of the 170 mm × 60 mm × 240 mm FOV.

11.4.2 Evaluation of Imaging Performance

The full device has been simulated using *GAMOS* [30], to assess the expected performance of the VIP PEM in terms of counting performance, spatial resolution, and image quality [68]. Though no dedicated protocols for the evaluation of organ dedicated PETs exist, the NEMA NU 4-2008 prescriptions are generally followed, with some modifications to adapt them to the case of a coplanar scanner.

A first test is performed to calculate the system sensitivity with a single point–like source placed in different positions to estimate the expected sensitivity and the point spread function (PSF) across the FOV as defined in the NEMA NU 4-2008 [43]. The sensitivity is defined as the number of collected coincidences divided by the total number of events. Values of the sensitivity for a point-like source placed in different positions along the z-axis are presented in Ref. [68]. As expected for a detector with sliding heads, the sensitivity is flat across most of the FOV and goes quickly to zero at the edges. The estimated average sensitivity is around 2 cps/kBq. The results are compatible with those obtained when evaluating crystal PEMs with analogous coplanar geometry and sliding heads [61].

The second test aims to evaluate the expected scatter and random coincidence fractions. The phantom described by the NEMA NU 4-2008 standards is the same high-density cylindrical phantom employed in the evaluation of the VIP-PET counting performance. Due to the excellent energy resolution of the CdTe, it is shown that the contamination from scatter events is negligible at any value of the activity [68]. The random fraction is negligible up to 10^6 Bq and peaks at around 10^8 Bq. Saturation effects become evident for activity bigger than 10^8 Bq, well above the level expected in a standard positron emission mammography ($\sim 10^6$ Bq).

A third test is performed to assess the PSF in order to measure the ability of the system to distinguish two closely emitting points. Simulation results indicate a spatial resolution around 1 mm FWHM regardless of the direction [68].

For the image quality evaluation, the NEMA NU 4-2008 image quality phantom is simulated in two different configurations, with the phantom axis perpendicular or parallel to the paddle surfaces in order to compare the quality of the in-plane and cross-plane images, respectively. The resulting images and the corresponding activity line profiles, as presented in Ref. [41], show that with both configurations, good images can be reconstructed with comparable image quality. The simulation results show that the VIP PEM scanner can achieve an excellent image contrast and easily distinguish objects down to 1 mm diameter in the absence of background activity. The conclusions are valid for both in-plane and cross-plane images.

As an extra proof of the excellent resolution of the VIP PEM scanner along the vertical axis due to the absence of the parallax error, a Derenzo phantom [69] is simulated [41]. Also in this case, two different scenarios, with the phantom axis placed parallel and perpendicular to the paddle surfaces, are simulated. The Derenzo phantom consists of hot rods with the increasing diameters of 1, 1.5, 2, 2.5, and 3 mm and a total of 10 million coincidences were collected for each scan. The reconstructed images in Figures 11.26 and 11.27 show that the VIP PEM can resolve all the hot rods down to 1 mm diameter with no significant difference between the in-plane and the cross-plane alignment.

According to the simulation, the VIP-PEM scanner is expected to detect lesions down to 1 mm diameter [41].

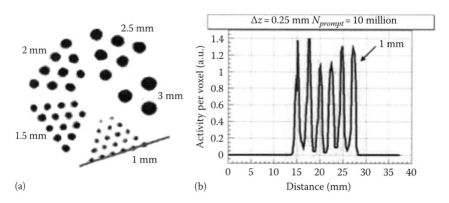

FIGURE 11.26 Results with Derenzo phantom aligned along the Y axis, as published in Ref. [41] (licensed under CC BY). (a) hot rods with diameter 1, 1.5, 2, 2.5 mm; (b) activity line profile of 1 mm rods.

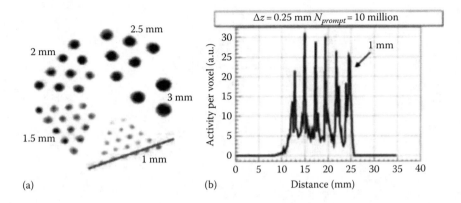

FIGURE 11.27 Results with Derenzo phantom aligned along the Z-axis, as published in Ref. [41] (licensed under CC BY). (a) hot rods with diameter 1, 1.5, 2, 2.5 mm; (b) activity line profile of 1 mm rods.

11.5 VIP COMPTON CAMERA

11.5.1 Introduction to SPECT and Compton Camera

11.5.1.1 SPECT

SPECT, like PET, produces tomographic images of the activity of radioactive tracers. However, whereas PET images are reconstructed from back-to-back gamma pairs produced by electron position annihilation, SPECT images are reconstructed from single gamma hits in the detector.

At present, SPECT devices most commonly employed in nuclear medicine are based on the *Anger camera* (Figure 11.28). The Anger camera consists of scintillator crystals coupled to photomultiplier tubes (PMTs), to detect mechanically collimated gammas coming from an active source. An advantage of SPECT is that different isotopes can be used simultaneously, with a large choice of radio tracers available that emit single or double gammas with different energies. Also, because SPECT detectors do not need to have full angular coverage, they are cheaper. A disadvantage of SPECT is the use of mechanical collimation to reject photons that do not travel along a path within a certain angular bin, which makes that SPECT has a significant lower sensitivity than PET by two orders of magnitude. A low sensitivity means that a higher radiotracer dose or a longer exposure time is necessary to obtain an image with sufficient quality, which is not in the advantage of the patient. Additionally, with SPECT to acquire complete tomographic images it is necessary to rotate the gamma camera and obtain projections at different angles during the rotation.

Depending on the type of object, the shape of the collimators can vary from parallel-hole to fan-beam (collecting converging rays) or cone-beam (collecting diverging rays) collimation. Better image resolution can be obtained by pinhole collimation. A pinhole camera works on the same principle as a camera obscura. It consists of a collimator in which a single small opening lets pass the gamma beam which

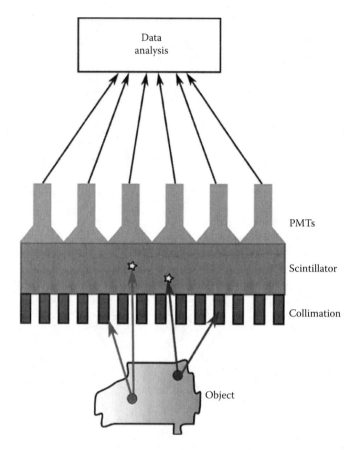

FIGURE 11.28 (See color insert.) Schematic depiction of an Anger SPECT camera.

is projected upside down onto the detector plane. A multi-pinhole SPECT camera is a cylinder containing various pinhole apertures. With multi-pinhole SPECT, one gains a better resolution with identical sensitivity compared to Anger SPECT cameras, but with the price of a smaller field of view. Typically, multi-pinhole SPECT cameras are used for small animal imaging and have an improved spatial resolution of less than 2 mm.

In recent years, much progress has been made in the development of SPECT systems for nuclear medicine applications, with advances both in hardware and software, to optimize dose and acquisition time [70–72]. The main advance in hardware is an increase in photon sensitivity, mainly achieved by new methods of collimation. By using the multiple pinhole principle and dedicated scanner geometry, the sensitivity can be significantly improved and allow simultaneous collection of photons from all angles without the need for camera rotation. Additionally, the use of semiconductors instead of scintillation crystals results in a better energy resolution and hence an improved rejection of scattered events. Advances in SPECT software include new image reconstruction algorithms using physical modeling of the collimators and detectors and compensation for resolution loss. The HICAM project [71] developed two Anger cameras with 5×5 cm² and 10×10 cm² FOV and with two different collimators. With the collimator with largest hole, they can achieve a sensitivity of 245 cpm/uCi (0.11 cps/kBq) with a planar source placed at 5 cm distance of the detector and an overall spatial resolution of ~2.5 mm. Another effort [72] on a new semiconductor SPECT camera reports a count sensitivity 10 times higher than that of conventional cameras. Their novel D-SPECT camera, optimized for cardiac imaging, achieves a sensitivity of 127 cpm/kBq.

11.5.1.2 Compton Camera

The application of Compton gamma cameras for nuclear medicine was first proposed in 1974 by Todd et al. [73], presenting electronic collimation as an alternative to mechanical collimation as used in SPECT. In the early 1980s, Singh and Doria analyzed in more detail a first working prototype for medical imaging [74,75]. Since then, many Compton camera designs have been presented, taking advantage of the advances in semi conductor technology.

Compton gamma cameras, instead of mechanical collimation, use the kinematics of Compton scattering to localize the radioactive source, as depicted in Figure 11.29. Gamma rays emitted by a radioactive source scatter inelastically (Compton scattering), depositing some of their energy, in the scatter detector and, subsequently, are absorbed in the absorber (photoelectric effect) where all of its remaining energy is completely deposited. Hence, the choice of the material for the scatter detector depends on the attenuation strength of Compton scattering in the detector material, whereas for the absorber, it depends on the attenuation strength of the photoelectric attenuation in the used material.

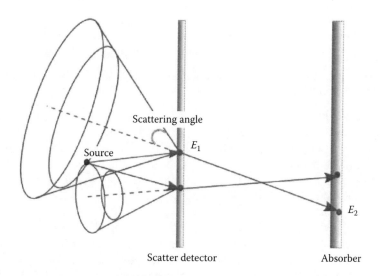

FIGURE 11.29 Schematic depiction of two Compton cones, reconstructed from two Compton scattering events in a Compton gamma camera.

The original gamma ray source is located on the surface of the Compton cone identified by the cone axis and apex, determined from the hit locations, and the scattering angle θ, obtained from the following equation:

$$\cos(\theta) = 1 - m_e c^2 \left\{ \frac{1}{E_\gamma - E_{scatter}} - \frac{1}{E_\gamma} \right\} \quad (11.3)$$

Where
E_γ is the known energy of the original gamma
$E_{scatter}$ is the energy deposited in the scatter detector.

However, because the momentum of the electron the gamma scatters from is unknown, Equation 11.3 is only approximately valid. This deviation of the relation between scattering angle and energy deposited in the scatter detector is known as the *Doppler broadening effect* [76]. Without knowledge of the momentum of the recoil electron, the Doppler broadening effect in addition to the energy resolution contributes to the smearing of the Compton scattering angle. In Ref. [76], the Doppler broadening effect is compared for different materials. More detailed information about the Doppler broadening effect in Compton cameras for nuclear medical imaging can be found in Ref. [77].

There are a number of advantages of Compton cameras over mechanically collimated SPECT cameras:

- Because there is no mechanical collimation, higher sensitivity can be achieved without the trade-off of worse spatial resolution. For nuclear medical imaging purposes, this means a reduction of the radiation dose or exposure time for the patient.
- Higher energy sources can be used without the need for thicker collimators.
- Three-dimensional tomographics images can be made without having to rotate the detector with complete angular coverage.

However, there are also various challenges of the Compton camera compared with traditional SPECT:

- A dominant contribution of the Doppler broadening effect to the spatial resolution, more pronounced for low-energy photons.
- The need to construct coincidences from two hits that can lead to falsely identified random hits.
- Image reconstruction for a Compton camera is challenging for a number of reasons, as explained in Section 11.2.

Compton cameras have applications not only in medical imaging, but also in homeland security and astronomy. Since the first proposal for its use in astronomy in the 1970s, various Compton cameras have been used in as telescopes at space observatories, for example, the Compton Telescope (COMPTEL) at the Compton Gamma Ray Observatory (CGRO) of NASA, detecting gamma light in a range from a few keV up till GeV [78,79]. Another report on a semiconductor

Compton camera [80] quotes a resolving power better than 3 mm for a 364 keV ^{131}I source at a distance of 30 mm of the detector and spatial resolutions of 6 and 18 mm for targets at distances of 30 and 150 mm, respectively. It also reports results on an extended source in a horseshoe shape. Another article from the same group [81] reports on further intents with a real Si/CdTe Compton camera to perform 3D imaging, on a sacrificed rat, injected with various radiotracers which were imaged simultaneously. Multiple cameras from different angles are used to improve the quality of the 3D images. They report a spatial resolution of 8 mm in the x and y axes and 10 mm in z, and an efficiency of 0.23 cps/MBq for a 356 keV source at 10 cm distance, which is still poor compared to PET and SPECT.

Compton cameras currently are still not used for medical applications because of the requirement of high energy resolution, high granularity and a large field of view, which is not possible with detectors made of scintillator material. Additionally, there is the challenging requirement to image reconstruction, especially in the presence of background activity.

11.5.2 VIP Compton Camera Design

The aim of the VIP project is to show that using pixelated semiconductor technology with a high voxel granularity for nuclear medicine detectors will improve both the energy and position resolution of the measurements with high sensitivity compared to state of the art crystal detectors. For this, a basic unit module has been designed that easily can be stacked to build up nuclear medicine detectors for applications like (PET), PEM or Compton gamma camera. The basic VIP unit module, as explained in Section 11.1.1, is made of 4 CdTe pixelated detectors of 20×10 voxels each (Figure 11.5). Each voxel has a size of $1 \times 1 \times 2$ mm^3 and is connected to its own independent readout channel for the energy, position, and arrival time. The chosen voxel size represents a good compromise between spatial resolution and fabrication cost.

A schematic view of the VIP Compton camera design is shown in Figure 11.30. The scatter detector and the absorption detector of the VIP Compton camera are made of pixelated Si and CdTe sensors, respectively. The distance between scatterer and absorber and the thickness of both detectors are optimized for the best compromise between spatial resolution and detection efficiency. Details about the optimization procedure are described in Section 11.5.3.1.

The absorption detector of the Compton camera has a parallelepiped shape with an active area of size $380 \times 540 \times 40$ mm^3. The scatter detector of the Compton camera has a rectangular parallelepiped shape with an active area of size $380 \times 540 \times 20$ mm^3, and is built following a similar modular design as the absorber, with some differences in the module geometry. The size of the voxels in the scatter detector is chosen to be 1 mm $\times$ 1 mm $\times$ 2 mm for consistency with the CdTe sensors, although Si can be made much thinner at acceptable cost.

The total number of independent readout channels of the VIP Compton camera design amounts to more than 3 million, which means an additional challenge for the image reconstruction, as explained in Section 11.2.

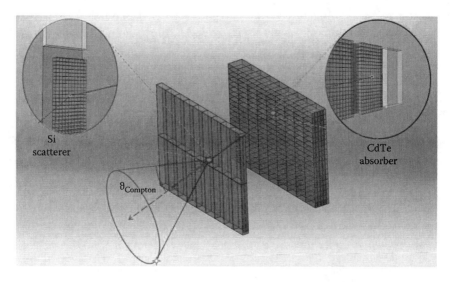

FIGURE 11.30 Schematic view of the VIP Compton camera. (Reprinted from Calderón Y. et al., Modeling, simulation, and evaluation of a Compton camera based on a pixelated solid-state detector, *IEEE NSS MIC Conf. Rec.*, 2011, pp. 2708–2715. Copyright 2011 IEEE with permission.)

The big advantage of the VIP design lies in the use of pixelated semiconductor materials for the scatterer (Si) and the absorber (CdTe). These advantages are

- Semiconductor detectors have small statistical variation of the pulse width of produced charge carriers. Hence, the energy resolution of semiconductors is by far superior to that of scintillation material [19].
- The good energy resolution improves the rejection of background noise and also the spatial resolution because of a more precise estimation of the Compton scattering angle.
- Small millimeter-sized voxels are possible with pixelated semiconductor detectors, resulting in a better spatial resolution.
- Semiconductor detectors, such as CdTe, CdZnTe, and HgI_2, are operational at room temperatures.
- Semiconductors can be operated in a magnetic field allowing for hybrid operation of SPECT and MRI.

The choice of silicon as semiconductor material for the scatter detector is justified by its optimal Compton cross section and a relatively small Doppler broadening effect compared to other semiconductor materials [76]. Additionally, the modular design of the VIP detector designs allows for easy portability and easy extension of the detector.

The choice of CdTe as semiconductor material for the absorber is justified by its excellent energy resolution and its high detection efficiency. Also, in the energy range up to 511 keV, which is the energy range of radiotracers used in medical imaging, the probability of photoelectric absorption is only sufficient for semiconductors with high Z. CdTe with a Z value in between 48 and 52, complies with this requirement.

11.5.3 VIP Compton Camera Performance

Before applying image reconstruction algorithms to the detector data, the Compton camera geometry should be optimized to guarantee an optimal behavior as far as sensitivity and angular resolution is concerned.

11.5.3.1 Optimization of Design

Using the phantom as specified by the NEMA-NU4 standard [43] for sensitivity measurements, the VIP Compton camera design was optimized for various design parameters. The isotopes that were used for these tests were ^{18}F (a positron emitting isotope, that will result in 511 keV gammas reaching the Compton camera) and ^{99m}Tc (a 141 keV gamma-emitting source).

The plot on the Figure 11.31a shows how the detector sensitivity changes with increasing thickness of the scatter detector. With ^{18}F, the optimal thickness is 3 cm Silicon and with ^{99m}Tc, the optimal thickness is 2 cm, with a sharp decrease for 3 cm. To have a good sensitivity for the entire energy range, a 2 cm Silicon thickness for the scatter detector is chosen. The plot on the Figure 11.31b shows how the detector sensitivity changes with increasing thickness of the absorber. With 4 cm thickness of CdTe, for a 511 keV gamma source, 95% of the maximum efficiency is achieved. With isotopes that emit gammas with lower energies, less thickness is necessary.

Note, however, that although the evaluation of the VIP Compton camera design is performed with 4 cm CdTe thickness, in fact other, more cost-efficient, solutions can be imagined. With a thickness of 2 cm, the cost of the CdTe detector would be half, while the loss in sensitivity would only amount to 11%. Alternatively, one could imagine, for the same price, to have a dual-headed Compton camera, with two orientations under a rotation of 90°, where the absorber would have 1 cm thickness but the loss of sensitivity would be compensated with a larger angular coverage and, additionally, an improvement in image quality (as explained in Section 11.5.4).

Figure 11.32 shows how the detector sensitivity depends on the distance between absorber and scatterer for both ^{99m}Tc and ^{18}F simulations. We found that the optimal distance between the back plane of the scatter detector and the front of the absorber is 10 cm. With smaller distances the angular resolution would begin to dominate the overall spatial resolution. With larger distances, the contribution of the angular resolution to the spatial resolution gets too small to compensate for the decrease of detector sensitivity. In Table 11.10, an overview is given of the optimized parameters for the geometry of the Compton camera.

Additional parameters are necessary to evaluate the system based on the performance of the readout and trigger electronics. The values of these are shown in

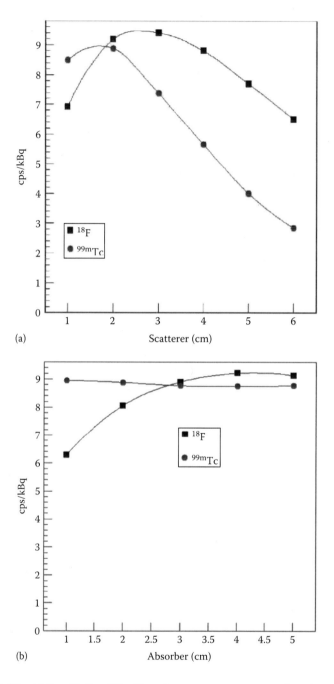

FIGURE 11.31 (a) Sensitivity of the Compton gamma camera as a function of Si thickness in the scatter detector for different isotopes, ^{18}F and ^{99m}Tc. (b) Sensitivity of the Compton gamma camera as a function of CdTe thickness in the absorber for different isotopes, ^{18}F and ^{99m}Tc.

Emerging Concept in Nuclear Medicine

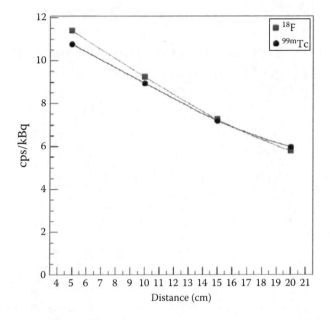

FIGURE 11.32 Efficiency of the Compton camera as a function of the distance between scatter detector and absorber for different isotopes ^{18}F and ^{99m}Tc.

TABLE 11.10
Optimized Parameters of the Compton Camera

Scatterer thickness (Si)	2 cm
Absorber thickness (CdTe)	4 cm
Scatterer to absorber distance	10 cm

Table 11.11. The details about the VIP readout system is described in Section 11.1.3. The readout output of the detector consists of a list of hit information including the hit position, energy deposition and timestamp. For a Compton camera, we are interested in coincidences, and hence, we have to compare pairs of hits, with one hit in the scatter detector and one hit in the absorber, that occur within a certain *coincidence time window*. The coincidence time should be large enough to compensate for smearing in the timestamps due to electronic jitter and signal delays and small enough in order to suppress random coincidences.

11.5.3.2 Detector Sensitivity

The dependence of the detector sensitivity on the source activity was studied with isotopes ^{99m}Tc and ^{18}F using a phantom defined by the NEMA-NU1 standard [83]. The phantom consists of a radioactive line source placed within a disk source holder filled with water, where the distance between the radioactive line source and the front plane of the scatter detector is 25 mm.

TABLE 11.11
Readout and Trigger Electronics Parameters of the Compton Camera

Scatterer trigger threshold (141 keV source)	10 keV
Scatterer trigger threshold (511 keV source)	20 keV
Absorber trigger threshold	20 keV
Dead time per voxel	130 µs
Measuring time per voxel	20 µs
Coincidence time window	20 ns

Notes: The trigger thresholds correspond to the minimum energy that should be deposited in a voxel to raise a trigger. The measuring time corresponds to the time, starting from the trigger, during which the semiconductor voxel is read out. Each voxel has its own individual readout and the energy of all hits in the voxel occurring during the measuring time are merged into one signal. The dead time corresponds to the time it takes to process the signal from a voxel. No new signals can be read during dead time.

Figure 11.33 shows for isotopes ^{99m}Tc and ^{18}F the total number of estimated coincidence events and the contributions from correctly classified coincidence events and background events. The results show that for activities of up to 10^7 Bq, the detector sensitivity is ~4.5 cps/kBq or more with ^{18}F and a detector sensitivity of ~3.2 cps/kBq can be obtained with ^{99m}Tc. with an optimized trigger threshold for the scatterer of 10 keV (instead of 20 keV).

Coincidence events are considered as correctly classified events when the original gamma undergoes Compton scattering in the scatter detector and gets absorbed by the photoelectric effect in the absorber. Incorrectly classified background events can be classified as *scattered* tracks, *random* coincidences and events that were caused by *other physics* processes. The results show that for the purity of the samples (i.e., the percentage of correctly classified coincidence events) 90% with a ^{18}F source and 77% with a ^{99m}Tc source can be achieved. Coincidence events are marked as "scattered" if the gamma has undergone additional scattering in the passive material of the source or detectors. The contribution of scattered events is of the order of 3% for ^{18}F and 10% for ^{99m}Tc. The relatively small contribution of scattered events to the overall coincidence event sample is because of the excellent energy resolution of the detector that allows for a small energy window around the target energy to reject events with missing energy due to scattering. The contribution of events from other physics processes is the largest source of background for low activities. The main contribution to these physics processes are events with additional hits in the detectors with energy deposits below the trigger threshold. The contribution of random events (i.e., hits from two different events, incorrectly assumed to be from the same coincidence event) is negligible with a ^{18}F source and only gets significant with ^{99m}Tc with activities higher than 10^7 Bq. Also, at such activities, the detector efficiency rapidly goes down because of the increasing chance to detect more than two hits in the same coincidence time window.

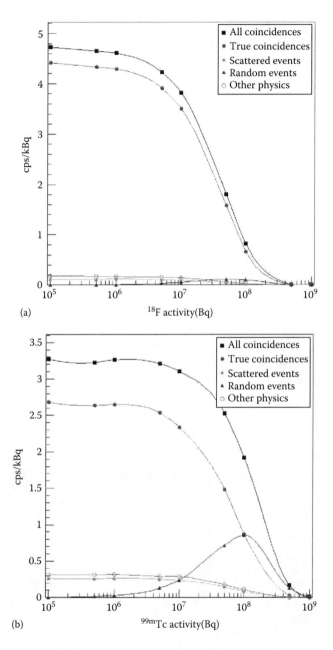

FIGURE 11.33 (See color insert.) Sensitivity of the Compton gamma camera as a function of source activity for different isotopes: (a) ^{18}F and (b) ^{99m}Tc.

11.5.3.3 Spatial Resolution

Image reconstruction algorithms applied to VIP Compton camera simulation data are explained in detail in Section 11.2. It is shown that with iterative image reconstruction algorithms as OE or LM-OSEM, a point spread function (PSF) of the order of 2 mm (FWHM) can be achieved with a single point source located at 100 mm distance from the scatterer plane.

11.5.4 VIP Compton Camera Image Reconstruction Results

Results for Compton camera with image reconstruction algorithms OE and LM-OSEM are reported in Refs. [84, 85].

Single Point Source: Using OE, a PSF smaller than 2 mm (FWHM) was found in the x- and y-direction and 2.4 mm in the z-direction. Using LM-OSEM, a PSF of about 1.9 mm was found in the x-direction and 2.9 mm in the y-direction. However, for LM-OSEM, the resolution in the direction perpendicular to the Compton camera is worse than in the lateral directions. To obtain a good resolution in the perpendicular z-direction with LM-OSEM, the Compton camera has to be rotated with 90° around the y-axis to obtain a PSF of 1.9 mm, as illustrated in Figure 11.34.

3D Cube phantom: It has been shown ([85], see Figures 11.35 through 11.37) that with LM-OSEM good images can be reconstructed with a phantom consisting of eight sphere-shaped sources on the corners of a cube, where the sources have a center-to-center distance of 5 mm. As explained before, to get a good resolution in the z-direction, the Compton camera should be rotated 90° around y.

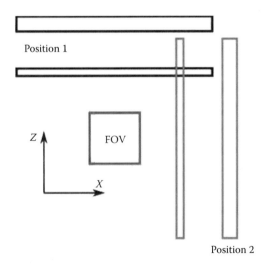

FIGURE 11.34 (**See color insert.**) Depiction of two orthogonal Compton cameras to obtain a good PSF in all direction. (From (Kolstein M. et al., *J. Instrum.*, 9, 2014), licensed under CC BY.)

Emerging Concept in Nuclear Medicine

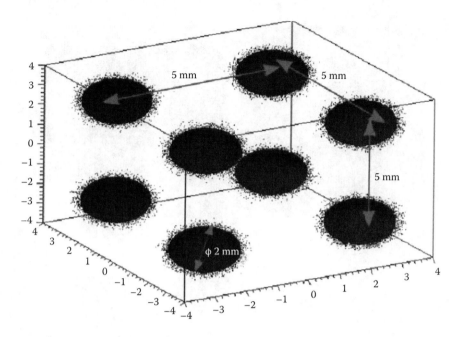

FIGURE 11.35 Depiction of the 3D cube phantom. (From (Kolstein M. *et al.*, *J. Instrum.*, 9, 2014), licensed under CC BY.)

With OE, it is not possible to get good results with sources that are separated 5 mm or less. Instead, it is possible to reconstruct a phantom with point-like sources on only 4 corners of the cube, and with a distance between the sources of 10 mm [84]. In this case, it is not necessary to rotate the Compton camera to obtain a good 3D image with a comparable resolution in all three directions.

Horseshoe phantom: An important requirement for nuclear medicine applications is the ability to correctly reproduce extended source objects, such as a horseshoe-shaped phantom. Figure 11.38 shows that very good results can be obtained with LM-OSEM and OE on Compton camera data with a horseshoe phantom ([85]).

Derenzo phantom: Another phantom with extended source objects is the Derenzo phantom [69], which serves as a measure of the spatial resolution of the detector. The Derenzo phantom consists of five segments, each containing rods with equal activity, of length 12 mm and with varying diameters and distances (left image in Figure 11.39). Figure 11.39 shows that very good results can be obtained with LM-OSEM ([85]). The rods with diameter 1.5 and 6 mm distance between the rod centers are distinctly reconstructed, as also can be seen from the line profile. With OE, it is not possible to produce an image of a Derenzo phantom with similar image quality from the same amount of Compton camera events.

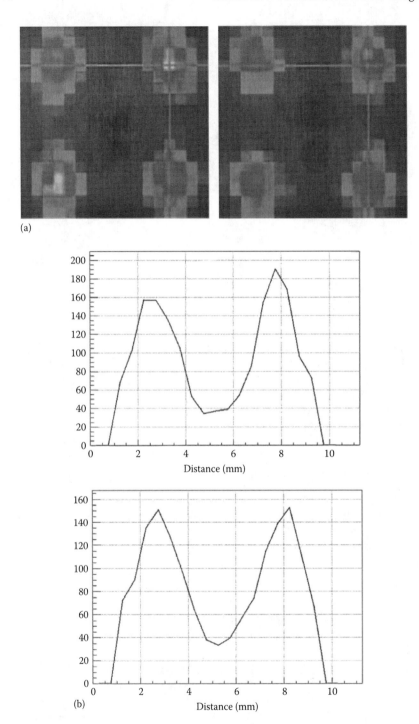

FIGURE 11.36 Results with a 3D cube phantom; showing the 2D image in the *xy* plane (a) and line profiles along *x* (b). *(Continued)*

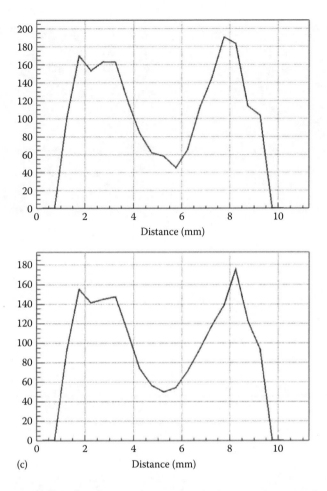

FIGURE 11.36 (*Continued*) Results with a 3D cube phantom; along *y* (c) for a slice with z = 2.5 mm (top) and z = −2.5 mm (bottom). (From (Kolstein M. et al., *J. Instrum.*, 9, 2014), licensed under CC BY.)

2D brain slice: Another phantom used to test the imaging capabilities of the Compton camera, was a 2D human brain slice [86]. For this, a DICOM 2D image file was used, containing information of a slice of a human brain. The original image was made based on a ^{18}F isotope with 105 Bq activity located in the grey matter of the brain slice. The LM-OSEM image reconstruction was used on 34 million coincidence events, with 20 iterations and a field of view of $1 \times 1 \times 5$ mm^3 divided in $220 \times 220 \times 1$ bins. Figure 11.40 shows that the grey matter structures in the image, as given by the DICOM file, can be clearly distinguished in the reconstructed image.

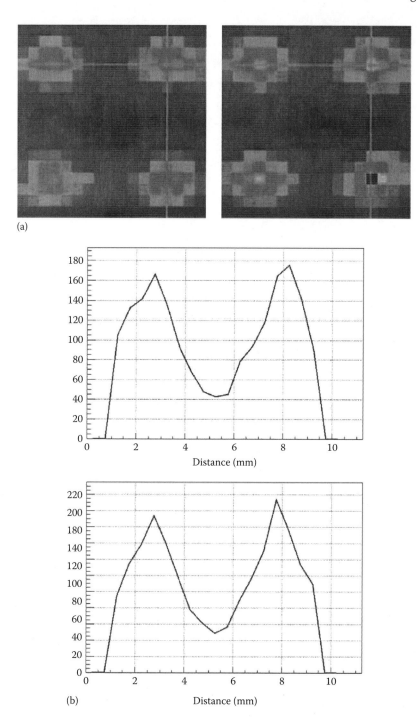

FIGURE 11.37 Results with a 3D cube phantom; showing the 2D image in the *yz* plane (a) and line profiles along *y* (b). (*Continued*)

Emerging Concept in Nuclear Medicine

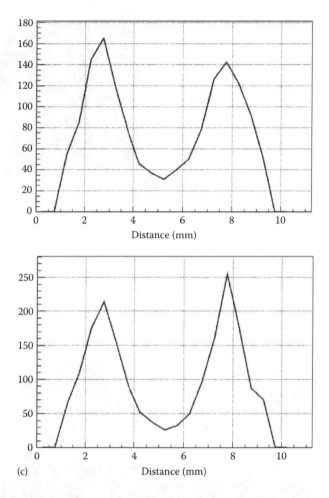

FIGURE 11.37 (*Continued*) Results with a 3D cube phantom; along z (c) for a slice with $x = 2.5$ mm (top) and $x = -2.5$ mm (bottom). (From (Kolstein M. et al., *J. Instrum.*, 9, 2014), licensed under CC By.)

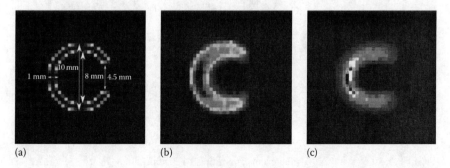

FIGURE 11.38 (a) depiction of the horseshoe phantom. Results on the horseshoe phantom with (b) LM-OSEM and (c) OE, as published in Ref. [85] (licensed under CC BY).

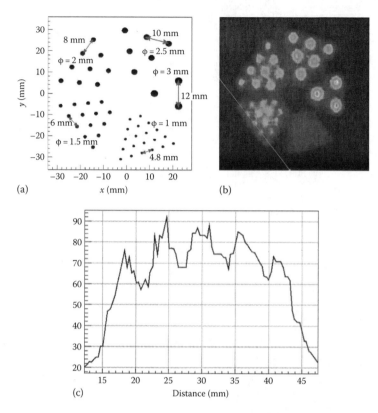

FIGURE 11.39 Derenzo phantom results as reported in Ref. [85] (licensed under CC BY). (a) Depiction of the Derenzo phantom in the x–y plane. (b) LM-OSEM result on 70 M Compton coincidences with a Derenzo phantom after applying a 3D median filter. (c) line-profile through the rods with 1.5 mm diameter. With the phantom at a distance of 100 mm from the scatterer and an activity of 2×10^8 Bq, the Compton camera sensitivity is 3.3 cps/kBq and it would take 1.8 min to get the 70 M coincidences used for this image.

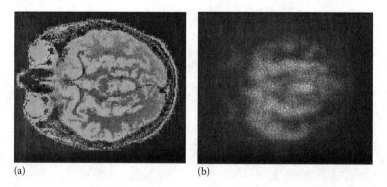

FIGURE 11.40 (a) DICOM image corresponding to a brain slice. (b) the reconstructed image using the LM-OSEM algorithm on Compton camera data. (From Calderón, Y. et al., Design, development, and modeling of a Compton camera tomographer based on room temperature solid state pixel detector, PhD thesis, UAB, Barcelona, Spain, 2014.).

Emerging Concept in Nuclear Medicine

11.6 CONCLUSION

With major advances in semiconductor detectors and frontend electronics (quantity, quality, and cost), employing high Z semiconductor detectors in the field of nuclear medicine imaging at the commercial level has become reality. GE is already commercializing the D-SPECT [72], which is a gamma camera made of CZT. Hitachi has already developed a full 3D brain PET using a CdTe detector [17]. Siemens has recently acquired Acrorad LTD, the largest producer of CdTe detectors, and this indicates an ambitious future plan by Siemens to use CdTe in devices that require high Z sensors such as CT, SPECT, and PET.

The novel design of the VIP detector is one-step forward on the same path and is ahead of the current development. It is meant to solve the intrinsic limitation of the current PET devices. The detector can have a trapezoidal or rectangular parallelepiped shape to make the design seamless. The design of the single unit detector makes it possible for the future nuclear medicine devices such as PET, Compton Camera, and PEM, to use a high granulated true 3D sensor made of CdTe/CZT or even TlBr detectors. Thus eliminating parallax and providing precise depth of interaction information to reconstruct images with high contrast and at low radiation dose. At any position the photon gets absorbed, one can measure its energy with high resolution. Removing the scattered events is now possible to achieve. In Table 11.12, one can see the performance of the VIP-PET when compared to current PET systems that are used for full body scan or dedicated for brain imaging. The VIP-PET even outperforms the small

TABLE 11.12
Comparison of 6 PET systems with the VIP-PET

	Gemini TF (Philips) [7]	Biograph mMR (Siemens) [88]	Discovery VCT (GE) [89]	HRRT (Siemens) [15]	G-PET (Phillips) [46]	CdTe-PET (Hitachi) [17]	VIP
Sensitivity (cps/kBq)	7.2	9.5	9.1	4.3	4.6	a	14.4
Transverse resol. at 1 cm (mm)	4.7	4.4	5.1	2.4	4.0	2.3	0.7
Transverse resol. at 10 cm (mm)	5.2	5.8	5.6	2.8	4.5	4.2	0.9
Axial resol. at 1 cm (mm)	4.7	4.5	5.6	2.5	5.0	4.8	1.3
Axial resol. at 10 cm (mm)	5.2	4.8	6.3	2.9	5.6	5.9	1.9
Scatter fraction (%)	31	36	38	46	34	37	4

Notes: The first 3 systems on the left are for whole body PET, the next 3 are brain PET. It is clear that the VIP-PET expected performance is significantly better than the current PET systems because it is a true 3D sensor that is converted into 6.3 million voxel, each has it is own readout electronics. The energy resolution of 1% at 511 keV shows its positive effect in the low percentage of the scatter fraction of the VIP-PET.

[a] The sensitivity in Ref. [17] is 25.9 cps/Bq/cm^3.

animal PET devices as one can see in the results shown in Figure 11.20. The excellent spatial resolution comes from two factors: first, the small size voxel, eliminating the DOI error, and, secondly, the excellent energy resolution that helps eliminating scattered photons and thus keeping only the good LORs. For both features, one has to use semiconductor detectors that can be subdivided into pixel sizes at the level of μm, something that is not possible to do with scintillator crystals. Additionally, only with detectors such as CdTe or CZT is it possible to achieve excellent energy resolution for 511 keV photons and have good mobilities for electrons and holes.

Stacking of the single unit module, as shown in Figure 11.5, will form a 3D sensor with voxels of size $1 \times 1 \times 2$ mm^3, which will make it possible to construct a large area Compton Camera. With the VIP detector one can define the Compton cone with high precision by using precise information about the interaction points in both the scatter and absorber detectors, and at the same time using precise measurement of the energy deposited in both detectors. Simulation shows that one can achieve a sensitivity of 4.4 cps/kBq and a resolution of 2 mm FWHM, which are comparable to the performance of the current state-of-the-art PET devices.

If one extends the VIP design to construct a PEM, comparable to Naviscan [61,87], the results show that the VIP-PEM can detect tumors with sizes down to 1 mm in a Derenzo phantom, with good SNR, as one can see in Figures 11.26.

With such a high number of independent channels, with about 6.3 million for PET and around 3 million channels for Compton Camera, the image reconstruction and processing is a major challenge to be solved. The VIP project developed and evaluated its own implementations of image reconstruction algorithms OE and LM-OSEM, which were especially needed for the VIP PEM and Compton camera and have achieved very good results as shown throughout this chapter.

The VIP project is developing future nuclear medicine diagnostic devices with today's technology. The very high cost of such development poses the question of the viability of marketing such expensive PET or Compton Camera. Although it is hard to argue on this issue, there is a lesson to learn from Thomas J. Watson, the chairman of IBM 1958, who said: "I think there is a world market for about 5 computers".

ACKNOWLEDGMENTS

The research leading to these results has received funding from the European Research Council under the European Union's Seventh Framework Programme (ER Grant Agreement 250207, "VIP"). We also acknowledge the support from the Spanish MINECO under the Severo Ochoa excellence program (grant SO-2012-0234).

REFERENCES

1. N.J. Patronas et al., Work in progress: [18F]Fluordeoxyglucose and positron emission tomography in the evaluation of radiation necrosis of the brain, *Radiology*, **144(4)**, (1982) 885–889.
2. T. Ido et al., Labeled 2-deoxy-D-glucos analogs 18F labeled 2-deoxy-2-fluoro-D-glucose, 2-deoxy-2-fluoro-D-mannose and 14C-2-deoxy-2fluoro-D-glucose, *J. Label Compd. Radiopharmacol.*, **14** (1978) 175–183.

3. K. Hamacher, H. Coenen, and G. Stocklin, Efficient stereospecific synthesis of no-carrier added 2-[18F]fluoro-2-deoxy-D-glucose using amino-polyether supported nucleophilic sub-stituation, *J. Nucl. Med.* **27** (1986) 235–238.
4. T. Beyer et al., A combined PET/CT scanner for clinical oncology, *J. Nucl. Med.*, **41**, (2000) 1369–1379.
5. Diagnostic Imaging Market by X-ray Systems (Digital, Analog, Portable), Computed Tomography, Ultrasound Imaging Systems (2D, 3D, 4D, Doppler), MRI Machines (Closed & Open), and Nuclear Imaging Systems (SPECT, PET, PET/CT)–Global Forecasts to 2018, http://www.marketsandmarkets.com/Market-Reports/diagnostic-imaging-market-411.html Markets and Markets, New Markers Reports.
6. W. Wong, N.A. Mullani, E.A. Philippe, R. Hartz, and K.L. Gould, image improvement and design optimization of the time-of-flight PET, *J. Nucl. Med.*, **24** (1983) 52–60.
7. S. Surti, S. Kuhn, M.E. Werner, A.E. Perkins, J. Kolthammer, and J.S. Karp, Performance of Philips Gemini TF PET/CT scanner with special consideration for its time-of-flight imaging capabilities, *J. Nucl. Med.*, **48** (2007) 471–480.
8. P. Dorenbos, Light output and energy resolution of Ce3+-doped scintillators, *Nucl. Instrum. Methods Phys. Res. A*, **486** (2002) 208–213.
9. D.L. Bailey, D.W. Townsend, P.E. Valk, and M.N. Maisey, *Positron Emission Tomography*: *Basic Sciences*, London: Springer-Verlag (2005).
10. G. B. Saha, *Basics of PET Imaging, Physics, Chemistry, and Regulations*, London: Springer-Verlag (2010).
11. W.W. Moses, Fundamental limits of spatial resolution in PET, *Nucl. Instrum. Methods Phys. Res. A*, **648(1)**, (August 21, 2011), S236–S240.
12. Guy A.A., Stefan P.S., John W. S., and Jick H.Y. What can be expected from high-Z semiconductor detectors? *IEEE Trans. Nucl. Sci.*, **24** (1977) 121–125.
13. Semiconductor Detector Material Properties, eV Products Inc., webpage: http://www.evproducts.com/pdf/material_prop.pdf. (Accessed 2014.)
14. W. Bencivelli et al., Evaluation of elemental and compound semiconductors for X-ray digital radiography, *Nucl. Instrum. Methods*, **A310**, (1991), 210–214.
15. K. Wienhard et al., The ECAT HRRT: Performance and first clinical application of the new high resolution research tomograph, *Trans. Nucl. Sci.*, **49** (2002) 104–110.
16. R. Nutt, The History of PET, *Mol. Imaging Biol.*, **4(1)** (2003) 11–26.
17. Y. Morimoto et al., Development of a 3D brain PET scanner using CdTe semiconductor detectors and its first clinical application, *IEEE Trans. Nucl. Sci.*, **58(5)** (2011) 2181–2189.
18. E. Mikhaylova et al., Simulation of the expected performance of a seamless scanner for brain PET based on highly pixelated CdTe detectors, *IEEE Trans. Med. Imaging*, **33(2)** (2014) 332–339.
19. G. Ariño et al., Energy and coincidence time resolution measurements of CdTe detectors for PET, *J. Instrum.*, **8**, (February 2013). doi:10.1088/1748-0221/8/02/C02015
20. J.C. Lund, R. Olsen, J.M. Van Scyoc, and R.B. James, The use of pulse processing technique to improve the performance of Cd1-xZnxTe gamma-ray spectrometers, *IEEE Trans. Nucl. Sci.*, **43(3)** (1996) 126–130.
21. Z. Mei, F. Hua, J. Glazer. Low temperature soldering, Twenty-First IEEE/CPMT International *Electronics Manufacturing Technology Symposium,* (1997), pp. 463–476.
22. J.-G. Macias-Montero et al., Toward VIP-PIX: A low noise readout ASIC for pixelated CdTe gamma-ray detectors for use in the next generation of PET scanners, *IEEE Trans. Nucl. Sci.*, **60(4)** (2013) 2898–2904.
23. J.-G. Macias-Montero et al., A 2D 4 × 4 channel readout ASIC for pixelated CdTe detectors for medical imaging applications *IEEE NSS MIC Conf. Rec.* Nuclear Science Symposium and Medical Imaging Conference (NSS/MIC), 2013 IEEE, (2013) 1–7.
24. H.M. Hudson et al., Accelerated image reconstruction using ordered subsets of projection data, *IEEE Trans. Med. Imaging.* **13(4)** (2011) 601–609.

25. S.J. Wilderman et al., Improved modeling of system response in list mode EM reconstruction of Compton scatter camera images, *IEEE Trans. Nucl. Sci.*, **48(1)** (2001) 111–116.
26. S.J. Wilderman et al., Fast algorithm for list mode back-projection of Compton scatter camera data, *IEEE Trans. Nucl. Sci.*, **45(3)** (1998) 957–962.
27. A. Sitek, Representation of photon limited data in emission tomography using origin ensembles, *Phys. Med. Biol.* **53(12)** (2008) 3201–3216.
28. A. Andreyev, A. Sitek, and A. Celler, Resolution recovery for Compton camera using origin ensemble algorithm, *IEEE NSS MIC Conf. Rec.*, Nuclear Science Symposium and Medical Imaging Conference (NSS/MIC), 2011 IEEE, (2011) 2774–2778.
29. C.X. Wang et al., Performance evaluation of filtered backprojection reconstruction and iterative reconstruction methods for PET images, *Comp. Biol. Med.*, **28** (1998) 13–25.
30. P. Arce et al., Gamos: A framework to do Geant4 simulations in different physics fields with an user-friendly interface, *Nucl. Instrum. Methods Phys. Res. A,* **735** (January 21, 2014) 304–313.
31. *STIR*, Software for tomographic image reconstruction, http://stir.sourceforge.net. (Accessed 2011–2014.)
32. A. Kishimoto et al., Development of a dual-sided readout DOI-PET module using large-area monolithic MPPC-arrays, *IEEE Trans. Nucl. Sci.*, **60(1)** (February 2013), 38–43.
33. F. Taghibakhsh et al., Detectors with dual-ended readout by silicon photomultipliers for high resolution positron emission mammography applications, *IEEE NSS MIC Conf. Rec.*, (2009) 2821–2826.
34. T. Mitsuhashi et al., 1 mm isotropic detector resolution achieved by X'tal cube detector, *IEEE NSS MIC Conf. Rec.*, (2010) 3093–3096.
35. H.W.A.M. de Jong et al. Performance evaluation of the ECAT HRRT: an LSO-LYSO double layer high resolution, high sensitivity scanner., *Phys. Med. Biol.*, **52** (2007) 1505.
36. G. Ariño et al., Characterization of a module with pixelated CdTe detectors for possible PET, PEM and Compton camera applications, *J. Instrum.*, **9** (2014).
37. T. Shiga et al., A new PET scanner with semiconductor detectors enables better identification of intratumoral inhomogeneity., *J. Nucl. Med.*, **50(1)** (2009) 148–155.
38. P. Vaska et al. Ultra-high resolution PET: A CZT-based scanner for the mouse brain., *J. Nucl. Med.*, **50(2)** (2009) 293.
39. P. Vaska et al. A prototype CZT-based PET scanner for high resolution mouse brain imaging, *IEEE NSS MIC Conf. Rec.*, 5 (2007) 3816–3819.
40. E. Mikhaylova et al., Simulation of pseudo-clinical conditions and image equality evaluation of PET scanner based on pixelated CdTe detector, *IEEE NSS MIC Conf. Rec.*, (2011) 2716–2722.
41. D. Uzun, G. De Lorenzo, M. Kolstein, and M. Chmeissani, Simulation and evaluation of a high resolution VIP PEM system with a dedicated LM-OSEM algorithm, *J. Instrum.*, **9** (2014).
42. NEMA standards publication NU 2-2001, *Performance Measurements of Positron Emission Tomographs*, Rosslyn, VA: National Electrical Manufacturers Association (2001).
43. NEMA standards publication NU 4-2008, *Performance Measurements for Small Animal Positron Emission Tomographs*, Rosslyn, VA: National Electrical Manufacturers Association (2008).
44. M.B. Sarrhini, Simultaneous attenuation and scatter corrections from the projections in small animal PET imaging, *Comp. Methods Prog. Biomed.,* **108(3)** (December 2012), 889–899.
45. J.S. Karp et al., Performance of a brain PET camera based on anger-logic gadolinium oxyorthosilicate detectors, *J. Nucl. Med.*, **44** (2003) 1340–1349.
46. S. Surti et al., Performance measurements for the GSO-based brain PET camera (G-PET), *IEEE NSS MIC Conf. Rec.*, **2** (2001) 1109–1114.

47. M. Cañadas, M. Embid, E. Lage, M. Desco, J.J. Vaquero, and J.M. Pérez, NEMA NU 4-2008 performance measurements of two commercial small-animal PET scanners: ClearPET and rPET-1, *IEEE Trans. Nucl. Sci.,* **58** (2011) 58–65.
48. Q. Bao, D. Newport, M. Chen, D.B. Stout, and A.F. Chatziioannou, Performance evaluation of the Inveon dedicated PET preclinical tomograph based on the NEMA NU-4 standards, *J. Nucl. Med. Vol.,* **50** (2009) 401–408.
49. R. Prasad, O. Ratib, and H. Zaidi, NEMA NU-04-based performance characteristics of the LabPET-8™ small animal PET scanner. *Phys Med Biol.,* **56(20)** (2011) 6649–6664.
50. H. Herzog, L. Tellmann, C. Hocke, U. Pietrzyk, M.E. Casey, and T. Kuwert, NEMA NU2-2001 guided performance evaluation of four Siemens ECAT PET scanners, *IEEE Trans. Nucl. Sci.,* **51** (2004) 2662–2669.
51. DICOM NEMA, Suite 900, medical.nema.org/dicom/geninfo/Brochure.pdf.
52. E. Mikhaylova et al., Optimization, evaluation, and comparison of standard algorithms for image reconstruction with the VIP-PET, *J. Instrum.* **9** (2014).
53. C.J. Thompson, K. Murthy, I.N. Weinberg, and F. Mako, Feasibility study for positron emission mammography, *Med. Phys.,* **21(4)** (1994) 529–537.
54. C.J. Thompson, K. Murthy, I.N. Weinberg, and F. Mako, Positron emission mammography (PEM): A promising technique for detecting breast cancer, *IEEE Trans. Nucl. Sci.,* **NS42** (1995) 1012–1017.
55. K. Shilling, The role of positron emission mammography in breast cancer imaging and management, *Appl. Radiol.* (2008) 26–36.
56. J.E. Kalinyak et al., PET-guided breast biopsy, *Breast J.,* **17(2)** (2011) 143–151.
57. M. Furuta et al., Basic evaluation of a C-shaped breast PET scanner, *IEEE NSS MIC Conf. Rec.,* (2009) 2548–2552.
58. L. MacDonald et al., Clinical imaging characteristics of the positron emission mammography camera: PEM Flex Solo II, *J Nucl. Med.,* **50** (2009) 1666–1675.
59. M.C. Abreu et al., Design and evaluation of the clearPEM scanner for positron emission mammography, *IEEE Trans. Nucl. Sci.,* **53(1)** (2006) 71–77.
60. K. Murthy et al., Results of preliminary clinical trials of the positron emission mammography system PEM-I: A dedicated breast imaging system producing glucose metabolic images using FDG, *J Nucl. Med.,* **41** (2000) 1851–1858.
61. W. Luo et al., Performance evaluation of a PEM scanner using the NEMA NU 4-2008 small animal PET standards, *IEEE Trans. Nucl. Sci.,* **57(1)** (2010) 94–103.
62. M. Iima et al., Clinical performance of 2 dedicated PET scanners for breast imaging: initial evaluation, *J Nucl. Med.,* **53** (2012) 1–9.
63. J.S. Eo et al., Imaging sensitivity of dedicated positron emission mammography in relation to tumor size, *The Breast,* Elsevier Ltd., **21** (2012) 66–71.
64. Diagosis of Breast Carcinoma: Characterization of Breast Lesions With CLEARPEMSONIC, Feasibility Study, Assistance Publique Hopitaux De Marseille, http://clinicaltrials.gov/ct2/show/study/NCT01569321.
65. C.W. Lerche et al., Dependency of energy-, position- and depth of interaction resolution on scintillation crystal coating and geometry, *IEEE Trans. Nucl. Sci.,* **55(3)** (2008) 1344–1351.
66. W.W. Moses and J. Qi, Instrumentation optimization for positron emission mammography, *Nucl. Instrum. Methods Phys. Res. A,* **527** (2004) 76–82.
67. *Breast Cancer Facts and Figures 2011–2012,* American Cancer Society, Atlanta: American Cancer Society, Inc. (2012).
68. G. De Lorenzo et al., Pixelated CdTe detectors to overcome intrinsic limitations of crystal based positron emission mammographs, *J. Instrum.,* **8** (2013).
69. T.F. Budinger, S.E. Derenzo, G.T. Gullberg, W.L. Greenberg, and R.H. Huesman, Emission computer assisted tomography with single-photon and positron annihilation photon emitters, *J. Comput. Assist. Tomogr.,* **1(1)** (1977) 131–145.

70. P.J. Slomka et al., Advances in nuclear cardiac instrumentation with a view towards reduced radiation exposure, *Curr. Cardiol. Rep.*, **14(2)** (2012) 208–216.
71. C. Fiorini et al. The HICAM gamma camera. Published in *IEEE Transactions on Nuclear Science*, **59(3)** (2012) 537–544.
72. S.S. Gambhir et al., A novel high-sensitivity rapid-acquisition single-photon cardiac imaging camera, *J. Nucl. Med.*, **50(4)** (April 2009) 635–643.
73. R. W. Todd, J. M. Nightingale, and D.B. Everett, A proposed γ camera, *Nature*, **251**, (1974) 132–134.
74. M. Singh, An electronically collimated gamma camera for single photon emission computed tomography. Part I: Theoretical considerations and design criteria, *Med. Phys.*, **421(10)** (1983) 421–427.
75. M. Singh and D. Doria, An electronically collimated gamma camera for single photon emission computed tomography. Part II: Image reconstruction and preliminary experimental measurements, *Med. Phys.* **428(10)** (1983) 428–435.
76. C.E. Ordonez, A. Bolozdynya, and W. Chang, Doppler broadening of energy spectra in compton cameras, *IEEE NSS Conf. Rec.*, **2** (1997) 1361–1365.
77. W.L. Rogers, N.H. Clinthorne, and A. Bolozdynya, Compton camera for nuclear medical imaging, Chapter from *Emission Tomography: The Fundamentals of SPECT and PET*, 1 st edn., M.N. Wernick and J.N. Aarsvold (eds.) London: Elsevier Academic Press, (2004) pp. 383–420.
78. V. Schönfelder et al., The imaging compton telescope comptel on the gamma ray observatory, *IEEE Trans. Nucl. Sci.*, **31** (1984) 766–770.
79. V. Schönfelder et al., COMPTEL overview: Achievements and expectations, *Astronomy Astrophys. Suppl.*, **120** (1996) 13–21.
80. S. Takeda et al., Experimental results of the gamma-ray imaging capability with a Si/CdTe semiconductor compton camera, *IEEE Trans. Nucl. Sci.*, **56** (2009) 783–790.
81. Y. Suzuki et al., Three-dimensional and multienergy gamma-ray simultaneous imaging by using a Si/CdTe compton camera, *Radiology*, **267(3)** (2013) 941–947.
82. Y. Calderón et al., Modeling, simulation, and evaluation of a Compton camera based on a pixelated solid-state detector, *IEEE NSS MIC Conf. Rec.*, (2011) 2708–2715.
83. NEMA Standards Publication NU 1–2007, *Performance Measurements of Gamma Cameras*, Rosslyn, VA: National Electrical Manufacturers Association (2007).
84. M. Kolstein et al., Evaluation of origin ensemble algorithm for image reconstruction for pixelated solid-state detectors with large number of channels, *J. Instrum.*, **8** (April 2013).
85. M. Kolstein et al., Evaluation of list-mode ordered subset expectation maximization image reconstruction for pixelated solid-state compton gamma camera with large number of channels. *J. Instrum.*, **9** (2014).
86. Y. Calderón et al., Design, development, and modeling of a compton camera tomographer based on room temperature solid state pixel detector, PhD thesis, UAB, Barcelona, Spain (2014).
87. Naviscan, www.naviscan.com.
88. G. Delso et al., Performance measurements of the siemens mMR integrated whole-body PET/MR scanner, *J. Nucl. Med.*, **52(12)** (2011) 1914–1922.
89. M. Teräs, T. Tolvanen, J.J. Johanson, J.J. Williams, and J. Knuuti. Performance of the new generation of whole-body PET/CT scanners: Discovery STE and Discovery VCT, *Eur J. Nucl. Med. Mol. Imaging*, **34(10)** (2007) 1683–1692.

Index

A

ADCs, *see* Analog to digital convertors (ADCs)
Amorphous selenium (a-Se) direct X-ray detector
   by ANRAD Corporation, 29
   carrier drift, 33
   carrier generation modeling, 33–34
   columnar recombination, 34
   geminate recombination, 34
   sensitivity reduction, 31
   trapping effects, 35
Amplification Technologies (New Jersey, United States), 121
Analog mammography, 233–235
Analog to digital convertors (ADCs), 56, 199
Anger gamma cameras
   components, 200–201
   LFOV, 196
   position logic circuit, 201
   transaxial tomography, 195
Application-specific imaging systems, 219–220
Application-specific integrated circuits (ASICs), 48, 81, 93–95
a-Se direct X-ray detector, *see* Amorphous selenium (a-Se) direct X-ray detector
Avalanche photodiode (APD) cell gain
   definition, 122–123
   FBK SiPM, 123–126
   Hamamatsu MPPC, 123–126
   SensL SiPM, 124–126

B

Barium fluoride (BaF_2) crystals, 196
Beer–Lambert law, 257
Bismuth germinate (BGO) crystals, 196
Block detectors, 202–203, 207–209, 214, 217
Blood pool agents, 164, 187
Bone mineral densitometry (BMD), 83, 92, 111
   ASICs, 94
   spectral BMD (*see* Spectral bone mineral densitometry)

C

CdZnTe and CdTe detectors, 211
   advantages, 1–2
   applications, 2
   atomic numbers, 2
   band gaps, 3

bulk resistivity, 4
carrier transport
   detrimental trapping, 11
   high photon flux (*see* High photon flux, carrier transport)
   nonuniform trapping, 11, 15
   recombination effect, 12
   residence time *vs.* transit time, 11
   shallow-level trapping, 11
   uniform trapping (*see* Uniform trapping)
carrier traps, 5
chemical potentials, 4
chemical properties, 4
densities, 2–3
DEXA, 2
dual-energy CT systems, 63
duality, 3
DxRay, Inc., 63
electrical compensation, 8–10
electron and hole mobility-lifetime products, 3
FWHM energy resolution, 62
Gamma-Medica Ideas, 62
grain boundaries, detrimental effects, 5
growth technologies
   annealing, 24
   atomic level perfection, 18
   defect interactions, 22–24
   directional solidification, 18
   melt-growth techniques, 18
   parasitic nucleation, 19–21
   physical defect generation, 21–22
   point defects, 18
   THM technique, 24–25
   VGF technique, 25
leakage current, 4
Medipix, 64–65
NEXIS, 61–62
NOVA R&D, Inc., 61
n-type conductivity, 3
point-defect structure, 6–8
p-type conductivity, 3
solubility, 5
Cesium fluoride (CsF) crystals, 196
Compton cameras
   advantages, 306
   challenges, 306–307
   cone axis and apex, 306
   designing, 307–309
   Doppler broadening effect, 306

performance
    design optimization, 309–311
    detector sensitivity, 311–313
    spatial resolution, 314
radioactive source, 305
reconstruction algorithms
    Derenzo phantom, 315, 320
    DICOM image, 317, 320
    horseshoe phantom, 315, 319
    single point source, 314
    3D cube phantom, 315–319
Compton scatter, 57, 92, 108, 200, 205, 258, 305
Computed radiography (CR) detectors
    advantages, 238
    manufacturers, 237
    phosphor plate readout process
        columnar photostimulable phosphor, 242
        components, 238–239
        dual-side reading approach, 239–241
        line-scan reading approach, 240–241
    PSL mechanism, 238
Converging collimator, 202
Corrections and compensation methods
    forward imaging process, 152
    material decomposition, 153
    PIECE, 153
    PIECE-1, 154–155
    pulse pileup, 154–155
    SRE and pileups, 152–153
CR detectors, *see* Computed radiography (CR) detectors
CsF crystals, *see* Cesium fluoride (CsF) crystals
CT angiography (CTA) scan, 86–87

D

Dark-field (DF) images, 263–264
Dead-time loss ratio (DLR), 157–158
2-Deoxy-2-(^{18}F)fluoro-D-glucose (^{18}F-FDG) radiotracer, 270, 277
Depth of interaction (DOI) detectors, 213–216
Detective quantum efficiency (DQE), 32, 41–43, 83, 181, 235
DEXA, *see* Dual-energy X-ray absorptiometry (DEXA)
DF images, *see* Dark-field (DF) images
Differential leading edge discriminator (DLED) method, 138
Digital Imaging and Communications in Medicine (DICOM) image, 298, 317, 320
Digital mammography
    advantages, 234–235
    characteristics, 235
    digital detectors, 235
    FFDM (*see* Full-field digital mammography (FFDM))
    vs. screen-film mammography, 234
Digital SiPM (dSiPM), 136–137, 141–142, 216
Direct conversion flat-panel detector
    hexagonal geometry, thin-film transistor, 246–248
    optical readout, 246–247
    schematic diagram, 245
    thin-film transistor readout, 246
Direct radiography systems, *see* Flat-panel (FP) detectors
Diverging collimator, 202
DLED method, *see* Differential leading edge discriminator (DLED) method
DLR, *see* Dead-time loss ratio (DLR)
DOI detectors, *see* Depth of interaction (DOI) detectors
Doppler broadening effect, 306
DQE, *see* Detective quantum efficiency (DQE)
dSiPM, *see* Digital SiPM (dSiPM)
Dual-energy X-ray absorptiometry (DEXA), 2, 85, 104
Dual-modality imaging systems, 220
Dynamic bow tie filter, 151, 183–184

E

ED-PC X-ray detectors, *see* Energy-dispersive photon-counting (ED-PC) X-ray detectors
EID-CT, *see* Energy integrating detector computed tomography (EID-CT)
EIDs, *see* Energy-integrating detectors (EIDs)
Electron–hole pairs (EHP) transport, 31
Emission computed tomography (ECT), *see* Positron emission tomography (PET); Single-photon emission computed tomography (SPECT)
Energy-dispersive photon-counting (ED-PC) X-ray detectors
    advantages, 86–87
    clinical radiology, 87
    dose efficiency, 86
    fabrication, 86
    SiPM-based direct conversion, 89
    SiPM-based indirect conversion, 88
    spectral CT, 97
    spectral radiology, 93
Energy integrating detector computed tomography (EID-CT), 154, 160–161, 163
Energy-integrating detectors (EIDs), 49–50, 56, 63, 67, 75, 84, 150
Excess noise factor (ENF), 133–134

F

FFDM, *see* Full-field digital mammography (FFDM)
Filtered back-projection (FBP), 282–283, 298–299
Flat-panel (FP) detectors
   direct conversion detector
      a-Se layer interaction, 245–246
      optical readout, 246–247
      TFT readout, 246
      TFT with hexagonal array, 246–248
   indirect conversion detector
      CMOS readout, 244–245
      efficiency and spatial resolution, 242
      scintillator phosphor, 242
      TFT readout, 243–244
      thallium-activated cesium iodide columnar phosphor, 243
   manufacturers, 237
Full-field digital mammography (FFDM)
   CR detectors
      advantages, 238
      description, 237
      manufacturers, 237
      phosphor plate readout process (*see* Phosphor plate readout process)
      PSL mechanism, 238
   current technologies, 236
   FP detectors (*see* Flat-panel (FP) detectors)
   slot-scan systems
      description, 237
      direct conversion approach, 249–251
      indirect conversion approach, 249
      manufacturers, 237
      soft copy workstation, 236
Full width at half maximum (FWHM), 62, 71, 73, 87, 99, 101, 106, 109, 133, 215, 218

G

Gamma-ray interaction tracking (GRIT) program, 38–40
Gamma-ray pinhole camera, 194–195
Gamma spectrometry
   662 keV gamma rays, energy spectrum of, 137
   linearity characteristics, 138, 140
   3 × 3 mm^2 MPPC, 137
   6 × 6 mm^2 MPPC, 137
   12 × 12 mm^2 MPPC, 137
   MPPC array, 138–139
   XP5212 PMTs, 138–139
   XP2020Q, 138–139
Geant4-based Architecture for Medicine-Oriented Simulations (GAMOS) package, 285, 288, 301

GE Healthcare's SIGNA PET/MR, 142–143
General image weighting (GIW) method
   CT projections, 52–53
   PCXCT imaging, 51–52
   signal-to-noise ratio, 50
   temporal weighting method, 51
Gold nanoparticles (GNPs), 187
Grating interferometry
   analyzer/absorption grating, 260, 262
   deflection angle, 262
   intensity datasets, 262–263
   limitations, 265
   microfocus X-ray source, 260
   optimization, 264–265
   phase grating, 260, 262
   reconstruction method, 263–264
   source grating, 260
   spherical wave, 261
GRIT program, *see* Gamma-ray interaction tracking (GRIT) program

H

Hamamatsu Photonics (Japan), 121
High photon flux, carrier transport
   drift mobility, 15
   dynamic equilibrium with photon field, 16
   native and impurity defects, 18
   space charge distribution, 16–17
   thermal emission of carriers, 17
   transit time of carriers, 16–18

I

Image-guided therapy, 220
Indirect conversion flat-panel detector
   CMOS readout, 244–245
   CsI:Tl scintillator, 242
   description, 242
   schematic diagram, 243
   thin-film transistor readout, 243–244
Input count rate (ICR), 86, 156, 181
In situ ingot annealing, 24

K

K-edge digital subtraction angiography (KEDSA), 84–85
K-edge filtration method, 74
K-edge imaging
   Au-HDL, 188
   contrast agent, 187
   dual-energy systems, 186
   gadolinium image, 186–187
   GNPs, 187
   intravascular pathologic epitopes, 188
   K-edge attenuation, 186

Index

metal nanoparticles, 187
PET and SPECT applications, 187
PET-CT and SPECT-CT, 187
"two-bin" energy configuration, 186
Ketek GmbH (Munich, Germany), 121

L

Large-field-of-view (LFOV) Anger gamma camera, 196
Longitudinal tomography, 195

M

MANTIS package, *see* Monte Carlo X-ray, electron transport imaging simulation (MANTIS) package
MC simulation, *see* Monte Carlo (MC) simulation
MDM system, *see* MicroDose mammography (MDM) system
Medipix2 detector, 64
Medipix3 detector, 64–65
Metastatic tumor imaging, 194
MicroDose mammography (MDM) system, 60–61
Micropixel APD, *see* Silicon photomultipliers (SiPMs)
Modular detector, 62, 219
Modulation transfer function (MTF), 32, 40, 285
Monte Carlo (MC) simulation
  ARTEMIS, 35
  charge generation models, 32–34
  columnar recombination, 34
  EHPs
    generation of, 38
    sample transport tracks, 38–39
    transport of, 38
  flow chart, 36
  geminate recombination, 34
  high-energy electron lose kinetic energy, 36–37
  indirect detectors, 38–40
  MANTIS package, 32
  monoenergetic X-rays tracking, 36–37
  trapping effects, 35
Monte Carlo X-ray, electron transport imaging simulation (MANTIS) package, 32, 40
MTF, *see* Modulation transfer function (MTF)
Multipixel Geiger-mode APD, *see* Silicon photomultipliers (SiPMs)
Multipixel photon counter (MPPC), *see* Silicon photomultipliers (SiPMs)
Multivariate normal distribution, 154
Multiwire proportional chamber (MWPC), 196

N

N-Energy X-ray Imaging System (NEXIS), 61–62

O

Ordered subset expectation maximization (OSEM), 283–286
Origin ensemble (OE) algorithm, 284–285

P

Parallel-hole collimator, 202
Particle transport, recombination, and trapping in semiconductor imaging simulation (ARTEMIS), 35
PC, *see* Photon counting (PC)
PCD-CT systems, *see* Photon-counting detector computed tomography (PCD-CT) systems
PCD models, *see* Photon-counting detector (PCD) models
PCXCT imaging, *see* Photon-counting X-ray/CT (PCXCT) imaging
PCXI, *see* Phase-contrast X-ray imaging (PCXI)
PEM, *see* Positron emission mammograph (PEM)
Penetration and energy loss of positrons and electrons (PENELOPE), 30–31
PET, *see* Positron emission tomography (PET)
Phase-contrast X-ray imaging (PCXI)
  advantages, 259
  Beer–Lambert law, 257
  complex refractive index, 258–259
  Compton scattering, 256–257
  grating interferometry
    analyzer/absorption grating, 260, 262
    deflection angle, 262
    intensity datasets, 262–263
    limitations, 265
    microfocus X-ray source, 260
    optimization, 264–265
    phase grating, 260, 262
    reconstruction method, 263–264
    source grating, 260
    spherical wave, 261
  human finger, 257–258
  pair production, 257
  photoelectric absorption, 256
PHO-CON scanner, 195
Phosphor plate readout process
  columnar photostimulable phosphor, 242
  components, 238–239
  dual-side reading approach, 239–241
  line-scan reading approach, 240–241

Index

Photodetectors, 212–213
Photomultiplier tubes (PMTs), 120
Photon counting (PC)
  candidate materials, 172–173
  clinical applications
    dose reduction, 190
    K-edge (*see* K-edge imaging)
    spatial resolution, 188–190
  CT systems, 176–177
  detector fabrication and signal formation
    advantages, 174
    CZT energy resolution, 175–176
    equipotential contours surrounding, 175
    pixilation of anodes, 174
    Shockley–Ramo theorem, 172
  direct conversion, 169–170
  dose efficiency, 180–181
  examination time
    bowtie filters, 183
    direct conversion sensors, 181–182
    DQE, 181, 183–184
    ICR, 181, 183–184
    peripheral points, 182
    pulse-counting electronics, 183
    trauma scanning, 181
  image quality, 179–180
  K-edge imaging, 178–179
  material properties, 170–171
  multienergy imaging, 185
  multirow detector systems, 178
  ring artifacts, 184
  X-ray medical imaging system, 178
Photon-counting breast CT, 74–75
Photon-counting detector computed tomography (PCD-CT) systems
  clinical merits and applications, 159–160
    beam-hardening artifacts, 162
    computer simulated XCAT phantom image, 157
    contrast-to-noise ratio, 161
    K-edge imaging, 162–163
    molecular CT imaging, 163–164
    quantitative CT and X-ray imaging, 162
    simultaneous multiagent imaging, 164
    spatial resolution, 161–162
    X-ray radiation and contrast agents, 161
  compensation methods, 153–155
  corrections methods, 152–153
  image reconstruction
    interior problem, 157–159
    spectral data, 159–160
  models (*see* Photon-counting detector (PCD) models)
  philosophical approaches (*see* Corrections and compensation methods)
  pulse pileup, 150
  SRE, 150
  X-ray beam-shaping filters, 150–151
Photon-counting detector (PCD) models
  cascaded model, 157
  DLR, 157–158
  EID-CT systems, 154
  Monte Carlo simulation, 157
  spectrum distortion, 156
  SRE and pulse pileup, 154–155, 157–158
Photon-counting X-ray/CT (PCXCT) imaging
  advantages, 48
    electronic noise rejection, 49–50
    GIW method (*see* General image weighting (GIW) method)
    material decomposition, 53–55
  ASIC electronics, 48
  charge sharing, 70–72
  count rate limitations, 65–66
  designs
    ADC, 56–57
    detector architecture, 55–56
    detector technologies (*see* CdZnTe and CdTe detectors; Si strip detectors)
    high demands, 55
    imaging configurations, 58–60
    material selection, 57–58
  gas-filled detectors, 49
  intensity-dependent line artifacts
    beam flattening filters, 69
    breast CT, 69
    flat field correction, 67–68
    high-Z material, 69
    nonuniform pixel response, 67–68
    1D pixel array, 68
    photon-counting CZT detector, 70
    pixel response *vs.* X-ray intensity, 68
    planar X-ray imaging, 67
  isotope emission imaging, 48
  low-energy tailing, 66–67
  microchannel plates, 49
  photon-counting breast CT, 74–75
  semiconductors, 49
  suboptimal energy resolution, 72–74
  transmission imaging, 48–49
  X-ray photons, 48
Photon detection efficiency (PDE), 124, 126–127
Photostimulated luminescence (PSL)
  mechanism, 238
Pinhole collimator, 202
Pixelated Geiger-mode avalanche photon detector, *see* Silicon photomultipliers (SiPMs)
Point spread function (PSF), 291–292, 301, 314
Positron emission mammograph (PEM)
  breast immobilization, 299–300
  definition, 299

parallel paddles, 300–301
partial volume effect, 300
unit detector, 301
Positron emission tomography (PET), 97, 119, 186
   BaF_2 crystals, 196
   BGO crystals, 196
   block detector modules, 202–203
   body and brain imaging, 196
   coincidence detection, 204–205
   CsF crystals, 196
   DOI detectors, 213–215
   image-guided therapy, 220
   MWPC, 196
   overview, 198–199
   pulse-height analysis, 199–200
   readout methods, 216–217
   scintillation detectors
      attenuation length, 208
      BaF_2, 208–209
      BGO-based block detectors, 209
      decay time, 208
      gamma-ray photon detection, 199
      light yield, 208
      L(Y)SO-based systems, 209
      NaI, 208
   silicon photomultipliers, 216
   spatial resolution
      annihilation, 206
      crystal size, 205
      crystal width, 207
      DOI blurring effect, 207
      empirical rule, 207
      factors affecting, 205–206
      photon noncollinearity, 206
   TOF detectors, 215–216
   waveform sampling, 217–219
Prototype PET scanner, 141–142
PSF, *see* Point spread function (PSF)
Pulse-height spectroscopy (PHS), 41

R

Radon transform, 281
Reconfigurable detector, 219
Rectilinear scanners, 194
Room temperature X-ray imaging arrays, 95

S

Scintillation cameras, *see* Anger gamma cameras
Scintillation detectors
   PET
      attenuation length, 208
      BaF_2, 208–209
      BGO-based block detectors, 209
      decay time, 208

      gamma-ray photon detection, 199
      light yield, 208
      L(Y)SO-based systems, 209
      NaI, 208
   SPECT, 199–200, 209–210
Scintillator detectors, 28
Screen-film mammography, 234
Semiconductor detectors, 210–212
Semiconductor X-ray detectors
   analytical methods, 32
   applications, 41–43
   a-Se detector, 29
   block diagram, 29–30
   EHP transport, 31
   MC simulation (*see* Monte Carlo (MC) simulation)
   photon–electron interactions, 30–31
   schematic diagram, 28–29
SensL (Ireland), 121
Shockley–Ramo theorem, 172
Signal-to-noise ratio (SNR), 120
Silicon photomultipliers (SiPMs), 87
   after-pulses, 126, 128–129
   commercial applications, 121
   dark-noise rate, 130–131
   dSiPM, 136–137
   ENF, 133–134
   fast timing, 138, 141
   gain, 122–126
   gamma spectrometry (*see* Gamma spectrometry)
   high-energy physics, 121
   $LaBr_3$, 120
   linearity of, 131–133
   LSO, 120
   LYSO, 120
   manufacturers, 121
   medical instrumentations, 141–143
   neutrino physics, 121
   optical crosstalks, 129–130
   PDE, 124, 126–127
   photodetector, 119
   PMTs, 120
   scintillation light, 119
   simplified electric structure, 121–122
   TOF-PET scanners, 120–121
   TTS, 133, 135–136
Single bow tie filters, 151
Single-photon avalanche diode array, *see* Silicon photomultipliers (SiPMs)
Single-photon emission computed tomography (SPECT), 97, 187
   Anger gamma cameras
      components, 200–201
      position logic circuit, 201
   converging collimator, 202
   diverging collimator, 202

Index

image-guided therapy, 220
overview, 197–198
parallel-hole collimator, 202
photodetectors, 212–213
pinhole collimator, 202
pulse-height analysis, 199–200
rotating-camera approach, 195
scintillation detectors, 199–200, 209–210
semiconductor detectors, 210–212
VIP, 303–305
SiPMs, *see* Silicon photomultipliers (SiPMs)
Si strip detectors, 60–61
Slot-scan systems
description, 237
direct conversion approach, 249–251
indirect conversion approach, 249
manufacturers, 237
Software for Tomographic Image Reconstruction (STIR) toolkit, 285
Solid-state photomultiplier, *see* Silicon photomultipliers (SiPMs)
SPECT, *see* Single-photon emission computed tomography (SPECT)
Spectral bone mineral densitometry
CdTe sensors, 105
DEXA scanner, 104–105
ED-PC BMD detector design, 105
ED-PC detectors, 104
EI detectors, 104
OCR, 106
osteoporosis diagnosis, 104
X-ray spectra, 107
Spectral clinical radiology
applications, 83
ASIC development, 93–95
cadmium telluride (CdTe), 81
cadmium zinc telluride (CdZnTe), 81
DEXA, 85
digital 2D projection image, 82
digital flat panel X-ray imaging arrays, 81
ED-PC detectors, 86–87
EI detectors, 84
indirect/direct conversion methods, 82–83
KEDSA, 84–85
panel X-ray imaging arrays, 81–82
semiconductor sensor development, direct conversion
ASIC devices, 88
CdTe, 89
CdTe and CdZnTe detectors, 92
CdZnTe, 89
cerium (Ce)-doped scintillators, 87–88
Compton effect, 92
ED-PC X-ray imaging arrays, 87
HgI_2, 89
high-Z semiconductors, 90
lower Z semiconductors, 90
PbI_2, 89
scintillating materials, 88
semiconductor materials, 89
silicon (Si), 89
SiPM devices, 87
SiPM element, 88
small pixel effect, 91
TlBr, 89
X-ray photon, 90
Si-based semiconductor, 81
single-kVp and dual-kVp, 85–86
spectral BMD
CdTe sensors, 105
DEXA scanner, 104–105
ED-PC detectors, 104–105
EI detectors, 104
OCR, 106
osteoporosis diagnosis, 104
X-ray spectra, 107
spectral computed tomography
Am spectrum, 99–101
CdTe and CdZnTe sensors, 97
CdTe crystals, 96–97
CdZnTe and CdTe detectors, 99
ED-PC CT module design, 96–97
energy spectra, 99–100
FWHM energy resolution, 101
mean *vs.* time, 102–103
OCR, 98–101
optimal energy weighting and material decomposition methods, 96
output count rate, 102
PET, 97
room temperature X-ray imaging arrays, 95
single-crystal semiconductor, 95
SPECT, 97
2D ASIC, 96–98
variance *vs.* mean, 103–104
spectral digital mammography
advantage, 108
applications, 109
ASICs, 109, 111
breast cancer screening, 107
ED-PC DM module design, 108
ED-PC X-ray imaging arrays, 110
EI X-ray imaging arrays, 110
flat field response, 109–110
high-Z semiconductors sensors, 110
iodine contrast injection, 108
limitations, 108
OCR, 109
SNR advantages, 107–108
2D projection image, 108
TFT array, 83
X-ray intensity, 81
X-ray photographic film, 81

Spectral computed tomography
  Am spectrum, 99–101
  CdTe and CdZnTe sensors, 97
  CdTe crystals, 96–97
  CdZnTe and CdTe detectors, 99
  ED-PC CT module design, 96–97
  energy spectra, 99–100
  FWHM energy resolution, 101
  mean *vs.* time, 102–103
  OCR, 98–101
  optimal energy weighting and material decomposition methods, 96
  output count rate, 102
  PET, 97
  room temperature X-ray imaging arrays, 95
  single-crystal semiconductor, 95
  SPECT, 97
  2D ASIC, 96–98
  variance *vs.* mean, 103–104
Spectral digital mammography
  advantage, 108
  applications, 109
  ASICs, 109, 111
  breast cancer screening, 107
  ED-PC DM module design, 108
  ED-PC X-ray imaging arrays, 110
  EI X-ray imaging arrays, 110
  flat field response, 109–110
  high-Z semiconductors sensors, 110
  iodine contrast injection, 108
  limitations, 108
  OCR, 109
  SNR advantages, 107–108
  2D projection image, 108
Spectral response effect (SRE), 150

T

Thallium bromide (TlBr), 89
Thin-film transistor (TFT) array, 83, 237, 244
Thyroid imaging, 194
Time jitter/transit time spread (TTS), 133, 135–136
Time-of-flight (TOF) detectors, 215–216
Time-of-flight PET (TOF-PET), 120
Transaxial tomography, 195
Traveling heater method (THM), 24–25

U

Under-Bump-Metal (UBM) layer, 279
Uniform trapping
  causes, 11
  deep-level defects trapping, 12–14
  shallow-level defects trapping, 14–15

V

Vertical gradient freeze (VGF) technique, 25
Voxel imaging PET (VIP) project
  CdTe detector
    BiSn and eutectic solder, 278
    mean free path, 277
    mobility-lifetime product, 274, 276
    511 keV photon, 274–275, 277
    rectangular shape, 275
    Schottky contact electrodes, 277
  Compton cameras (*see* Compton cameras)
  counting performance, 289–291
  CT scan, 270–271
  detection efficiency, 273–274
  ECAT-HRRT, 274
  evaluation
    Derenzo phantom, 302–303
    NEMA NU 2-2001, 288
    NEMA NU 4-2008, 288, 302
    PSF, 302
    scatter and random fraction, 302
    system sensitivity, 301
  ^{18}F-FDG radiotracer, 270
  image quality
    calculated coefficients, 293–294
    NEMA NU 4-2008, 292
    recovery coefficients, 293–294
    test conditions, 292
  limitations, 286–287
  line of response (LOR), 271, 273
  minimum detectable lesion
    cold and hot spheres, 296–297
    HR + PET scanner, 296–297
    NEMA NU 2-2001, 295–296
    tumoral and normal tissues, 295
  MRI scan, 271–272
  PEM
    breast immobilization, 299–300
    definition, 299
    drawbacks, 300
    paddles touching, 300
    partial volume effect, 300
    unit detector, 301
  reconstruction
    algorithms, 298–299
    FBP, 282–283
    origin ensemble algorithm, 284–285
    OSEM, 283–284
    quality, 285
    realistic 3-D human brain phantom, 298
    simple back-projection, 281–282
    software, 285–286

Index

resolution, 291
scatter fraction, 288–289
scintillator crystals, 271–273
semiconductor radiation detectors, 272, 274
SPECT, 303–305
VIP-PIX chip
   pixel-channel properties, 278–279
   processing unit, 279–280
   UBM, 279–280

W

Wafer-level annealing, 24

X

X-ray detectors
   advantages, 27
   applications, 27
   scintillator detectors, 28
   semiconductor-based detectors
      (*see* Semiconductor X-ray detectors)

Z

Zecotek Photonics (Singapore), 121